# Simulation and Testing for Vehicle Technology

Clemens Gühmann · Jens Riese
Klaus von Rüden
Editors

# Simulation and Testing for Vehicle Technology

## 7th Conference, Berlin, May 12–13, 2016

Springer

*Editors*
Clemens Gühmann
TU Berlin
Berlin
Germany

Jens Riese
IAV GmbH
Berlin
Germany

Klaus von Rüden
IAV GmbH
Berlin
Germany

ISBN 978-3-319-32344-2 ISBN 978-3-319-32345-9 (eBook)
DOI 10.1007/978-3-319-32345-9

Library of Congress Control Number: 2016936428



Printed on acid-free paper

This Springer imprint is published by Springer Nature
The registered company is Springer International Publishing AG Switzerland

# Preface

The potential provided by innovative transmission systems, supercharged combustion engines, and hybrid or electric drive concepts for reducing fuel consumption and emissions is far from exhausted. Developments in this regard however, are limited to the motor vehicle system. Intelligent mobility, on the other hand, also involves the human element, other road users and the environment. As a result, modern driver assistance systems offer considerable scope for broadening the potential in reducing fuel consumption and exhaust emissions.

Modern automotive engineering is inconceivable without virtual development processes. The use of validated simulation and test methods is necessary to ensure that the decisions taken in the early concept phase are justifiable and future-proof. Constantly rising numbers of electrical components in vehicles increase the complexity of processes, models, and methods and call for new strategies to control complete systems and boost efficiency in the development process.

Just a few years ago, the use of simulation was concentrated on the area of ECU testing using HiL simulation, the current environment uses a wide range of simulation methods at all stages of the V-model development process. Car2Car communication and driver assistance systems that have developed to the point of autonomous driving demand further development methods to validate complete systems, the aim being to achieve end-to-end, model-based processes that integrate a virtual application process alongside hardware and software development.

This objective, however, can only be met with validated methods in data-based and physically-oriented modeling. Complex, non-linear, and dynamic system effects must also be modeled with finite effort if growing numbers of ECU functions are to be efficiently developed with the help of simulation in the future. The advantages of simulation will only bear fruit, however, when tests, applications and optimization can be carried out automatically in dry runs.

Organized jointly by the Chair of Electronic Measurement and Diagnostic Technology (Technische Universität Berlin) and Ingenieurgesellschaft Auto und Verkehr (IAV GmbH), the seventh conference on simulation and testing for vehicle technology in May 2016 picks up on current trends in modeling, simulation, and control unit testing in connection with developing automotive electronics. Competent experts from industry and universities will be reporting on current research and development projects on modeling and simulating driver assistance systems, internal combustion engines, transmission systems, and e-vehicles. Besides modeling and simulation, the conference will devote equal attention to the subject of "testing". In addition to examining new testing concepts and test methods up to system testing level, test case generation and testing tools will also be discussed in detail.

We hope the presentations delivered will give all conference documentation readers and conference participants' insight and ideas for them to use in their own development and research projects.

Our special thanks go to the authors for their topical and interesting contributions; to Krystina Boettcher and Patrick Paska from IAV GmbH and to Springer International Publishing for being so cooperative.

Berlin
May 2016

Clemens Gühmann
Klaus von Rüden
Jens Riese

# Contents

# Part I
# XiL-Simulation

# Real-Time Hardware-in-the-Loop Simulation of Multiphase DC/DC Converters

**Sebastian Schulz and Hendrik Gerth**

**Abstract** In the course of the development of fuel cell electric vehicles, "high voltage" DC/DC converters—known from power split hybrid electric vehicles—are going to be a new component for European automotive engineering. These converters couple high voltage nets with different voltages of more than 60 V. The power throughput ranges from several 10 of kW to over 100 kW. The corresponding converters are usually designed as galvanic coupled boost and buck/boost converters. Signal-based hardware-in-the-loop (HiL) tests are an efficient solution for the functional development and software test. This work shows how to implement such a signal-based HiL real-time test bench with an analog simulation of the power unit. For this purpose, it is shown how the power section of the real DC/DC converter can be scaled so that an equivalent signal behavior with currents in the milliampere range and voltages up to 5 V are achieved. Thus, it is demonstrated that transient effects, such as load dumps or short circuits, can be simulated.

## 1 Introduction

### *1.1 Hardware-in-the-Loop Simulation of a DC/DC Converter*

Similar to the hardware-in-the-loop simulations of electric machines for the testing of power inverters, there are two different concepts involved [1]. On the one hand, it is possible to carry these tests out using powerful simulators which reproduce the load and thermal conditions of the vehicle exactly. This means that the test bench has to provide the respective power in- and output, high voltages have to be applied

---

S. Schulz (✉) · H. Gerth
IAV GmbH, Hochvolt-Antriebsstrang Konzepte, Gifhorn, Germany
e-mail: Sebastian06.Schulz@iav.de

H. Gerth
e-mail: Hendrik.Gerth@iav.de

 
C. Gühmann et al. (eds.), *Simulation and Testing for Vehicle Technology*,
DOI 10.1007/978-3-319-32345-9_1

and the cooling has to be implemented. Though this kind of simulation is able to provide a very detailed environment, the effort is very high.

On the other hand, so-called signal-based simulations are possible. In this case, the controller board of the components is separated from the power module. The simulator evaluates all signals from the controller board and reproduces the signals from the power module to the controller board (Fig. 1). These are the signals to drive the power switches ([s*] Fig. 1), and all signals for the measurement values, that is, for the voltages at the input and output ([$u_{\text{in}}$] and [$u_{\text{out}}$], respectively, in Fig. 1), the inductor current ([$i_{\text{L}}$] in Fig. 1), the currents at the input and output side and temperatures, for example, of the choke core or the semiconductors ([$\vartheta$] in Fig. 1). The advantage is that respective tests can be carried out at the desk without the need for any cooling of the device under test (DUT) and with safe voltages. Though it is, of course, impossible to test the power module, which is simulated by the test bench, it is an efficient way to carry out software development, and software and diagnosis tests of the controller board.

It is necessary to fulfill a number of requirements for the HiL simulator described here. At first, the simulator should allow the emulation of all possible topologies (see, for example, [2–5]), those are:

- boost converters with unidirectional current flow from the fuel cell to the drive in the fuel cell electric vehicle (FCEV), where the battery/traction drive voltage is always larger than the fuel cell voltage;
- buck/boost converters with unidirectional current flow in FCEVs, where the fuel cell voltage and the battery voltage range overlap;
- boost converters with bidirectional current flow to supply an electric drive from a battery with lower voltage, such as in power split hybrid electric vehicles (HEVs) (Fig. 1, top); and
- buck/boost converters with bidirectional current flow for so-called twin battery concepts (Fig. 1, bottom).

Secondly, as in any real-time simulation, the real system dynamics of the buck- and boost-converter have to be preserved. This is especially important with respect to the fast control of such converters.

Furthermore, the simulation will take disturbances and the impedance of the traction network into account. The voltage ripple caused by the inverter of the electric drive of several volts could particularly lead to impaired regulation. The impedance of the fuel cell, battery, traction network or the inverters also affect the control. The internal resistance of the battery, which reaches several Ohms at low temperatures, can be especially critical. A DC/DC converter, which supplies the traction network under these conditions, has to cope with a significant feedback of its output current on the associated output voltage. This reaction may also lead to problems regarding the control of the converter. Therefore, the DC/DC HiL allows the simulation of a simple battery model.

Finally, the simulator should be modular as a component in an FCEV is build up in several single phases to achieve the respective output power and to use the

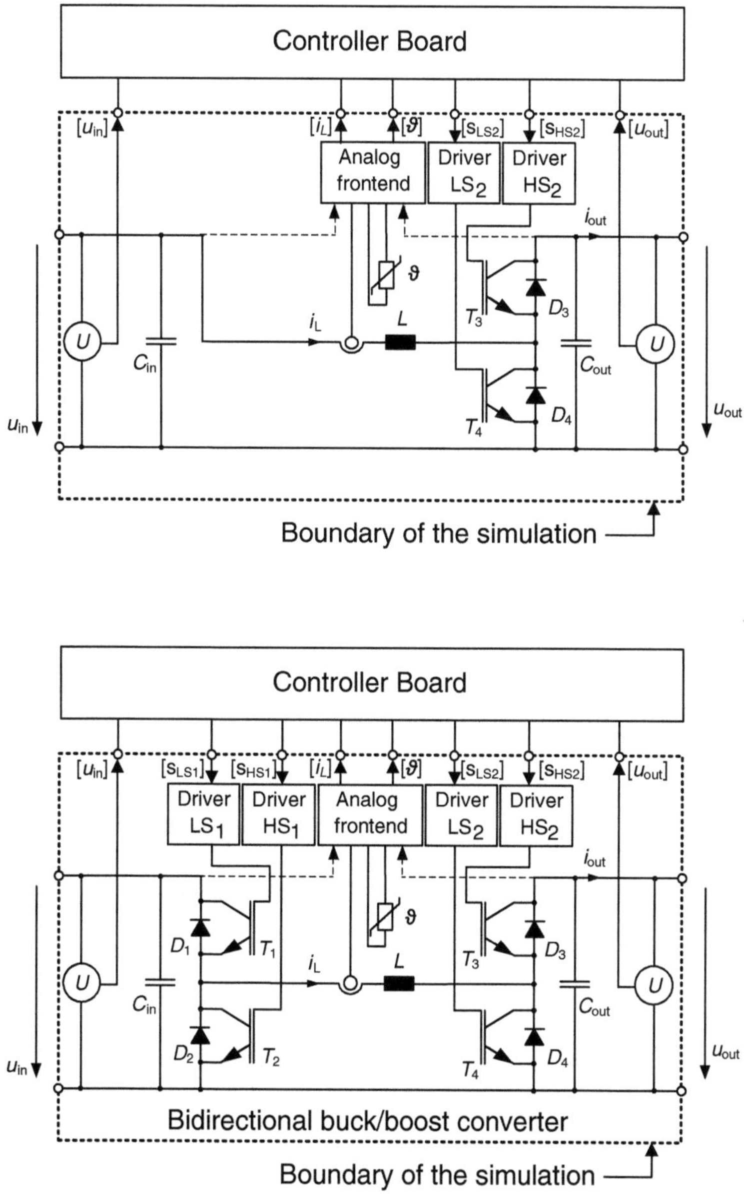


**Fig. 1** Examples of two topologies the simulator has to cover, i.e. a boost converter with $u_{out} > u_{in}$ with bidirectional current flow (*top*) and a buck/boost converter with bidirectional current flow

advantages of interleafed switching. Thus, real multiphase converters can also be simulated and not only single phase converters. Therefore, an inexpensive solution for the simulation of a single phase is essential.

## 2 Setup

The phase currents in HiL simulations of electric machines are more or less sinusoidal and the ripple due to the switching of the inverter is only a more or less disturbing artifact for the controller. The ripple of the inductor current may be essential for DC/DC converters, depending on the control strategy. Considering switching frequencies of 50 kHz or more, simulations on FPGAs even with a step size of only 200 ns [6] may not provide a sufficient resolution when evaluating the pulse width modulation driving the power switches.

Therefore, the simulator is based on an analog computer due to the necessity of tracking the time variation of the inductor current. The advantage of this principle is that it provides the real dynamics without any delays caused by a discrete-time signal processing. However, this makes it necessary to adapt the simulator to a specific DUT by modification of the simulator hardware.

The HiL model of the single phase is essentially constructed in the classical buck/boost topology (Fig. 2). The power switches, which are insulated-gate bipolar transistors (IGBTs) and diodes in reality, are replaced by metal-oxide-semiconductor field-effect transistors (MOSFETs), which have a very small voltage drop in the activated state.

### 2.1 *Scaling Currents and Voltages*

The dynamic behavior of the real system of the DC/DC converter has to be preserved. Real currents and voltages have to be reproduced on a secure and easily workable level. Two scale factors are introduced for this purpose. With these factors, the voltages and currents of the real system ($u_{\mathrm{real}}$ and $i_{\mathrm{real}}$) can be scaled to sizes of the simulator ($u_{\mathrm{sim}}$ and $i_{\mathrm{sim}}$) by

$$u_{\mathrm{sim}} = u_{\mathrm{real}} \cdot V_{\mathrm{u}} \tag{1}$$

$$i_{\mathrm{sim}} = i_{\mathrm{real}} \cdot V_{\mathrm{i}}. \tag{2}$$

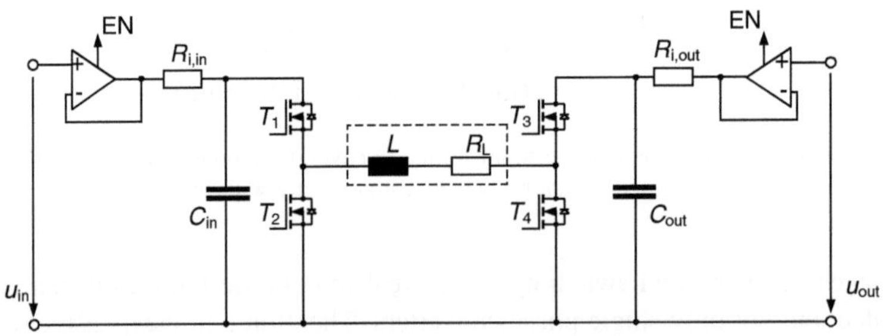


**Fig. 2** Fundamental schematic configuration of one phase of the simulator

These scaling factors $V_u$ and $V_i$ also allow the scaling of the other component values (inductance of the storage inductor, and capacity of the DC link capacitors and resistors).

The inductance of the real system is given by

$$u_{L,real} = L_{real} \cdot \frac{di_{L,real}}{dt}. \tag{3}$$

The inductance in the simulator is described by

$$u_{L,sim} = L_{sim} \cdot \frac{di_{L,sim}}{dt}. \tag{4}$$

Dividing Eq. (3) by Eq. (4) and using $V_u$ and $V_i$ leads to

$$L_{sim} = L_{real} \cdot \frac{V_u}{V_i}. \tag{5}$$

Similarly, capacities can be transformed using

$$u_{C,real} = \frac{1}{C_{real}} \int i_{C,real} dt \tag{6}$$

$$u_{C,sim} = \frac{1}{C_{sim}} \int i_{C,sim} dt \tag{7}$$

$$C_{sim} = C_{real} \cdot \frac{V_i}{V_u} \tag{8}$$

and resistances can be transformed by using

$$R_{sim} = R_{real} \cdot \frac{V_u}{V_i}. \tag{9}$$

By doing so, the relevant system dynamics are preserved.

The possibility of scaling the real power module is limited by the voltage drop over the silicon IGBTs and diodes, which reach between 1.2 and 1.8 V. With a typical scaling factor of $V_u = 0.01$, a switch with a voltage drop of 12 up to 18 mV is necessary to emulate this mathematically correctly scaled form in simulation by the HiL.

In this way, it is not realizable. However, the constant voltage drop is less than one percent in maximum of the voltage supplied in high voltage applications. By using MOSFETs—as quasi-ideal switches—the deviation can be neglected. However, if the converter uses diodes, such as in the unidirectional boost converter, the respective MOSFETs in the simulator have to be turned on in a similar way as those in a synchronous rectifier. The voltage drop of the body diode of the MOSFET of, for example, 700 mV would otherwise be equivalent to 70 V in reality with a $V_u$

of 0.01. This means that the MOSFET representing a diode has to be turned off when the current reaches zero if the converter is operated in discontinuous mode.

The factors which are used in this simulation are $V_u = 0.01$ and $V_i = 0.001$. Thus, voltages in the range of 0…500 V will be scaled down to 0…5 V. The usual values of the single phase current will be scaled from 0…100 A down to 0…100 mA.

## 2.2 *Modeling the Electric Nets on the Input and Output Side*

At first, it should be possible to simulate the voltage ripple caused by the converter and also the internal resistance of the fuel cell or battery by modeling the traction network. The voltage ripple can be simulated using an arbitrary function generator as a signal input source and a linear amplifier (Fig. 3). The amplifier has to possess sufficiently high dynamics and has to be stable with a high capacitive load as the simulated DC link capacitors of the output stage reach up to several microfarads. At the same time, the output current has to be sufficiently large, so that it is suitable for the simulation and provides a sufficient signal to noise ratio. Therefore, at least 100 mA seem to be a good choice.

Respective integrated linear amplifiers are available. As they tend to have a certain voltage drop at higher currents and a quite large offset voltage, they can be placed inside a feedback loop to provide a voltage source with a low output resistance (Fig. 3). The (purely resistive) internal resistance of the network is simply realized by a scaled resistor in series with the amplifier ($R_{Bat,in}$ in Fig. 3).

In order to simulate a load dump, there has to be an opportunity to disconnect the voltage source from the converter, for example, when the battery opens its contactors or the inverter of a drive shuts down immediately. This is achieved by using the two MOSFETs $T_1$ and $T_2$. Their $R_{DS,on}$ has to be subtracted from the simulated internal resistance of the battery. Furthermore, the switch $T_3$ in Fig. 3 allows the simulation of a short circuit on the traction network. The resistance of $R_{SC}$ has to be as small as possible to guarantee the survival of $T_3$ when the capacitors of the simulated power board are rapidly discharged.

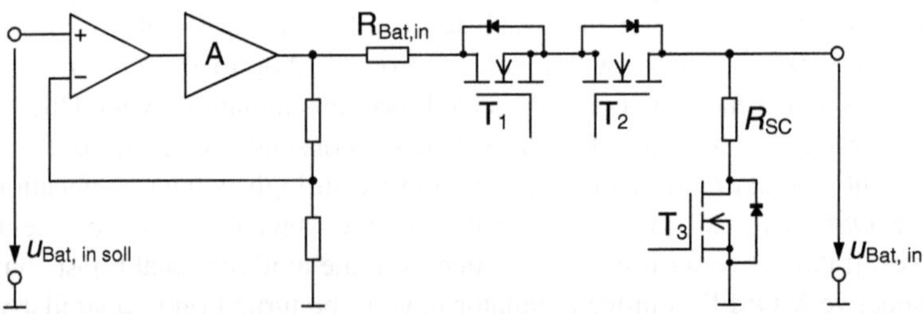


**Fig. 3** Emulation of the traction net, here the net for a simulated battery. The switch combination of $T_1$ and $T_2$ allows for the generation of load dumps, and $T_3$ for the switching of short circuits. The internal resistance of the battery is modeled by $R_{Bat,in}$

## 2.3 Single-Phase Simulator

Figure 4 shows the general layout of the power module of the HiL representation of the DC/DC converter. Different measurement points are provided depending on the actual implementation in the simulated power module.

Actually, the current measurement is done using shunts, i.e. $R_{m,in}$ and $R_{m,out}$ for the input and output current, respectively, and $R_m$ for measuring the inductor current. The frequency response of the measurement system of the real power board has to be implemented in the simulator especially if the real converter uses current sensors.

The inductance has to be scaled according to Eq. 5. It is also possible to reproduce the saturation behavior by an appropriate design of the magnetic core. Therefore, the saturation current of the simulator choke and the saturation current of the real DUT also have to be scaled by $V_i$. It is, furthermore, helpful if the same material is used. With it, the required values of the magnetic conductance $A_L$ and the minimum cross section $A_{min}$ of the toroidal core can be determined [7] using

$$B_{sat} = \frac{L_{sim} \cdot i_{sat,sim}}{N \cdot A_{min}}. \tag{10}$$

According to the Eq. (12), the appropriate number of turns is given by

$$N = \sqrt{\frac{L_{sim}}{A_L}} \tag{11}$$

and the $A_L$ value by

$$A_L = \mu \frac{A}{l}. \tag{12}$$

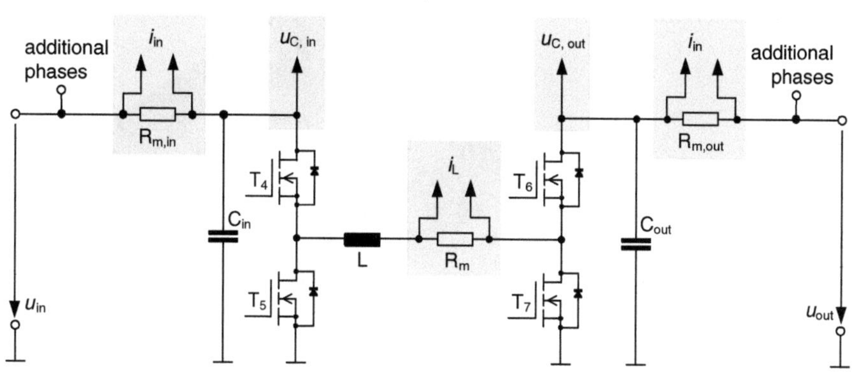


**Fig. 4** Emulation of the power module of the DC/DC converter

Figure 5 shows the simulator for one single phase. This board includes the model of the power module, the models for the sources and nets on the input and output side, and all relevant auxiliary components, such as the internal power supply, the power unit and the measurement technology required with the associated signal conditioning.

## 2.4 *Multiphase Setup*

Several of these single phase simulator boards can be individually controlled by the controller boards and connected in parallel in order to simulate a multiphase system. In this work, a three-phase system was chosen. Figure 6 shows these three coupled single phase modules. As shown in this picture, it is possible to stack multiple boards to a multiphase converter [8].

# 3 Simulation Examples

Different operating conditions are demonstrated in the following. In this case, the simulator is used to develop and test software for a DC/DC converter and its control before the real power module is ready for real world tests. This allows the development and test of software parallel to the development of the hardware. In this application, the DUT is a boost converter with bidirectional current flow, similar to that at the bottom of Fig. 1, which will be connected to a battery on the input side and the traction net with the battery and the electric traction drive on the output side of the converter. Critical error conditions are a load dump, when the drive has an emergency shutdown and the converter continues to drive current into the output side and a short circuit on the output side. The converter is operated in current control on its input side.

**Fig. 5** Simulator board for a single phase of a DC/DC converter

**Fig. 6** Combination of three single phase boards for a three-phase converter

## 3.1 *Normal Operation*

In normal operation, the individual controllers of the three phases of the converter get the same target values. The three phases are controlled individually by a single controller board and they are synchronized so that the switches can be operated in phase interleaf. That means that the phases are switching using a center-aligned pulse width modulation, which helps to reduce the ripple of the resulting input and output current of the converter significantly.

Figure 7 shows the simulated step response of the converter to a change of the desired value. As can be seen, the input voltage of the converter drops due to the load on the battery and gives a feedback to the current control.

The slightly different values of the inductor currents are caused by different measurement errors of the measurement circuits of the DUT.

## 3.2 *Load-Dump*

The simulator allows for a realistic test of the correct behavior in case of a load dump (Fig. 8). Here, it is possible to analyze if the overvoltage shutdown is working correctly and to analyze the behavior of the controller due to the dramatic change in the impedance of the traction net on the output side of the converter.

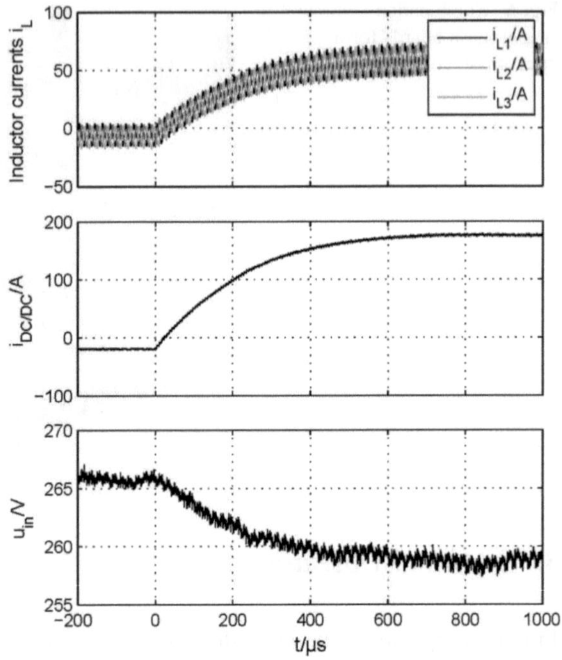


**Fig. 7** Normal operation of the three-phase converter with a simulated change of the desired value

## 3.3 *Short Circuit*

The short circuit on the output side shows how time critical the simulation is. The over current shutdown takes place only about 30 μs after initiating the short circuit (Fig. 9). The HiL simulation here allows the precise emulation of the behavior inside a real vehicle in order to analyze the behavior of the DUT without the risk of damaging it.

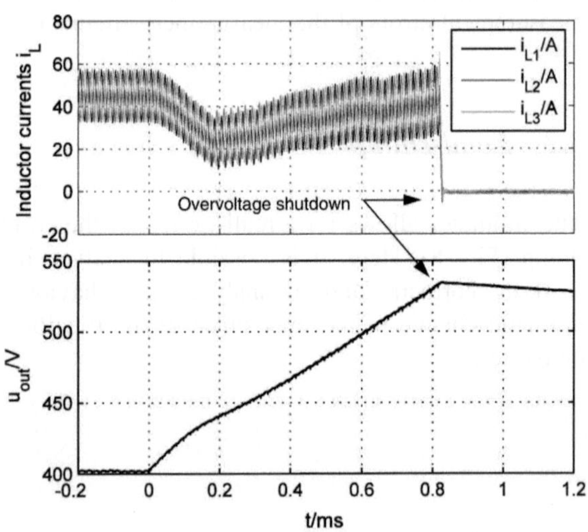


**Fig. 8** Simulated load dump for the three-phase simulator

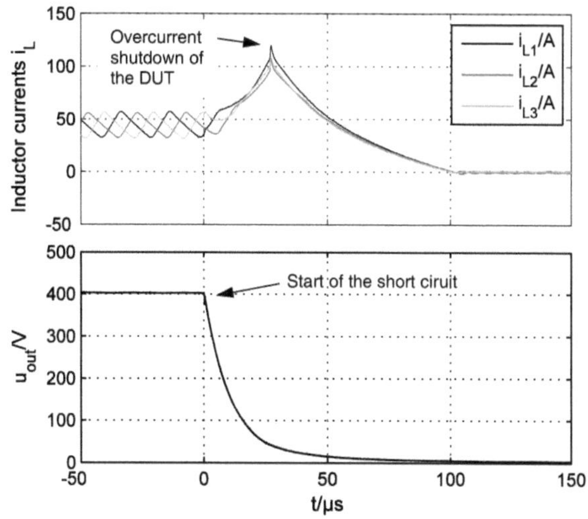


**Fig. 9** Currents of the three inductors and output voltage of the simulated DC/DC converter in case of a short circuit in the traction net connected to the output side of the converter

## 4 Conclusions

This paper presents a concept for a HiL simulator for a multiphase DC/DC converter designed as an analog computer. This simulator allows a precise simulation of the dynamic behavior of real existent DC/DC converters. As in any signal-based simulation, the real currents and voltages are represented by voltages of a secure and easy to handle level.

The simulator allows the support of the development and calibration of the software of DC/DC converters and to run functional and diagnostic tests with low effort. It is, therefore, a helpful tool for the development of galvanically coupled DC/DC converters.

## References

1. Bouscayrol, A.: Different types of hardware-in-the-loop simulation for electric drives. Proc. IEEE Int. Symp. Ind. Electron. 2146–2151 (2008)
2. Tendillo, L.A., Idiarte, E.V., Altés, J.M., Moncus, J.M., Blav, H.V.: Design and control of a bidirectional DC/DC converter for an electric vehicle. EPE/PEMC 2012, pp. LS4d.2–1–LS4d.2–5 (2012)
3. Young, C.M., Cheng, Y.S., Peng, B.R., Chi, S.H., Yang, Z.: Design and implementation of a high-efficiency bidirectional DC-DC converter. Future energy electronics conference (IFEEC), 2015 IEEE 2nd International, Taipei, ISBN 978-1-4799-7655-3, pp. 1–5, 1–4 Nov 2015
4. Schmidt, W.: DCDC converter for hybrid vehicle applications. Integrated power systems (CIPS), 2008 5th International Conference on, Nuremberg, ISBN:978-3-8007-3089-6, pp. 1–3, 11–13 March 2008

5. Kim, B.S., Kim, H.J., Jin, C., Huh, D.Y.: A digital controlled DC-DC converter for electric vehicle applications. 2011 international conference on electrical machines and systems (ICEMS), pp. 1–5
6. Herrera, L., Wang, J.: Fpga based detailed real-time simulation of power converters and electric machines for ev hil applications. Energy conversion congress and exposition (ECCE), 2013 IEEE, pp. 1759–1764
7. Zach, F.: Leistungselektronik—Ein Handbuch. Springer Verlag, ISBN: 978-3-658-04898-3, pp. 1847ff (2015)
8. Xu, H., Wen, X., Qiao, E., Guo, X., Kong, L.: High power interleaved boost converter in fuel cell hybrid electric vehicle. Proc. IEEE IEMDC 1814–1819 (2005)

# Integrating a SiL into a HiL Test Platform

**I. Matheis, T. Dörsam and W. Hoffmann**

**Abstract** At Daimler, a HiL for the high power net of an electrical vehicle was built. The rest-bus simulation was planned to be implemented as a script that simply pushed constant values to the bus. Later on, the script was to be extended by further logic to become more realistic. At the same time for another team of Daimler, a SiL was developed to carry out pretesting of code and data changes of some ECUs. The SiL contained three virtual ECUs that used the real in-house software, a plant model and of course some rest-bus simulation. By chance an engineer of the HiL team met and discussed about their current work. A synergy was detected: when the SiL would be put into HiL context, it would be a perfect rest-bus simulation since the original ECU code could be used to calculate all signals needed. Also the rest-bus simulation was coming almost for free because the SiL was ready to use. The virtual ECU tool Silver was then integrated into the HiL and further synergy effects were found. For the price of a powerful desktop PC, the plant model of the SiL, which was more detailed than needed, could be reused for the HiL. At last, the Silver could be extended to remotely control the hardware of the HiL. Thus, the SiL could even drive test scripts on the HiL. Integrating the SiL into the HiL improved its quality and sped up its build. Last but not least, the SiL helped to reduce costs because the three virtual ECUs replaced the three real ones for which special HiL hardware would have been required as well.

I. Matheis (✉)
QTronic GmbH, Kreuznacher Str. 62, 70372 Stuttgart, Germany
e-mail: ingo.matheis@qtronic.de

T. Dörsam · W. Hoffmann
Daimler AG, Hanns-Klemm-Straße 45, 71034 Böblingen, Germany
e-mail: thomas.doersam@daimler.com

W. Hoffmann
e-mail: werner.hoffmann@daimler.com



C. Gühmann et al. (eds.), *Simulation and Testing for Vehicle Technology*,
DOI 10.1007/978-3-319-32345-9_2

# 1 Planning Targets of the High Voltage Vehicle Power Net HiL

Due to high voltages and power, the high voltage power net of a hybrid or electric vehicle needs to meet special safety requirement. Apart from the usual functionalities like boost charging of the battery or running strong electric motors, safety has to be guaranteed. To examine how the components of the high voltage power net fulfill their specification, manifold tests with the target hardware must be carried out.

The HiL of the high voltage power net was used for:

- Testing prototypes during development of the hybrid drive,
- Verifying component tests,
- Administering specific standardized component tests,
- Analyzing sensitivity: what affects the system reaction the most,
- Simulating incidents measured in the vehicle and
- Developing system tests for the high voltage power net of the vehicle.

The design of the HiL was such that it was as close as possible to the vehicle layout (Fig. 1).

The HiL was modularized so that single components as well as subsets of the setup above could be tested. E.g. the high voltage battery could be replaced by a programmable power supply unit.

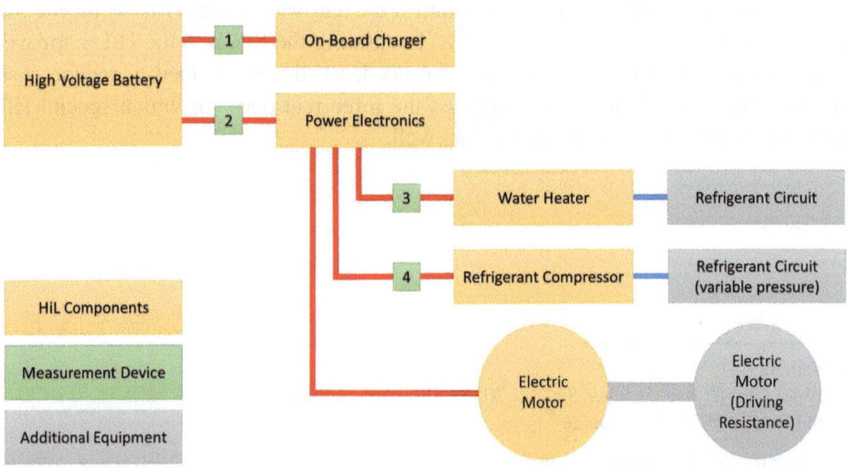


**Fig. 1** The HiL setup

## 2 Powertrain-SiLs at Daimler

Since many years Daimler relies on virtual development solutions by QTronic [1–4]. Development and application engineers are able to test their changes within minutes on a standard Windows PC. This fast development cycle improves the quality of the code and the data before it is passed on to the build process of the real ECU. Thus, the quality of the product improves and costs can be reduced. At Daimler there are also highly accurate plant models available. Putting this together results in closed-loop simulations that are well accepted by the engineers.

The SiLs used in this project was a hybrid drive containing three virtual ECUs: the engine control unit (ECU), the transmission control unit (TCU) and the central powertrain coordinator (CPC). The virtual ECUs were built out of original C-code and original data. Any change could be applied and tested in less than 5 min.

The components of the SiL were integrated using Silver by QTronic. Because Silver provides an emulation layer for XCP and CAN, the engineers could reuse their application and measurement software (e.g. CANape or INCA) which they normally use for the car or the HiL. In particular, Silver was also able to control real hardware [5] running in a real-time mode, which turned out to be useful for integrating the SiL into the HiL. The SiLs were updated on a daily basis.

## 3 Reutilization of the Existing SiL in the HiL

Like any HiL test platform, the HiL for the high voltage power net needed a plant model and a rest-bus simulation. The initial plan for the rest-bus was to send constant values to the CAN bus. Later on, this was planned to grow in complexity in order to get closer to the performance of the real vehicle. The plant model of the HiL was planned to be implemented in Simulink starting with a simple model and enhancing the quality with every iteration.

Because the SiL was validated already and used by many engineers, it could provide the features of the rest-bus simulation. Since the SiL contained the real code and real data of the vehicle, the rest-bus was very realistic from the start. The amount of work to connect the SiL with a CAN bus of the HiL was a matter of less than one day's work. Furthermore it was not necessary to implement a plant model because the SiL already contained one.

At the beginning of the project the Modelica based plant model was too slow in some situations. This problem was solved by running the SiL on a high-end computer with a water cooling system. Silver then was set to slow down the SiL so that it followed real time.

However, further enhancement opportunities were detected. Using python scripts in Silver and the package PyVISA, Silver could control the HiL hardware. In order to connect to the HiL hardware connections such as RS232, GPIB (via GPIB bus or Ethernet), FTP, CAN (CANopen and Automotive CAN) and USB-IO were implemented.

One challenge was that the HiL hardware did not always accurately respond to the python programs. There were delays and sometimes the hardware simply failed to reach the target state. The problem of the delays was solved by moving any control code to python threads that ran in the background while Silver continued to act as the rest-vehicle. Retry-counters covered the case when the hardware returned with some error code. In the end, the python test scripts running in Silver became so robust that they could run several hours without a human in the loop.

Silver became one of the central components of the HiL.

## 4 Conclusion

The integration of the SiL into the HiL influenced the progress of the project in three ways:

1. Fast set-up of the HiL: using a high-end computer the SiL could run in real time. Integrating it into the HiL was done in less than a day. The SiLs were updated on a daily basis and the SiL covered the needs for a rest-bus and a plant model. Furthermore, at the time the HiL was setup, there were no real ECUs available. Integrating the virtual ECUs was the only way to start the project this early.
2. Improved quality of the HiL: because the rest-bus was based on three virtual ECUs made of original code and data, the rest-bus was nearly perfect. Furthermore, the well tested and highly accurate plant model was much better than needed.
3. Reducing the costs of the HiL: integrating three ECU in the HiL would have needed several HiL racks. Instead, the SiL needed a single high-end computer with a water cooling system. Furthermore the human resources for maintaining the HiL, programming the rest-bus and the plant model could be saved.

## References

1. Gloss, et al.: Systematic validation of over 200 transmission variants. *ATZelektronik,* 04/2013
2. Tatar, M., Schaich, R., Breitinger, T.: Automated test of the AMG Speedshift DCT control software. In: Proceedings 9th International CTI Symposium Innovative Automotive Transmissions (2010)
3. Brückmann, H., Strenkert, J., Keller, U., Wiesner, B., Junghanns, A.: Model-based development of a dual-clutch transmission using rapid prototyping and SiL. In: International VDI Congress Transmissions in Vehicles 2009, Friedrichshafen, Germany, 30.06.-01-07.2009. http://qtronic.de/doc/DCT_2009.pdf

4. Rink, A., Chrisofakis, E., Tatar, M.: Automating test of control software—method for automatic test generation. ATZ elektronik 6/2009, Dec 2009. http://qtronic.de/doc/ATZe_2009_en.pdf
5. Liebezeit, T., Bazarsuren, U., Beilharz, J., Mees, A., Serway, R.: Rapid control prototyping for frontloaded. In: Vehicle testing of transmission control system algorithms by Laptop, 5. Tagung Simulation und Test für die Automobilelektronik, 15.-16. Mai 2014, Berlin

# RPCsim: Model-Based Analysis Within the Calibration-Process

**Steve Laux, Sven Freitag and Frank Geschner**

**Abstract** RPCsim (*R*apid *P*rototyping *C*alibration: *sim*ulation) provides an easy to use and fast interface between Simulink®, parameter-files and measurement-data. By creating individual calculation-routines in the graphical way provided by Simulink, the user may perform meaningful calibration- and analysis-tasks, which are readable and easily adaptable by others. With the built-in functionalities, it is possible to simulate previously recorded measurement-data with different adjustments and instantly compare the corresponding results.

**Keywords** Simulink · Simulation · Automotive · Calibration · Data-analysis

## 1 Introduction

For several decades now, electronic components have become indispensable in nearly every branch of technology and engineering. This trend of course also includes the automotive industry. In particular the digital engine-control-unit (ECU) is becoming increasingly important. It allows, among others, an improvement in fuel consumption, reducing the impact on the environment and providing a consistent engine performance under all conditions. The increasing demands regarding fuel-economy and the stricter emission-legislation are consequently leading to innovative engine components and complex control systems. This means that the automobile manufacturers are increasingly faced with new challenges, while on the other hand development-time and -budgets are constantly being reduced, mostly due to competitive reasons.

S. Laux (✉) · S. Freitag · F. Geschner
IAV GmbH, Chemnitz, Germany
e-mail: steve.laux@iav.de

S. Freitag
e-mail: sven.freitag@iav.de

F. Geschner
e-mail: frank.geschner@iav.de

 
C. Gühmann et al. (eds.), *Simulation and Testing for Vehicle Technology*,
DOI 10.1007/978-3-319-32345-9_3

## 1.1 Motivation

The situation described above inevitably leads to the fact, that the periods for the execution of calibration-tasks by conventional means are now no longer acceptable. Engineers are confronted with the necessity to adjust a highly complex system with several thousand parameters, often by means of trial and error and experience. It also requires a large amount of time-consuming and repeating data evaluations (see Fig. 1). The sheer number and complexity of given tasks for a single engineer are not manageable anymore. The increase in efficiency is an important element to ensure an appropriate handling of project-topics in the future. The main objective would be to save time while maintaining quality, which leads to the question how to achieve all substantial requirements?

One way could be the implementation of reliable and highly automated methods. Relieve the employee from inconvenient analysis-tasks, which would be better accomplished by a machine altogether (see Fig. 2). The development of appropriate tools for series-applications is nowadays not only helpful but essential. But what would calibration-engineers ask for?

## 1.2 Tool Requirements

The declared goal during the development-process was the creation of a powerful offline simulation-environment for the analysis and evaluation of a huge amount of measurement data. The items in the following list have therefore been defined as fundamental requirements:

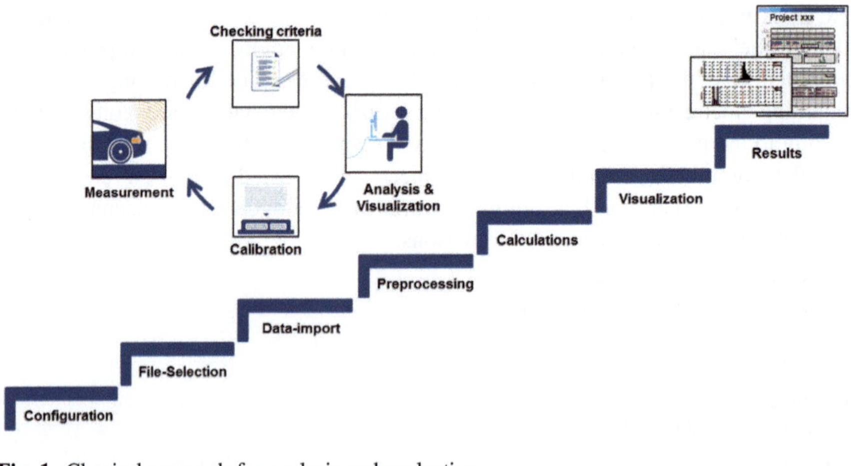


**Fig. 1** Classical approach for analysis and evaluation

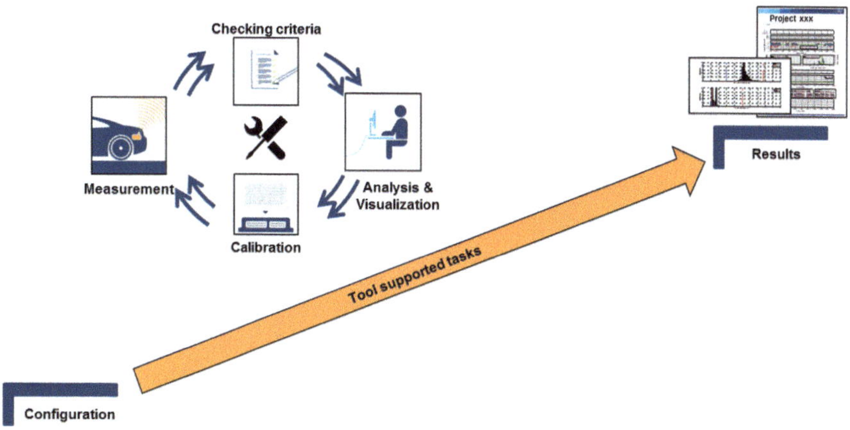


**Fig. 2** Faster procedure using tool-support

- Implementation of any given calculation-formula (as model)
- Possibility to edit models at any time directly by the user
- Direct linking of measurement-data and simulation-models
- Processing of any number of measurement-files simultaneously
- Adjustability of model-parameters (manually or automated)
- All of this as transparent and user-friendly as possible

Especially for the first two items, it means that we had to pay closer attention to the way the user can enter his formulas into the tool. This led to some basic design decisions during the development-phase and had a major influence on the final layout of the simulation-interface.

## 2 Simulation-Environment RPCsim

This section will give a brief outline of the model-based simulation-tool RPCsim; *R*apid *P*rototyping *C*alibration and *Sim*ulation. Since the beginning of development back in 2005, the appearance and handling of the tool underwent some major changes. Starting with an interface in the form of a text-based editor, the basic idea behind it always remained the same. The user should have the possibility to create his individual calculation-routines, although the actual simulation-process, including different parameters and measurement-data, is automatically performed in the background. Countless development-iterations later and in close networking with the calibration-engineers, the graphical user-interface illustrated in Fig. 3 has been accomplished.

The interface is divided into four major sections. This includes the model-library on the left, the parameter-block in the middle and the management of measurement-data on the right side. Those parts are completed by the

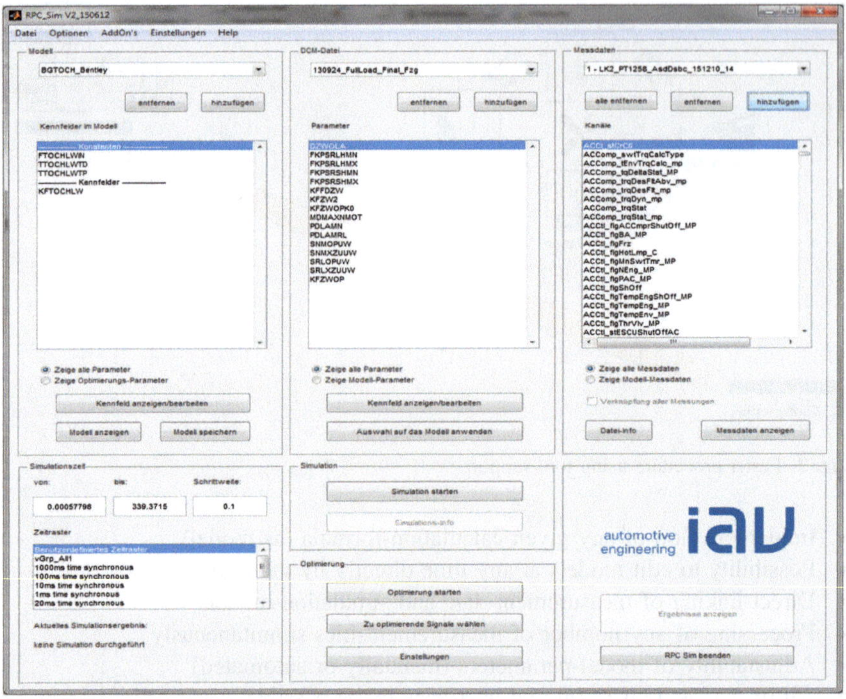


**Fig. 3** Graphical user-interface for RPCsim

simulation-section at the bottom. Going through the appropriate configuration-steps and selection-dialogs in the right order, the user is easily guided through the whole simulation-process. He only needs a few clicks to achieve his personal analysis-report, of course providing the existence of the needed calculation-routines.

In the beginning the evaluation had to be written in the form of scripted files. However this approach did not last for long, because not every automotive-engineer has sufficient programming-skills and not every programmer has the deep understandings for the context in concern. After the implementation of a text-based formula-editor, which was no more successful than the previous attempt, we finally agreed on the use of Simulink made by The MathWorks® [1]. This crucial decision was the solution of the previously mentioned contradiction between programmer and user. And beyond that it comes with more suitable advantages:

- Using an established computing-software for engineers and scientists
- Availability in a network-based and affordable licensing-model
- Large scope of implemented tool-boxes for specific tasks (e.g. signal-processing)
- Easy-to-use graphical design-language with a high recognition-value
- Small limitations on construction and layout of various applications
- Simple versioning and portability of models

Through the graphical specification of calculation-algorithms, the expert for the relevant working-package is directly involved into the development-process. The programmer on the other side can focus his efforts on core-functionalities like data-handling and visualization. In essence the analysis of waveforms, specific data-conversion or any arbitrary computation of signals remains in the hands of the user. He is supported by standardized routines for file-exchange, data-manipulation, pre-selection and pre-processing as well as the fast presentation of results. In conclusion we achieved the successful development of an efficient simulation-environment, which easily can find its way into the everyday work of any calibration-engineer.

# 3 Model-Based Calibration and Analysis

Model-based simulations, as part of our offline calibration-process, can be divided into two major categories: *technical optimization* and *model-based analysis.*

## *3.1 Technical Optimization*

The first application is used for an automated adaptation of certain model-properties. In this case it means the modification of the parameters of the calculation-formulas, in order to achieve the desired system behavior. In most cases the model is used in a loop, repeatedly called by a suitable kind of numerical optimization-algorithm to perform multiple simulations with modified parameters (see Fig. 4). The input-vector would be identical for each single simulation, and the respective output-vector will then be compared to given target-values. The resulting deviation, e.g. the mean-squared-error [2], between target and simulation-result is then used to estimate the quality of the current parameter-set. The model which produces the least error will then represent the desired result of this optimization.

Use-cases for this procedure are the determination of parameters for neural-networks, an initial calibration for ECU-functions or the matching of variables for physical based engine-models for the subsequent use in further simulations (e.g. the THEMOS-library). These special cases are listed as examples and are not described here in more detail. At this point it should be noted, that the simulation-environment RPCsim is also designed and suitable for such applications.

## *3.2 Model-Based Analysis*

The second part which is explained a bit further in detail is the model-based analysis of very large amounts of data. Especially with regards to the achievement of a faster

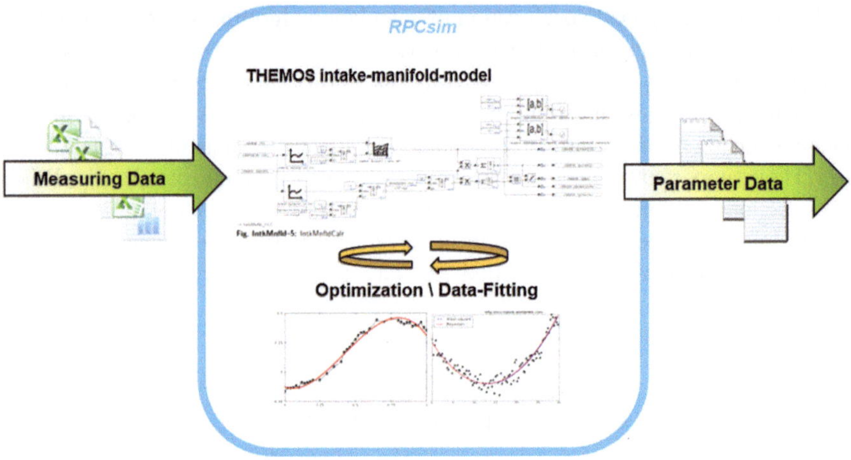


**Fig. 4** Model-based "in-the-loop" optimization

calibration process, this is an important feature of our standardized simulation-environment. It enables the user to design individual analysis-algorithms, in an easy-to-read graphical way. Those models can then be integrated into a common model-library, thus made available for other employees facing similar problems (see Fig. 5).

All the previously mentioned time-consuming operations during the analysis are performed as an automated part of the tool-chain. This includes import routines with different file-formats, any particularly needed steps of pre-processing, the

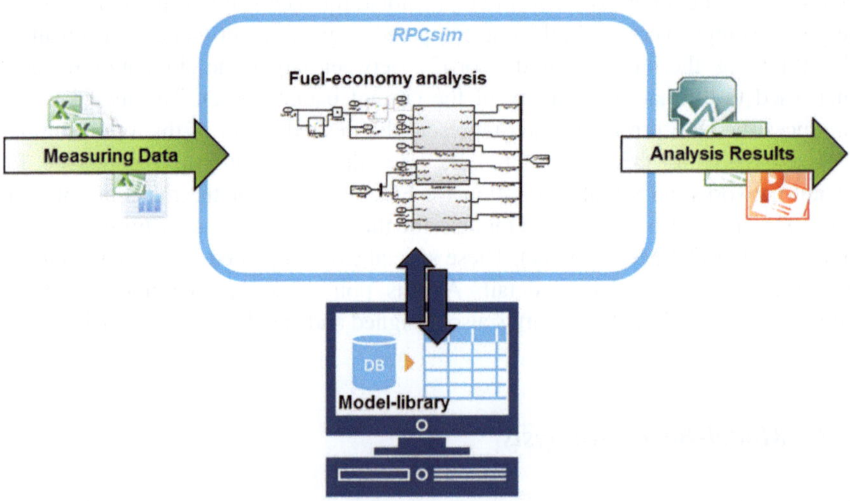


**Fig. 5** Model-based analysis principle

simulation itself and the final graphical preparation of the result and its visualization. This supports the engineer with the creation of standardized output-formats and reports, rapidly speeding up the whole procedure and delivering the opportunity to spend his time with more important calibration-tasks. Some examples for this application are the evaluation of drivability-related characteristics, the classification of potentials in fuel-economy and the interpretation of measured outcomes of durability-trials. The last two points are described in detail in the following section.

# 4 Examples for Model-Based Analysis

The two most popular use-cases to be discussed more detailed, are the *evaluation of fuel-economy* and the *analysis of durability-runs*. Both applications were originally designed by single engineers and are now an integral component of the simulation-library. Besides their own simulation-package, they share the necessity of some pre-processing operations as well as some special output-formats.

## *4.1 Evaluation of Potentials in Fuel-Economy*

This method basically consists of the virtual transfer of different driving-cycles, including specific test-patterns like the NEDC, onto the testbed. This does not mean to actually drive the test-cycle on the testbed, but to repeat the measurement by simulation as if it was under optimal conditions (see Fig. 6). Using this procedure, we can identify the driving-states where the actual cycle differs from the expected optimum and, within certain limits, also find the reason for it. Most of the time this would reveal a loss in fuel-efficiency caused by effects normally not investigated on the testbed, like interventions related to driveability, emission-optimization or diagnoses.

As a requirement to obtain meaningful results, the influences caused by friction or gas-exchange-losses have to be eliminated. Therefore the fuel-consumption has to be simulated in terms of engine-speed and indicated mean-effective cylinder-pressure (IMEP). This means that the vehicle measurements have to be performed using cylinder-pressure-indication, mostly by means of measuring spark-plugs, in order to be directly comparable to the testbed-results. The acquired data first has to be sorted by cylinder and firing-order. Then it has to be synchronized with the parallel recording of the ECU-signals. This gives the ability to draw the right conclusions from the various driving-maneuvers for each single combustion event. The pre-processing is performed before the simulation as part of the data-import routines, exclusively connected to the fuel-economy-model. Apart from a few necessary configuration steps, the engineer himself is not further interfered with the whole process of data-management.

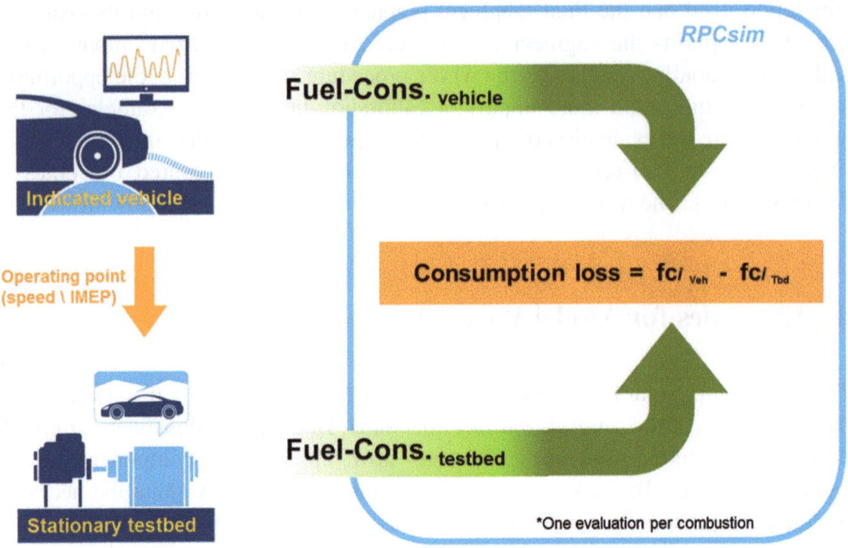


**Fig. 6** Principle of fuel-consumption evaluation

The simulation-model itself is divided into several blocks, each of it calculating different aspects of influences on fuel-economy. This separates the overall consumption-deviation into individual parts related to ignition-timing, lambda-deviation, camshaft-positioning and other impacts. By the implementation of additionally measured correlations and efficiency-curves in the form of certain lookup-tables, the final magnitude of potentials in fuel-consumption can be estimated. The results are reconditioned and exported as standardized plots, showing for example the cumulated consumption-loss over the entire test-cycle or the actual distribution over different operational points. This helps the engineer to identify several possibilities for refinement in only one single simulation and often leads to an improved calibration. Figure 7 shows an overview of this whole process.

The given example shows the application of the simulation-environment with the analysis of single measurement-files in connection to related testbed-data. The following topic shows the simultaneous simulation of huge amounts of data, which was recorded in different files but will be merged together into one report.

## *4.2 OBD-Analysis of Durability-Trials*

One special application is the analysis and evaluation of durability-measurements with respect to the mandatory monitoring in accordance with legal emission-requirements (e.g. OBDII). The actual analysis of diverse driving-conditions is again realized with the help of certain Simulink-models. These models are created by the engineer responsible for the calibration of the diagnosis-functionality or can be

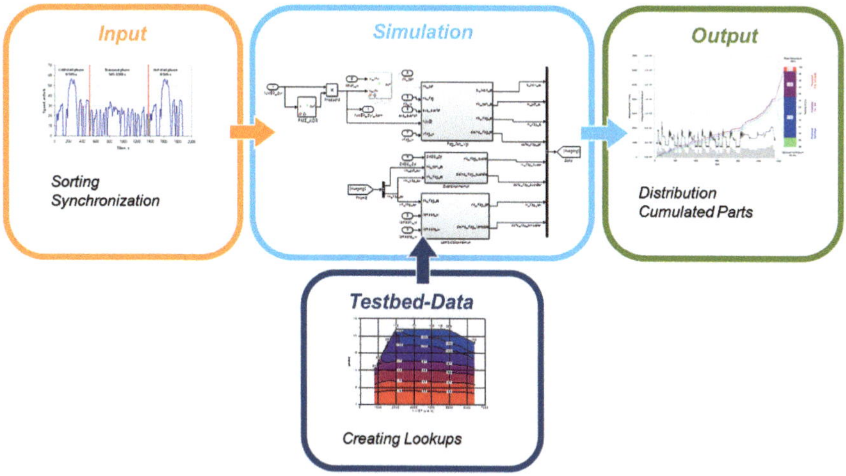


**Fig. 7** Simulation of driving-cycles under optimal conditions

adopted from previous projects with similar prerequisites. This is easily possible because the applied principles, founded by comparable legal requirements, are often identical. Therefore all the previously built models are collected in one single library, where all affected users have access to. In most cases the whole process is coordinated by an OBD-calibration-pilot. He assembles all the necessary routines into one common model, where only the interface and some parameters have to be adapted to the current project. The principle is illustrated in Fig. 8.

The routines that are used also include some unique data-processing algorithms, which are part of the whole simulation-environment. The first part is a script-based pre-selection of suitable measurements. The ability to search and parse a large number of data-files for specific events is an integral part in the evaluation of durability-trials or the numerous recordings of vehicle test-trips. In this way, it is possible to narrow down the effort on simulation only for those files actually containing relevant cases, with expected reactions of a diagnostic-function. On the other hand the user can specify certain driving-states or operating-points which he wants to examine in more detail. This includes defined targets for vehicle-speed and engine-load or various ambient-conditions, such as altitude and temperature thresholds. Different conditions can be linked directly to one another and the search can be limited to a pre-determined time-period, like for example a very specific winter-test. By reducing the huge amount of available measurements to a reasonable number, this automated file-selection helps to save time and allows for a rapid robustness-analysis of diagnosis-calibrations.

Another important capability, which is implemented through special Simulink-Blocks, provides the continuous possibility to save the entire output-vector as a single "shot" at designated simulation-steps. These steps can be triggered by certain events, such as the successful completion of a diagnosis-cycle, even at several times during the same simulation. This enables the user to collect the

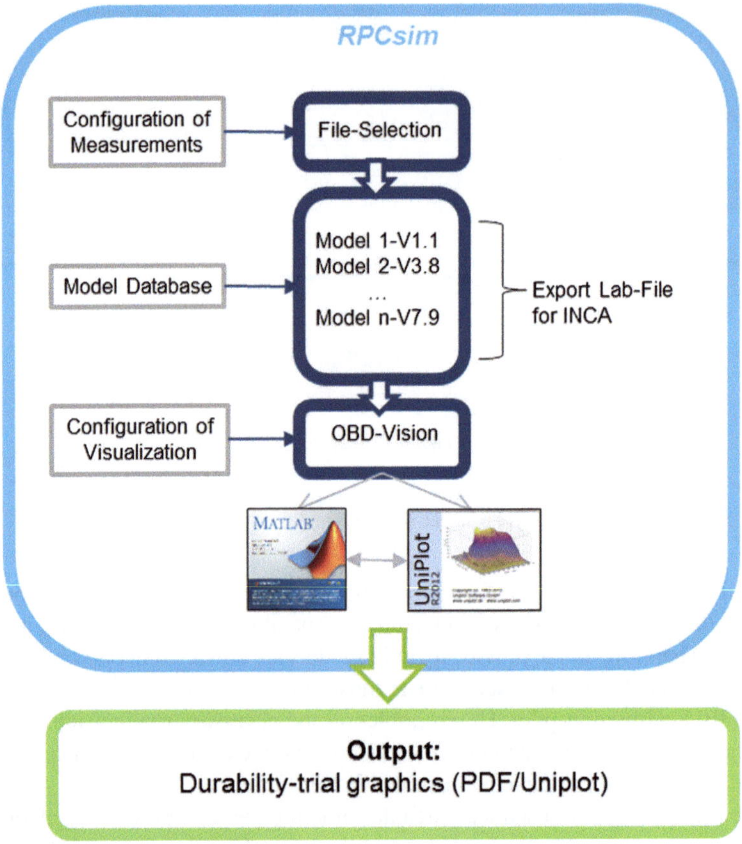


**Fig. 8** Sequence of durability-trial evaluation

analysis results as an extraction at multiple time-sections, which then could be classified and merged together into one robustness-report. This eliminates the need to search through the measurement-files to identify such events and document the results manually.

Figure 9 shows an excerpt of a standardized analysis-report of the canister-purge-diagnosis. Some critical parts of the ECU-functionalities were emulated during the simulation. These parts are expanded by the calculation of certain missing signals and specific signal-ratios to allow an extensive evaluation. These ratios correspond to the actual distance of the varying diagnosis-results to the appropriate fault-thresholds. A system with the default hardware, i.e. without an introduced malfunction, should return a ratio of around one. In the same way particularly small or high ratios indicate an existing fault or a potential for improvement. As described above, these ratios are collected from all the relevant measurement-files and are classified to show the overall percentage distribution. The bottom of the illustration shows, that the majority of the results demonstrate a

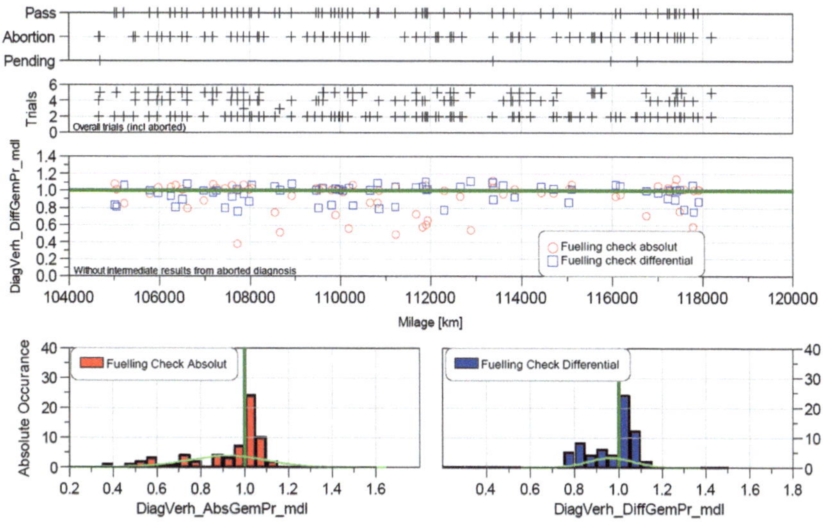


**Fig. 9** Example-report for the diagnosis of tank-ventilation

significantly robust calibration in this example. Additionally the analysis provides an overview of the total number of diagnosis-sequences and furthermore the successful and aborted attempts. The top of Fig. 9 shows corresponding results plotted versus mileage.

## 5 Summary

The request of IAV's calibration-engineers for an offline simulation and analysis of certain measurement-data, led to the development of an extensive and standardized simulation-environment. The detailed calculation-routines can be created directly by the users, without the need of profound programming-skills. By using Simulink the models can be easily build and implemented into the simulation-process and are readable by others. The biggest advantages are the possibility for a quick self-designed evaluation, the use of standard formats for import and export, and a common model-library where everybody can participate and simply adapt the existing routines to their own requirements. The ability to use characteristic curves and maps as parameters, which are directly linked to the graphical-user-interface, leads to a significant improvement of calibration- and analysis-tasks. Using the built-in functions for the manipulation of parameter-content, gives the possibility to simulate previously recorded measurement-data with different parameter-sets and instantly compare the results with respect to the applied adjustments. Besides the direct automated calibration of model-parameters, the main focus is currently on the analysis of big data, often recorded in a large number of measurements-files. The given examples for that process are the evaluation of fuel-economy compared to the expected optimum and the analysis of durability-trials to assess the robustness of OBD-related functionalities. This demonstrates the diverse application possibilities and the adaptability of the simulation-environment RPCsim.

## References

1. The MathWorks, Inc.: http://www.mathworks.com (2016)
2. Walz, G.: Mittlere Quadratische Abweichung. In: Lexikon der Mathematik. 1 Auflage. Spektrum Akademischer Verlag, Mannheim/Heidelberg (2000)

# Continuous Development of Highly Automated Driving Functions with Vehicle-in-the-Loop Using the Example of Euro NCAP Scenarios

**Raphael Pfeffer and Tobias Leichsenring**

**Abstract** Highly automated driving functions of the type currently being targeted for the near future are based on the interaction of several assistance systems. The actions of these assistance functions rely on an analysis of data from the virtual environment. Consequently, efficient development of individual advanced driver assistance systems (ADAS) that operate safely provides the basis for highly automated driving. The development and testing of such systems poses an enormous challenge to automobile manufacturers and automotive suppliers as the system complexity in the whole vehicle keeps growing. This includes the highly developed technical functions and increased interaction of the individual systems, as well as the growing complexity of the test scenarios in the field of advanced driver assistance systems. In the development process, this feat can only be mastered with assistance by virtual test driving. These factors are additionally intensified by external boundary conditions and regulations such as those specified and issued by legislators or testing organizations (for the European area, this is the European New Car Assessment Program—Euro NCAP). The seamless utilization of the CarMaker open integration and test platform in the field of advanced driver assistance systems provides a solution to meet the challenge of consistently growing testing requirements. This paper is intended to show how ADAS can be reproducibly tested and validated across the entire development process (MIL, SIL, HIL, and VIL). The utilization of the MIL method in particular makes it possible to easily test challenging maneuvers such as collision scenarios toward the end of the development cycle by embedding a real-word vehicle in a virtual environment. The proposal below will present the development work made possible by utilizing the VIL method as illustrated by the specific use case of defined Euro NCAP test scenarios.

R. Pfeffer (✉)
Dipl.-Wi.-Ing./Product Manager Test Systems, IPG Automotive GmbH,
Bannwaldallee 60, 76185 Karlsruhe, Germany
e-mail: raphael.pfeffer@ipg.de

T. Leichsenring
B.Eng./Test Systems & Engineering, IPG Automotive GmbH,
Bannwaldallee 60, 76185 Karlsruhe, Germany
e-mail: tobias.leichsenring@ipg.de



C. Gühmann et al. (eds.), *Simulation and Testing for Vehicle Technology*,
DOI 10.1007/978-3-319-32345-9_4

# 1 Motivation

Advanced driver assistance systems provide the basis for automated driving/ autonomous vehicles on various levels of automation [1]. They are designed to ease the driver's burden (for instance pulling into tight parking spaces) or, in the case of highly automated driving/autonomous vehicles, perform the driver's tasks (both longitudinal and lateral control) completely. Therefore, the development of advanced driver assistance systems plays a central part in vehicle development. The objective is to achieve faultless functions which, due to high system maturity, execute their tasks within the total system at any time and without any problems. At the same time, development and testing have to take validation requirements into account that have heavily increased in recent years. This growing complexity can be explained by the following factors, among other things:

- Greater diversity of vehicle variants.
- Shorter development cycles.
- Increasing interlinking of systems (interlinking of sensors, electronic control units, operator controls and actuators).
- Road environment data from several sensors providing the basis for information used by a single system (sensor data fusion).
- Greater variety of scenarios including complex environment factors.

When taking all of these factors into account, it soon becomes clear that an early real-world test of this scope is no longer practicable or economically feasible [2]. The scenarios are too complex to be set up over and over in the real world—here various approaches exist as well, such as using dummies on guide rails. While these variants can cover at least partial aspects of testing environment-based assistance functions, the number of scenarios to be tested, as well as their reproducibility, is limited. For comprehensive and particularly for consistent tests, virtual test driving provides a viable method. In addition to established techniques used in simulation, new test methods have to be employed as well. Therefore, IPG Automotive has extended the existing development process from the model- to software- and through to the hardware-in-the-loop stages by Vehicle-in-the-Loop tests.

# 2 Seamless Development in Virtual Test Driving

Virtual test driving can be used to test all vehicle components (engine, steering, ESC, ACC, etc.). In this context, virtual test driving should be understood as a means of modeling real-world tests, containing the same elements as a real-world road test (vehicle, driver, road, other road users, etc.). This approach can be used at any stage of the development cycle. Whereas in MIL and SIL all the elements are exclusively simulated, the HIL stage allows hardware prototypes (e.g. individual electronic control units) to be tested on a test rig. In this case, real-world electronic

| | MIL | SIL | ECU HIL | System HIL | PT HIL/ Stat. VIL | Dynamic VIL | Real World |
|---|---|---|---|---|---|---|---|
| ECU-Code / Function | V | R | R | R | R | R | R |
| ECU | V | V | R | R | R | R | R |
| System (e.g. Steering gear) | V | V | V | R | R | R | R |
| Vehicle | V | V | V | V | R | R | R |
| Road and static environment | V | V | V | V | V R | R | R |
| Vehicle dynamics | V | V | V | V | V | R | R |
| Driver | V | V | V | V | V | V R | R |
| System experience | V | V | V | V | V | R | R |
| Surrounding traffic | V | V | V | V | V | V | R |

V Virtual world R Real world

**Fig. 1** Closing the integration gap between HIL and real world tests

control units (ECUs) are integrated into a virtual vehicle. The ECU cannot tell that it has not been installed in a real-world vehicle, as it is being supplied with realistic signals via the HIL simulator, just like it would in a physical vehicle. The VIL method has been developed for testing advanced driver assistance systems particularly for the purpose of making it possible to combine real-world road tests and virtual test driving, which is achieved by fusing the real-world vehicle and the virtual environment (see Fig. 1 for the evaluation of the different integration levels), as shown in Chap. 3.

This approach allows vehicles to be repeatedly tested in an identical setting at varying depths of modeling according to their development stage. Initially, a consistent test description and definition of evaluation criteria can be established and, in the course of the development process, continually tested on the models as the depth of their development progresses. This allows early changes to be made according to the results obtained, and the process can easily be complemented by additional test scenarios as well. The utilization of virtual test driving offers the following advantages:

- Early development even without hardware prototypes (frontloading).
- Cost benefits as a result of hardware and manpower savings.
- No risk of personal injury or hardware damage.
- Re-usability of the test cases.
- Test automation = increased efficiency.
- Time savings due to availability of the test cases in multiple real time.

In summary, it is thus possible to check and compare the conclusiveness of the tests and test results at any stage of the development process. However, in recent years, highly specific requirements have been emerging in conjunction with component development, which have to be considered in the total vehicle context at an early development stage. To be observed in this context are regulations issued by legislators or organizations (for further details regarding the various requirement levels and corresponding test protocols, see Chap. 5). The new, most recently created testing methods are heavily focused on accident-avoiding assistance systems. Testing them requires interactions with third parties that may be involved in an accident. Therefore, making full vehicle tests experienceable in early stages, while excluding risks to pedestrians and hardware, is a necessary objective to be pursued in the development process. The Vehicle-in-the-Loop method is intended to close this gap and to meet the corresponding demands made on early, yet conclusive tests.

## 3 Vehicle-in-the-Loop Method

When using the Vehicle-in-the-Loop (VIL) method, a full real-world vehicle is embedded in a total vehicle simulation environment such as CarMaker. The objective is to combine elements of the simulation, such as the environment, with the real-world vehicle. Essentially, from a technical perspective, there are two aspects to be observed in this context as discussed below:

- Transitioning the position, orientation and motion of the real-world vehicle from the real world into simulation.
- Transferring the environment scenarios calculated in the simulation (such as distances from other road users or traffic objects) to suitable sensor interfaces in the real-world vehicle.

High-end, conventional measuring technology, which, for example, is used in real-world vehicle (dynamics) testing as well, lends itself to determining the vehicle's kinematics. Modern inertial navigation systems (INS) utilized in this context combine inertial sensors, such as acceleration sensors and gyroscope sensors, for a relative determination of the position and position changes with the absolute position data from GPS and DGPS, allowing accuracies of up to 2 cm to be achieved. The update rate of the sensors may reach a level of up to 1000 Hz, depending on the system used. To transfer the raw data from the INS into simulation, the data is filtered and (if necessary) interpolated in order to achieve the cycle time of the simulation (1 ms in the case of CarMaker). The following interpolation is applied for this purpose:

$$\vec{v}_{Ext}(t) = \vec{v}_{Ext}(t-1) + \left( \vec{a}_{Meas} + \frac{\vec{ds} * k}{dt^2} \right) * dt \tag{1}$$

$$\vec{s}_{Ext}(t) = \vec{s}_{Ext}(t-1)\left(\vec{v}_{Ext} + \frac{\overrightarrow{ds} * k}{dt}\right) * dt \tag{2}$$

$$\overrightarrow{ds} = \vec{s}_{Meas} - \vec{s}_{Ext} \tag{3}$$

With the measured accelerations $\vec{a}_{Meas}$ continuous signal curves can be calculated due to the extrapolation. Position variances $\overrightarrow{ds}$ are minimized by means of the filter constants $k$. To improve the result even further, other vehicle data, such as wheel speed or yaw rate, may be read from the vehicle's bus system and fed in as well.

In the simulation itself, the actual vehicle model is deactivated and replaced with measured data, while the simulation core runs on a real-time computer in the vehicle (e.g. RoadBox as a hardware component from IPG Automotive). The data is imported via a CAN interface. Now, vehicle dynamics in the simulation correspond to the real-world (measured) data from the real-world vehicle.

In order to be able to transfer environment scenarios (such as traffic objects or pedestrians for AEB testing) to the real-world vehicle, they have to initially be defined in the simulation platform and the actual tests determined. The following criteria, illustrated by the example of traffic objects here, should be met by the simulation environment:

- Flexible definition of objects in the virtual world (absolute or relative to the ego vehicle).
- Various types of objects (diverse vehicles, pedestrian models, etc.).
- Flexible object characteristics such as surface contours.
- Definable and reproducible behavior of the objects which, ideally, is oriented to definitions of maneuver-based tests.
- Triggering of actions taken by the objects according to optional criteria.
- Dynamically adjustable behavior of the objects in relation to the movement of the ego vehicle.

In the next stage, a closer look has to be taken at the respective sensor interface. There are various possibilities to emulate the sensor data, depending on the sensor technology used:

**Physical** Some of the environment sensors or environment sensor data can be physically reproduced or emulated. For instance, IPG Automotive's VIL demonstrator presented back in 2013 shows what a solution for ultrasonic sensors, used for automatic parking assist for example, could look like [3]. In this case, matching ultrasonic emitters are affixed to the original ultrasonic sensors. They emit exactly the same echo signals that are calculated in the simulation from the given distance information. Similar methods are also possible for camera-based systems by using a monitor HIL approach. In that case, the real-world camera, which may be located behind the windshield for example, is installed in the vehicle's trunk where it is oriented toward a monitor with artificial visualization that simultaneously displays

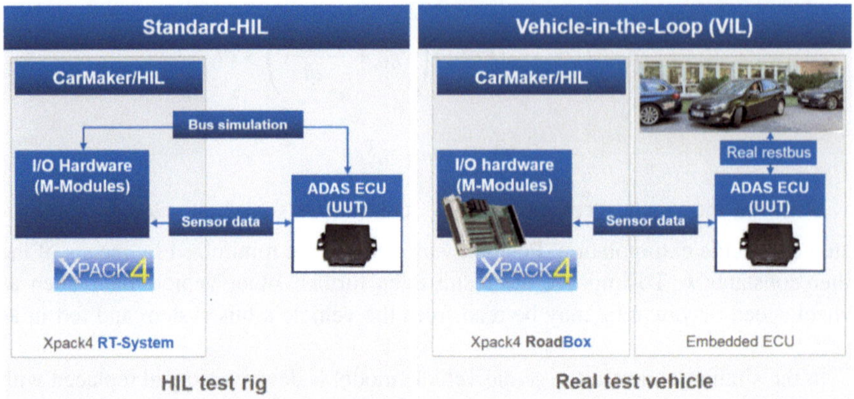


**Fig. 2** Comparison between an HIL and VIL schematic

exactly the signs, road markings or traffic objects that have been defined in the simulation.

**Virtual** For some types of sensors a physical sensor interface, due to technical boundary conditions such as limited assembly space, is no option. In this case, it would be appropriate to exclude the sensor from the investigation and, instead, to directly provide the simulation values to the respective ECU for example. Solutions on the level of object lists, such as those for radar sensors, are equally possible as injecting video data on the image sensor by using a video interface box (video injection) in order to circumvent the disadvantages of the monitor HIL approach mentioned in the previous passage).

A large number of sensors already exists in the vehicle and continues to provide real-world sensor data (e.g. wheel speed, temperature, steering angle sensor, etc.) which is available either directly or indirectly via the vehicle bus. These sensors do not have to be emulated. Here, the VIL method significantly differs from an HIL (component) test rig. Schematically, a generalized setup for the VIL method can be depicted as follows in Fig. 2 and distinguished from the conventional HIL approach:

## 4 Requirements Set by Institutional Regulations

In the case of passenger cars, many safety systems are still optional, even though they are becoming increasingly prevalent, and a growing number of newly registered vehicles have active safety functions installed in them. Safety, which is already on a very high level today, is consistently seeing further improvement as a result of actions taken by independent, global testing entities (e.g. ADAC, NCAP organizations, IIHS, etc.). In addition, institutions such as the U.S. National Highway Traffic Safety Administration (NHTSA) have a major influence on the testing procedures for new cars.

For automobile manufacturers, however, this is an additional aspect that drives up the scope of testing. For the European area, Euro NCAP (European New Car Assessment Program), which has been conducting various crash tests and classifying the vehicles accordingly since 1997, is an important organization. The safety requirements established by Euro NCAP usually go far beyond statutory standards, as consumer protection is an important aim pursued by the organization. Therefore, exacting requirements are defined in the field of safety, which vehicle manufacturers seek to comply with in the interest of receiving a positive rating. In the course of time, the criteria have consistently been revised and extended. Since the beginning of 2016, for instance, emergency brake assist systems have been mandatory for rating new vehicles. The systems have to detect pedestrians (AEB Pedestrian) and other vulnerable road users such as cyclists (AEB VRU), and initiate a braking event early enough. In the United States, emergency brake assist tests have been included in the NHTSA testing program as well. This safety technology will be tested as part of the U.S. NCAP rating criteria starting in 2018. Testing procedures are precisely specified and documented by Euro NCAP in the relevant protocols. For instance, the "Euro NCAP AEB VRU Test Protocol" [4] for testing emergency brake assist for pedestrians prescribes in detail what tests have to be performed and their respective variations (e.g. speed), what demands are made on measuring technology, and how the components (e.g. targets—dummy pedestrians) or VUT (vehicle under test) have to be set up and dimensioned. The scenarios are described as follows:

- Scenario A: An adult pedestrian starts crossing the street on his or her own side (i.e. coming from the right for right-hand traffic).
- Scenario B: An adult pedestrian starts crossing the street from the opposite lane (i.e. from the left).
- Scenario C: A pedestrian, a child in this case, is visually obstructed by a vehicle parked on the side of the street and starts running across the street from his/her own side.

The relevant traffic situations, i.e. collisions with pedestrians or cyclists, can be investigated in real-world road tests only to a limited extent or with a considerable investment of resources. However, with the CarMaker open integration and test platform, advanced driver assistance systems can be seamlessly developed and tested.

## 5 Potential of VIL in the Test Procedure

The reproducibility of the test scenarios for all test cases and rating criteria by using the same maneuvers, conditions and criteria enables efficient, seamless testing (MIL, SIL, HIL, VIL) across the entire development process. The Vehicle-in-the-Loop method in particular makes it possible to circumvent the difficulties addressed in this paper toward the end of the development process. By means of commonly used methods from the HIL environment (e.g. video injection) real-world collision objects are replaced with virtual targets and supplied to the real-world vehicle. Due to this combination of conventional road testing and simulation, advanced driver assistance systems can be reproducibly tested in the real-world vehicle without risk. Using the augmented reality approach, additional virtual environment information (e.g. other road users or instructions) can be provided to the driver via a head-mounted display in the driver's glasses. The driver can respond to them so that the relevant functions can be tested in the real world.

Conventional ADAS testing procedures are typically oriented to real-world road tests and are characterized by continually growing complexity and diversity of scenarios. In addition, an increasing number of approaches are used to standardize the procedures. This ensures that the VUT test results and score achieved become comparable and thus conclusive. Compared with the VIL method, these test procedures exhibit the following weaknesses:

- The realization of the targets becomes increasingly complex and difficult due to a variety of factors. For one, rigid objects are no longer sufficient for use as targets. The Euro NCAP Protocol, for instance, prescribes the movement of the pedestrian's legs as well. This will become an even more exacting requirement for future systems when, aside from additional degrees of freedom with respect to the body, the recognition of facial expressions, gestures and intentions may become relevant as well.
- Controlling the targets will become increasingly complex. In addition to dynamic trajectories, various movement patterns, such as "walking" versus "running," must be implemented.
- The impact must be timed within certain tolerances and, according to the protocol, take place at certain points of contact, and be reproducible.

Hazards for humans and hardware: Particularly when testing AEB functions, collisions cannot always be avoided. In many cases, the criterion is not if but at what speed the collision occurs. For this event, it must be ensured that, for one, the driver is not exposed to any risk of injury and for the other, that there will be no excessive damage to hardware (reusability of the targets) (Fig. 3).

In addition to these aspects, the combination with simulation in the VIL approach offers further advantages. As well as the criteria, the test cases can almost fully be adopted from earlier development stages. This reduces the time before, between and after actual test driving events. Spontaneous changes to the test setup can be made faster as well because there are clearly fewer mechanical components involved. Finally, even the driver model from the simulation [5] can be used for the drive tests, which significantly increases the reproducibility and accuracy of the results, particularly in more complex closed-loop tests.

When applying the quotient from the number of driven test cases and manpower invested as an efficiency criterion, individual investigations reveal a multiple increase when using the VIL method compared with road tests performed according to the "Euro NCAP AEB VRU Test Protocol."

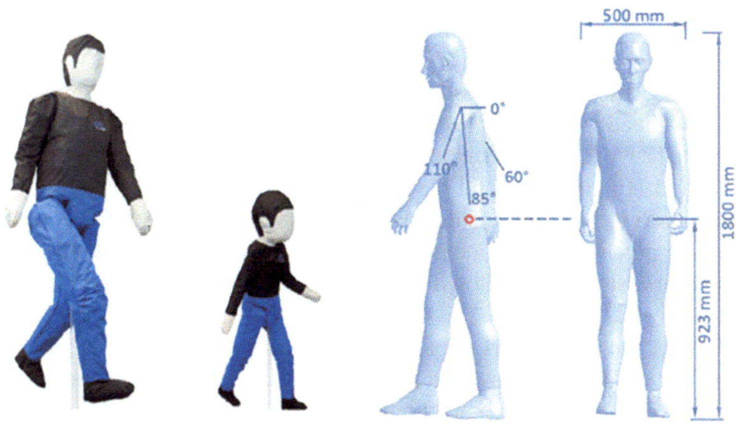


**Fig. 3** Euro NCAP pedestrian targets with dimensions as shown in the Euro NCAP test protocol [4]

## 6 Early Experienceability of Systems

The success, or at least the benefits, of a large number of advanced driver assistance systems heavily depend on driver acceptance. As a result, advanced driver assistance systems significantly differ from conventional control systems such as ESC, the use of which serves to protect vehicle occupants without active driver intervention and in many cases is not even consciously perceived. Conversely, this means that advanced driver assistance systems have to be tested at an early stage of the development process for acceptance and usability in order to effectively prevent a development from heading in the wrong direction.

The VIL method can assist in this respect as well. For instance, according to the setup described above, systems can be evaluated in studies with test subjects without incurring any risk whatsoever, as crashes involving potential pedestrians occur only in the virtual world. In addition, linking of the simulation platform enables testing of ADAS functions or algorithms as early as in the MIL or SIL stage. The ADAS model can be linked directly with the simulation environment and its performance modeled in the real-world vehicle.

This presupposes that the respective driving situation can be experienced as realistically as possible. To achieve this, the driver or test subject is equipped with a so-called head-mounted display. Here, augmented reality glasses with optical see-through technology haven proven their viability. Similar to a head-up display, they make it possible to lay the virtual world, for instance other road users, on a pair of otherwise completely clear glasses. As a result, the driver or test subject sees the real world as they would through normal glasses with an overlay of the virtual objects. A special challenge in this regard is posed by the system's latency which, for one, is caused by the necessary head tracking in order to determine head pose and, for the other, by the calculation of the projected image.

## 7 Outlook

As systems become more and more complex, they require an immensely increasing testing and validation effort. Conventional testing systems soon reach their limits in this context. The VIL method marks another step in complementing the established MIL, SIL and HIL methods and reducing development resources. Initial fields of application with various automobile manufacturers show the need for as well as the wide-ranging potential of this method. The next few years will show in what areas VIL can be of assistance and how the method can be anchored in the development process.

In addition, the current development and trends in the field of augmented and virtual reality speak for the fact that utilization of the VIL method may provide an alternative to conventional driving simulators.

## References

1. The German Association of the Automotive Industry (VDA), for instance, divides the various levels of automation into level 0 (no ADAS) to level 5 (no driver required)
2. Hakuli, S., Krug, M.: Virtuelle integration. Kapitel 8. In: Winner, Hermann; Hakuli, Stephan; Lotz, Felix; Singer, Christina: Handbuch Fahrerassistenzsysteme—Grundlagen, Komponenten und Systeme für aktive Sicherheit und Komfort. 3. Auflage. S. 125–138. Springer Fachmedien Wiesbaden, Germany (2015)
3. Schwab, S., Leichsenring, T., Zofka, M.R., Bär, T.: Durchgängige Testmethode für Fahrerassistenzsysteme. In: Jahrgang, S. (ed.) ATZ 01/2015, vol. 116, pp. 54–59 (2015)
4. Euro NCAP. http://www.euroncap.com/en/for-engineers/protocols/pedestrian-protection/ (2016). Accessed 29 Feb 2016
5. Pfeffer, R., Schwab, S., Leichsenring, T., Zofka, M.-R.: A real-world vehicle in a virtual environment: vehicle-in-the-Loop for reproducible real-time-capable testing of advanced driver assistance systems. In: 2nd International ATZ Conference on Advanced Driver Assistance Systems, Springer, Germany (2015)

# Finding Coupling Strategies of a Real-Time Capable Fourier-Transformation-Based Engine Model on a HIL-Simulator

**Aras Mirfendreski, Andreas Schmid, Michael Grill and Michael Bargende**

**Abstract** For the analysis of electronic control units (ECU) before series production, hardware-in-the-loop (HIL) simulators are used for model-based function diagnosis such that possible errors due to false applications can be detected at an early stage. Due to required real-time capability of engine models, they only can be represented in a highly simplified manner which leads to loss in predictability. Many specific function tests, therefore, cannot be carried out. In a first step, this paper presents a novel modeling concept for a 1D-engine-model with the Fourier transformation (FT) method, based on a mean value model providing the prerequisites for real-time capability, which has been developed at AUDI AG in cooperation with the Institute of Internal Combustion Engines and Automotive Engineering at the University of Stuttgart. By means of the mathematical FT-concept, it is possible to represent pressure pulsations emerging inside the engine air path, and so, to improve the standard of a mean-value model. Pressure pulsations inside the air path have a strong influence on the engine's internal behavior such as for the gas exchange of the cylinder, the responsive performance of the turbo charger, etc. On the basis of this model, various coupling strategies with a HIL-simulator have been examined. With a focus on the real-time capability, further essential simulation speed potentials were identified such that a higher ratio of the processor power can be harnessed in case of an optimal coupling. The ECU

A. Mirfendreski (✉) · M. Bargende
Institut für Verbrennungsmotoren und Kraftfahrwesen (IVK),
der Universität Stuttgart, Stuttgart, Germany
e-mail: Aras.Mirfendreski@ivk.uni-stuttgart.de; extern.aras.mirfendreski@audi.de

M. Bargende
e-mail: Michael.Bargende@fkfs.de

A. Schmid
AUDI AG, Neckarsulm, Germany
e-mail: Andreas1.Schmid@audi.de

M. Grill
Forschungsinstitut für Kraftfahrwesen und Fahrzeugmotoren Stuttgart (FKFS),
Stuttgart, Germany
e-mail: Michael.Grill@fkfs.de

 
C. Gühmann et al. (eds.), *Simulation and Testing for Vehicle Technology*,
DOI 10.1007/978-3-319-32345-9_5

and the engine simulation model both have individually configurable sampling times. When coupling the engine model with a HIL-simulator, both sampling times need to be synchronized such that a signal transfer can be executed at steady, discrete time intervals. The global sample time (GST) is usually determined by the ECU, since it has requirements which have to be fulfilled—the engine model, therefore, complies with the GST by either adapting it or taking a common divisor of it. By this method, the engine model loses speed potentials, since it is forced to adjust its sample rate to an invariable subset. Investigations have proven that it is beneficial to select a GST, with respect to the coupled engine model, that brings the most advantageous "synergy" between ECU and engine model. For this, first, the highest possible "sampling time limit" of the engine model (just stably executable by the model solver) across the whole engine map needs to be determined—the GST for the ECU, afterwards, is set with this or a common multiple value. With the presented method of a variably designable sampling rate between the ECU and the engine model, the system speed can be increased up to 22 %, which in conclusion means that additional processor performance can be deallocated, which in turn can be used to boost up the model with more Physics and more details.

## 1 Introduction

According to the Fourier Theorem, any periodic signal $f(t)$ with a periodic time $T$ can be expressed by a constant component and the sum of its infinite harmonic signals $h_n(t)$ and angular frequencies $\omega_n$, differing in their amplitudes $A_n$ and phases $\varphi_{n.}$ The angular frequencies of the harmonics are multiples of the basic angular frequency $\omega_0 = 2\pi/\mathrm{T}$. The total sum is also called the trigonometric sequence, or Fourier sequence, and is given by the following equation:

$$f(t) = c_0 + \sum_{n=1}^{\infty} [a_n \cos(n\omega_0 t) + b_n \sin(n\omega_0 t)] \quad (1)$$

The variables $c_0$, $a_n$ and $b_n$ are known as Fourier coefficients. $c_0$ represents the average value (mean value) of the signal $f(t)$. If $f(t)$ is represented by an oscillating pressure signal $p(t)$, the variable $c_0$ represents the mean pressure of the pressure signal. Equation (1) can be simplified by using the following correlations:

$$a_n \cos(n\omega_0\, t) + b_n \sin(n\omega_0\, t) = c_n \cos(n\omega_0\, t + \varphi_n) \quad (2)$$

with

$$c_n = \sqrt{a_n^2 + b_n^2} \quad (3)$$

and

$$\varphi_n = \arctan(a_n/b_n) \tag{4}$$

Equation (1) now can be transformed into the so-called spectral representation of the Fourier sequence:

$$f(t) = c_0 + \sum_{n=1}^{\infty} c_n \cos(n\omega_0 t + \varphi_n) \tag{5}$$

After executing the Fourier Analysis, a periodic signal $f(t)$ can be expressed by the following parameters [1]:

| | |
|---|---|
| $c_0$ | average value (mean value of signal $f(t)$) |
| $c_n = c_n(n\omega_0)$ | amplitude spectrum |
| $\varphi_n = \varphi_n(n\omega_0)$ | phase spectrum. |

In an integral and complex form for a periodic signal $f(t)$, the forward transformation from time domain into spectral domain can be performed with the following equation:

$$F(\omega) = \int_{-\frac{T}{2}}^{+\frac{T}{2}} f(t)e^{-i\omega t}\,dt \tag{6}$$

The back transformation from the spectral into time domain:

$$f(t) = \frac{1}{2\pi} \int_{-\frac{T}{2}}^{+\frac{T}{2}} F(\omega)e^{+i\omega t}\,d\omega \tag{7}$$

Often, however, a continuous function is not known but can be tapped N discrete times $t_k = k \cdot \Delta t$. In this case, the discrete Fourier transformation (DFT) is used. The DFT makes the assumption that $f(t)$ is periodic continuous beyond the chosen interval. The Fourier coefficients can be gained by the following equation:

$$F_n = \frac{1}{N}\sum_{n=0}^{N-1} f_n\, e^{\frac{-2\pi i}{N}} \tag{8}$$

The reverse discrete Fourier transformation (RDFT) can be carried out with:

$$f_n = \sum_{n=0}^{N-1} F_n\, e^{\frac{+2\pi i}{N}} \tag{9}$$

According to [2], a new algorithm was developed which was able to reduce the amount of complex computing operation required for the calculation of the spectral lines of the DFT by a factor of $\frac{N}{\ln N}$. Due to a lower computing time and, thus, a faster calculating process, the numerically favorable algorithm of the DFT is called fast Fourier transformation (FFT) [3].

## 2 Pressure Pulsations of a V6-TDI

The following investigations are based on a current Audi V6-TDI engine. First, pressure pulsations inside the engine air path are examined, using a calibrated, detailed 1D engine model. The pressure inside the exhaust manifold directly behind cylinder 1 is tapped and split into its spectral components by applying the Fourier Transformation. Afterwards, the gained Fourier coefficients (amplitude, phase and order) are superimposed with the reverse discrete fast Fourier transform (RDFT) method. The result of it is shown in Fig. 1. The left column, starting from the top with a mean pressure signal, lists all harmonic pressure signals with decreasing

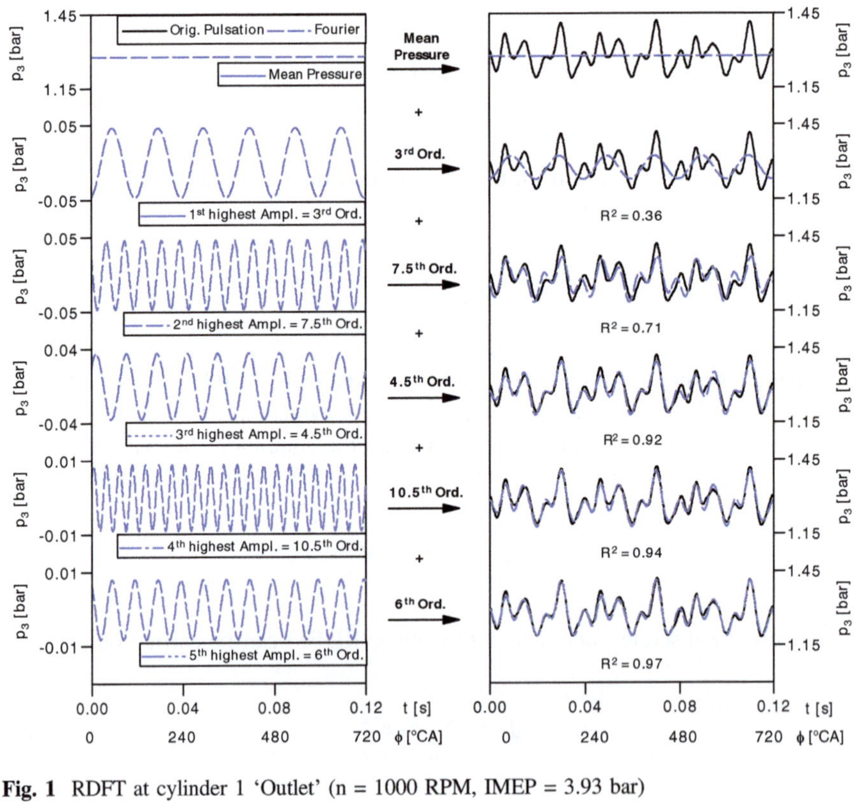


**Fig. 1** RDFT at cylinder 1 'Outlet' (n = 1000 RPM, IMEP = 3.93 bar)

amplitude height, gained from the FFT. The right column shows the results of the RDFT process. With each additionally superimposed harmonic pressure wave, the original pressure is approached. With five orders already, the original pressure pulsation can be reproduced up to a coefficient of determination $R^2$ of 97 %.

Pressure pulsations inside the engine air path have a strong influence on the gas exchange and hence on the combustion. Furthermore, the turbocharger shows different behavior when being applied with pressure pulsations instead of a constant pressure. In order to be able to take these relevant effects into account, four nodes at which a pulsating pressure brings advantages in terms of higher realistic results are defined.

The nodes 'Cylinder In' and 'Cylinder Out' will ensure a correct calculation of the cylinder gas exchange. The nodes 'Compressor Out' and 'Turbine In' will lead to calculation of a much more accurate balance of the fluid dynamics outside the turbocharger boundaries.

By means of four chosen operating points, Fig. 2 gives a good overview of how the coefficient of determination $R^2$ correlates with the number of superimposed harmonics for all four defined nodes.

Figure 2 shows that the number of orders needed to obtain a certain value for the coefficient of determination clearly differs from node to node. "Cylinder In" is a conspicuous node, at which obviously a higher number of orders is required for obtaining result quality similar to the other nodes. The pressure pulsation inside the air path gets overlaid by for- and backward pressure feeds, which strongly depend on the engine geometry. A long air path, flow splits, filling and tangential intake

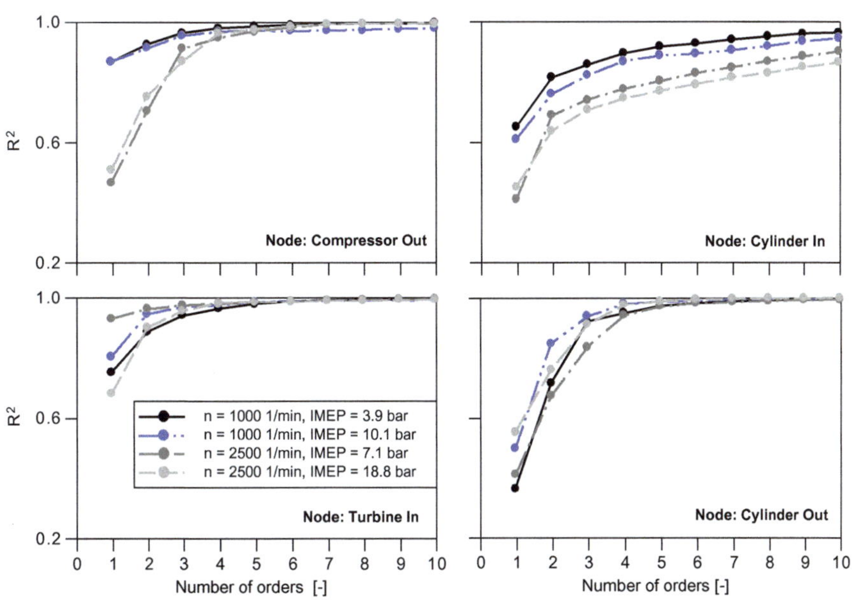


**Fig. 2** Coefficient of determination versus number of orders for four predefined nodes

manifolds, etc. cause many reflections inside the intake system. In contrast, the pressure pulsations at the other nodes can be quite well reproduced with 4 of the most important orders already. Additional harmonics here lead only to minor increase in quality.

For a quantification of the benefit a pulsating pressure brings compared to a non-pulsating pressure, the following examinations are presented. By way of example, Fig. 3 (top left) shows the gas exchange loop for the operating point n = 2500 RPM, IMEP = 18.83 bar. When applying a mean pressure on both sides (intake and exhaust), deviations in cylinder pressure up to 350 mbar result at discrete times. With five applied orders of harmonic pressure signals, the gas exchange loop can be well reproduced. Figure 3 (bottom left) demonstrates the resulting error in volumetric efficiency after the gas exchange cycle which attains a value of 4 %. The operating point n = 2500 RPM at IMEP = 18.83 bar even shows that an insufficiently accurate pulsation (1st order) can have an adverse effect on the gas exchange and, thus, on the volumetric efficiency. Incorrect pressure pulsations during valve opening times principally change the cylinder charge. By superimposing further orders, the error can gradually be reduced. The graph on the bottom right shows that the volumetric efficiency directly affects the maximal cylinder pressure during the combustion.

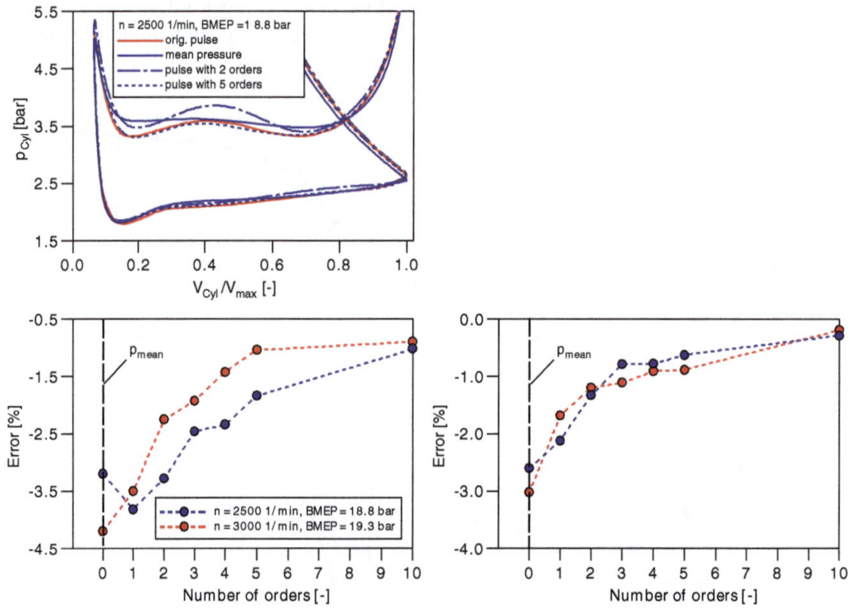


**Fig. 3** Model quality examinations (Internal engine): volumetric efficiency (*bottom left*), Max. pressure (*bottom right*)

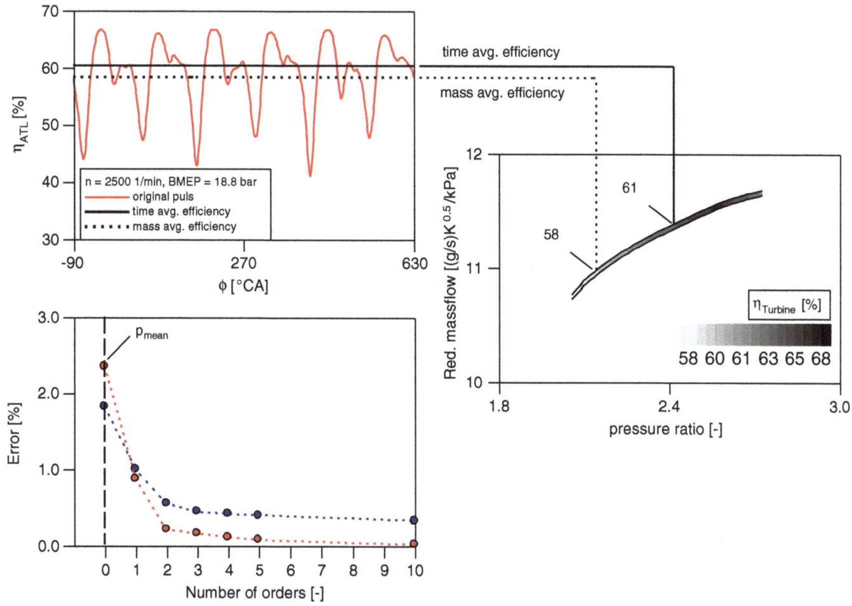


**Fig. 4** Model quality examinations: Turbocharger

The same quantitative analyses are carried out on the turbocharger. Figure 4 (top left) shows the crank angle resolved turbocharger efficiency for n = 2500 RPM at IMEP = 18.83 bar. The efficiency of this operating point fluctuates in a maximum range of 26 %. The mass averaged efficiency is approximately 3 % lower than the time averaged efficiency. This indicates that a turbocharger applied by a mean pressure results in a higher mass flow. A different operating point is consequently tapped from a turbocharger map as can be seen in Fig. 4 (top right). Hence, the turbocharger performance is overestimated. In the graph on the bottom left, it is evident again that the error in turbocharger efficiency can be eliminated using the FT-method [4].

## 3 Model Architecture

The FT-model can be generated based on a detailed 1D engine model or based on low pressure indication measurements from an engine test bench. This way, an automated measurement and model feeding will be possible.

The following investigations within the framework of this paper are carried out on the basis of a 1D-engine model. In Fig. 5 (top) pressure pulsations are displayed that are tapped right behind cylinder 1 and cylinder 4 (see schematic on the right). With respect to the firing order of the V6-TDI (1-4-3-6-2-5), the intake air of cylinder 4 takes place 120 °CA after the intake of cylinder 1.

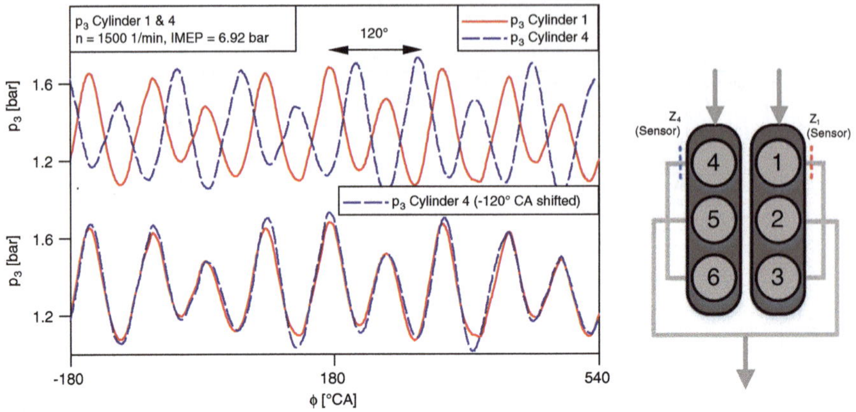


**Fig. 5** Offset of the air path

By offsetting the exhaust pressure pulsation $p_3$ of cylinder 1, the exhaust pressure $p_3$ of cylinder 4 can be gained (see Fig. 5 bottom). Slight differences in pressure peaks, however, may result due to deviation of the individual intake geometries. Moreover, the individuality of each cylinder combustion (injection timing, injection mode etc.) cannot be taking into consideration when applying this strategy. For gasoline engines, in addition, disregarding cyclic fluctuations might cause some further deviations in the results.

Likewise, on the intake and the exhaust side, the pressure pulsations of all other cylinders behave in the same manner. Therefore, the surrounding pressure pulsations of all other cylinders can be formed with a respective phase offset of the pressure pulsations of cylinder 1.

This step is only optional and might be very helpful, in case a real-time capability cannot be achieved due to a lack of processor capacity.

The FT-model was designed for the software GT-POWER and MATLAB/Simulink that are commercially used in engine development areas. According to the modeling architecture, there are some small differences to be considered. In this paper, the modeling method for GT-POWER will be presented.

Figure 6 (left) shows an engine model reduced in its detailed air path. An explicit calculation of the fluid dynamics along the stream line, therefore, will not be performed here. Next, all cylinders are replaced by a representative single cylinder as exhibited in Fig. 6 (right). Pressure pulsation on the intake side as well as on the exhaust side for cylinder 2–6 are copied and applied to the eliminated cylinder ports with respect to their firing angles. The output torque of cylinder 1 applied to the crank train is analogously taken into consideration.

At the earlier predefined nodes, the model is split into several subsystems (Fig. 7, left). Between the subsystems, an RDFT process with the FT-coefficients gained in a previous step can be performed (Fig. 7, right).

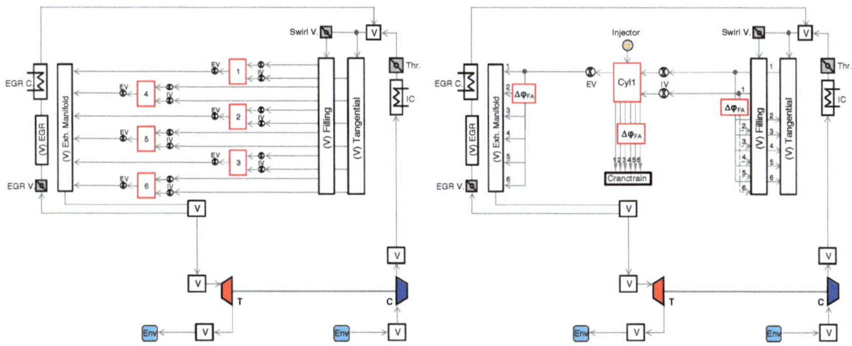


**Fig. 6** Volume-based engine model (*left*), cylinder reduction (*right*)

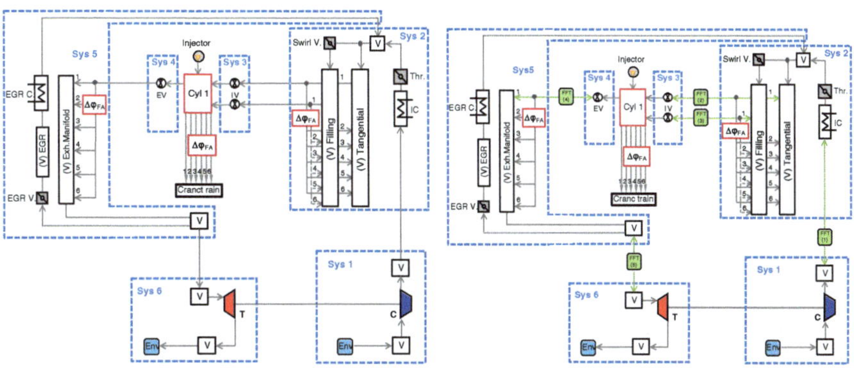


**Fig. 7** Model split into subsystems (*left*), application of Fourier coefficients at four predefined nodes (*right*)

Further details for the FT-blocks and the connecting procedure of the subsystems can be found in [5].

The final system architecture of the FT-model is presented in Fig. 8 (left). By means of this applied method a further advantage arises. Each subsystem is now available as an independent system. The numeric solution of each subsystem can, therefore, be chosen individually. For an optimization of the overall computation time, an optimal combination of implicit and explicit solvers can be chosen. By clustering, the potential of the computation time can be used more efficiently (Fig. 8, right), which brings an increase of up to 400 % compared to a solution with a single explicit solver.

A model of this same kind has also been developed in a MATLAB/Simulink environment. Unlike the modeling method in GT-POWER, splitting the model into subsystems is not necessary in MATLAB/Simulink, since a flow solver such as Navier Stokes does not exist.

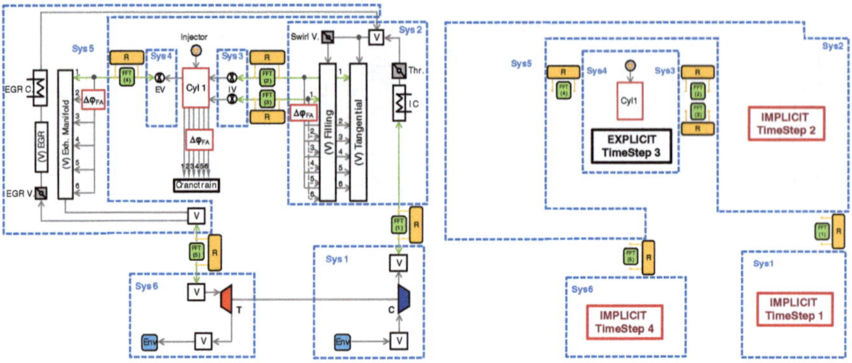


**Fig. 8** Connection of the air path by physical controllers (*left*), selection of individual solver types for each subsystem (*right*)

## 4 Stationary Behavior

Figure 9 shows the pressures for all four nodes for the stationary operating point n = 2000 RPM, IMEP = 10 bar. Theoretically, any number of orders can be stored in the model. However, an increasing number of computing operations increases the computation time. In order to minimize the computation load, the number of orders for each node is being limited so that for all operating points of the engine map a minimum coefficient of determination of 95 % is always guaranteed. Node 'Cylinder In' constitutes an absolute maximum number of orders in its extreme engine map range of 19.

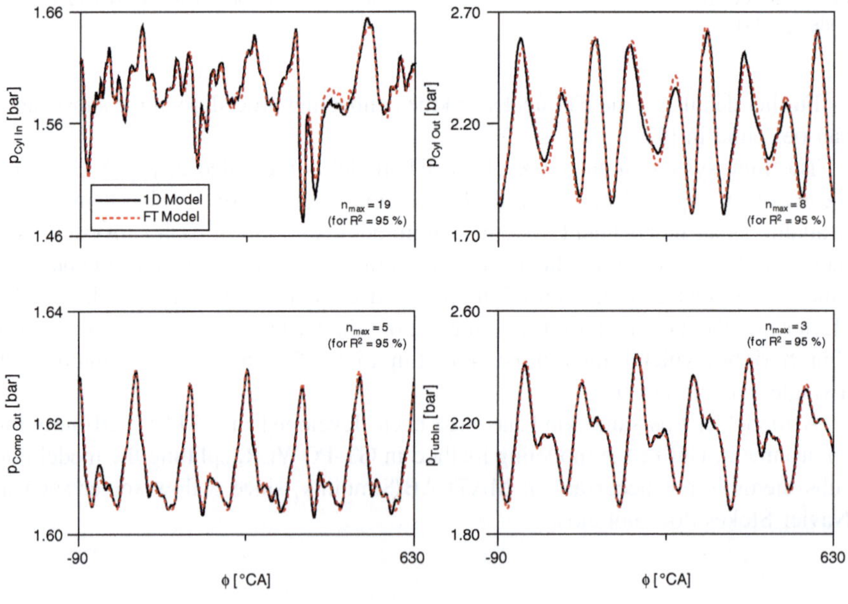


**Fig. 9** Model results at n = 2000 1/min, IMEP = 10 bar

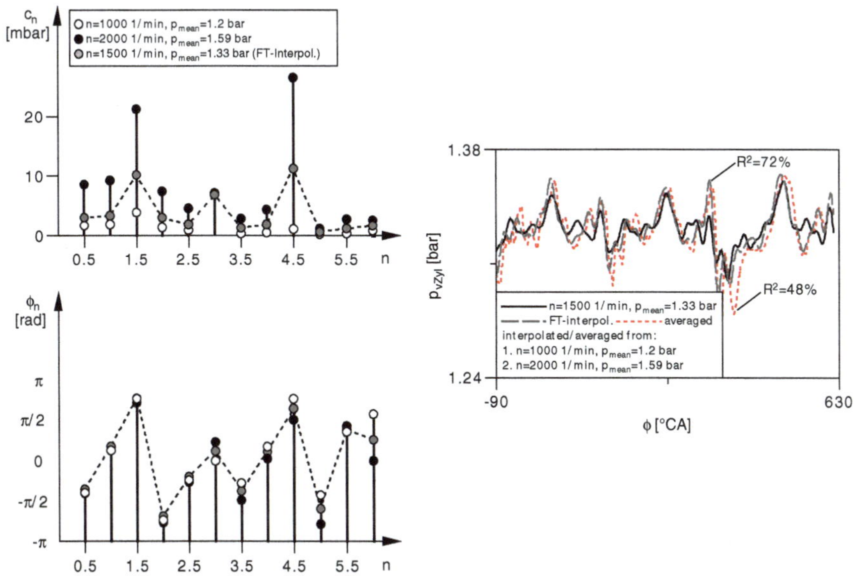


**Fig. 10** Interpolation quality of Fourier coefficients

For the calculation of a stationary operating point inside or outside the range of embedded data points or for transient simulations, the determination of the Fourier coefficients will be done by a two-dimensional interpolation of the input parameters (engine speed and adjacent mean pressure).

For an evaluation of the interpolation quality, two exemplary operating points for a complex pressure pulsation of the cylinder intake pressure with a gap of 1000 RPM in engine speed and 390 mbar in mean pressure were chosen. The two left graphs in Fig. 10 show the interpolation for the determination of phase and amplitude information for a number of orders.

The result in pressure pulsation from the interpolation of the Fourier coefficients is shown in the right graph. As evidenced, this methodology demonstrates an agreement in results compared to the original pressure curve at 1500 RPM, see dashed line.

The coefficient of determination is at $R^2$ = 72 %. As a comparison, the dotted line represents the average of the two pressure curves of the boundary operating points, that leads to a much lower determination coefficient of $R^2$ = 48 %.

In conclusion, it can be stated that due to its physical character, an interpolation of the Fourier coefficients leads to good quality.[1]

[1]Storing a pressure curve over crank angle (gained either from an engine test bench or from a 1D-engine model) would require a high data volume (comparison for one operating point: **pressure curve (PC)**: With a resolution of one data point per crank angle = 720 data points, **Fourier-Method (FM)**: amplitude + phase = 2 data points per order; ratio for 5 orders: PC/FM = 720/10 = 72.

## 5 Transient Behavior in the Time Domain

The transient behavior with particular attention to a changing engine speed represents a challenge for the FT-engine model in the time domain.[2]

With a steady[3] change in speed among a cycle, the angular frequency changes continuously. The following problems arise:

The DFT only transforms a finite number of N samples, which by definition serves as a tool for periodically and synchronously sampling. For a function in spectral domain, no temporal resolution is made. This means that the Fourier transform only provides interpretable results as long as the spectral composition remains constant over time [6].

With changing engine speed in transient mode, the periodic duration of a cycle is not foreseeable. A signal, calculated by the IDFT, containing spectral compositions gained from a periodic signal (orders of constant angular frequencies) would get clipped at a wrong position where the function has not yet reached or has already passed its full periodic length. The function, therefore, would become unsteady.

In practice, nonetheless, there are different methods to turn non-periodic functions into periodic ones of a finite window length.

Harris [7] gives a good overview of window functions for discrete Fourier transforms. They allow non-periodic functions to flush with the edges of the period by multiplication with pre-defined orthogonal functions. A disadvantage of this method is that potentially important orders may get eliminated which might distort the pressure around its edges and take a negative impact on the current pressure gradient.

Another method often applied to create a periodic continuation of a function is called "zero padding". The number of samples $N$ with the sampling time $\Delta t_1 = \omega/N$ can get transferred into a higher sample number $M > N$ with the sampling time $\Delta t_2 = \omega/M$. This essentially means there are $M$-$N$ zeros added on $N$ samplings in the frequency domain [8]. Hence, in the time domain, the periodic function gets stretched.

The subject area of non-stationary classes provides a third alternative. Those procedures are applied to add a temporal resolution to spectral components such that a "spectral dynamic" can be considered. The spectral representation is thereby formed into a function which, in addition to the angular frequency, also represents a temporal change of the output quantity. Methods such as short-time Fourier

[2]For an angle-based DFT calculation the respective problem does not occur.

[3]A jump in speed, for example starting from the characteristic frequency to a multiple of this frequency, would still let all spectral components finish the period. Within the framework of this application, this case would never take place due to material inertias.

transformation (STFT) [9, 10], modification of the Fourier integral [11], wavelet-transformation [9, 12–18] or filter bank analyses [19–22] are commonly used.

For all three variants, the period length of the signal to be transformed needs to be known in advance. Besides this, some additionally required modeling steps would make the application more difficult and affect the already critical calculation speed in transient mode adversely. For the third variant another problem arises, since a transient speed in a work cycle can never be predicted—it can have an infinite number of shapes.

In the following, this paper proposes an approach that fulfills the transient demands satisfactorily.

Considering a changing engine speed during a work cycle, an equivalent, time-specific cycle duration over 720 °CA can be calculated by the following integral:

$$T_{cycle_t} = \int_0^{\varphi=720} \frac{1}{6 \cdot n(\varphi)} d\varphi \tag{10}$$

On the assumption of a linear speed gradient starting from $n_0$, the time-specific, average period length is expressed with:

$$\overline{T}_{cycle_t} = \int_0^{\varphi=720} \frac{1}{6\left(n_0 + \frac{\Delta n}{720}\varphi\right)} d\varphi \tag{11}$$

Solving the integral for a full cycle, the solution is:

$$\overline{T}_{cycle_t} = \frac{120}{\Delta n} \ln\left(1 + \frac{\Delta n \cdot \varphi}{720 \cdot n_0}\right)\Bigg|_{\varphi=720} = \frac{120}{\Delta n} \ln\left(1 + \frac{\Delta n}{n_0}\right) \tag{12}$$

A respective, time-specific, mean engine speed over a cycle is obtained by:

$$\overline{n}_{cycle_t} = \frac{120}{\overline{T}_{ASP_t}} = \frac{\Delta n}{\ln\left(1 + \frac{\Delta n}{n_0}\right)} \tag{13}$$

The question that arises initially is: To which extent do changes in engine speed among a cycle generally take place during a drive movement?

By way of example, the WLTC has been run in a longitudinal model environment. The acceleration section in Fig. 11 shows that a maximum change in speed of Δn/cycle = 100 1/(min · cycle) during up-shift events may take place. This happens

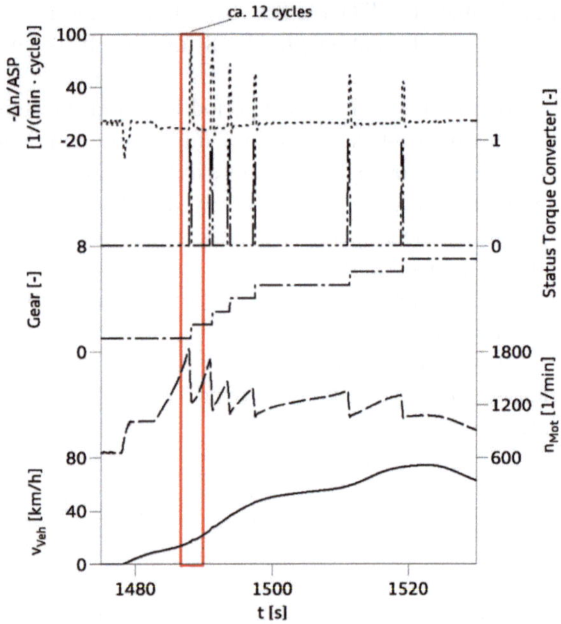


**Fig. 11** Acceleration section in WLTC

while the torque converter is actively regulating the speed difference between drive and output. Among the examinations, the order of 100 has proven to be independent from any driven cycle and therefore serves a good estimation in general.[4]

The logarithm from Eq. 13 is not legitimate for the argument zero, hence, for a cycle with no speed gradient. Applying a series expansion with the following formula

$$\ln(1+x) = \sum_{i=1}^{\infty} \frac{(-1)^{i+1}}{i} \cdot x^i \tag{14}$$

provides the continuous function:

$$\bar{n}_{cycle_t} = \frac{\Delta n}{\ln\left(1 + \frac{\Delta n}{n_0}\right)} = \frac{\Delta n}{\frac{\Delta n}{n_0} - \frac{1}{2}\left(\frac{\Delta n}{n_0}\right)^2 + \frac{1}{3}\left(\frac{\Delta n}{n_0}\right)^3 \ldots - \frac{1}{n}\left(\frac{\Delta n}{n_0}\right)^n} \tag{15}$$

The series expansion now needs to be limited to a certain order. Examination has proven a number of only 2 orders to sufficiently render results.

As an alternative, an angle-specific, average speed can be determined when defining a speed, starting from $n_0$ with a linear gradient $\frac{dn_0}{d\varphi}$ and tapping its value at 360 °CA.

[4]This value may deviate upwards in case of high differences between gear transmission rations.

$$\bar{n}_{cycle_\varphi} = n_0 + \frac{dn_0}{d\varphi} \cdot 360 \tag{16}$$

With the estimation for low speeds $\frac{dn_0}{d\varphi}$ and a maximum change in speed $\Delta n_{max} = n_{ASP_{WLTC,max}}$, the following approximation can be made:

$$\bar{n}_{cycle_t}\left(n_{0,min}, \Delta n_{max}\right) = \frac{\Delta n_{max}}{\ln\left(1 + \frac{\Delta n_{max}}{n_{0,min}}\right)} \sim \bar{n}_{cycle_\varphi} \quad \textit{for small}\, \Delta \mathrm{n} \tag{17}$$

Depending on $\Delta n$, Fig. 12 displays $\bar{n}_{ASP_t}$, $\bar{n}_{ASP_\varphi}$ and the respective period lengths $\bar{T}_{ASP_t}$, $\bar{T}_{ASP_\varphi}$ for different $n_0$. The marking $(\Delta n/cycle)_{WLTC,max}$ exhibits that the approximation in Eq. 17 is legitimate for the engine relevant $\Delta n/cycle$ area.

Taking a constant, representative speed during a work cycle, there will no longer be any concern that the speed function might become unsteady at its boundaries. A transient speed, therefore, is formed by a sequence of single, stationary cycles of constant speeds $n_{cycle_\varphi}$.

**Fig. 12** Temporal- and angle-specific, mean period time and speed

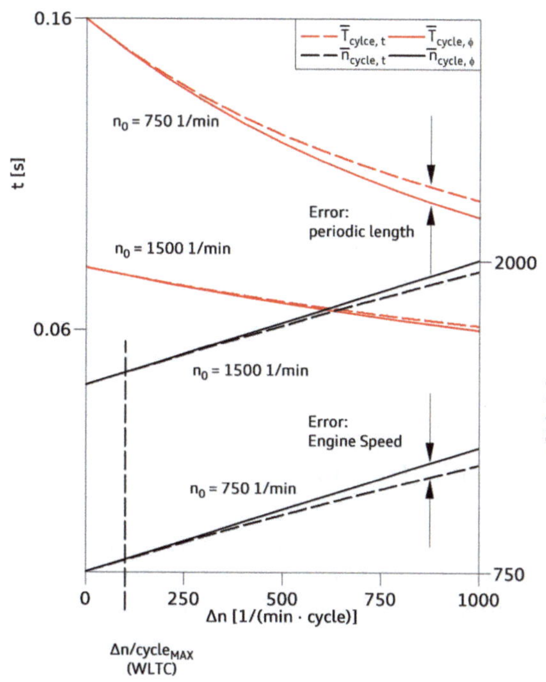

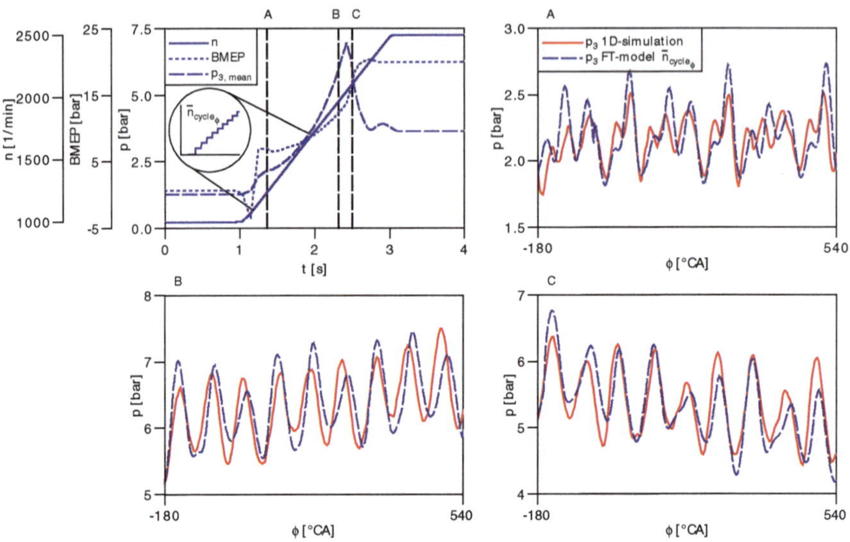


**Fig. 13** Transient behavior (change in speed and load)

For a validation of this method, a transient simulation has been performed. Both engine speed and load are increased at the same time. The resulting exhaust pressure $p_3$ is compared to the one from a 1D-simulation. Figure 13 (top left) presents the boundaries for the transient simulation. At three different marked times, the respective cycles are tapped.

Reaching full intake load ($p_3$ const., $n$ increasing)
$p_3$ increasing, $n$ increasing
$p_3$ decreasing, $n$ increasing

The angle-resolved figures (A–C) show that the pressure pulsation calculated with the FT-model exhibit very good agreement with the 1D-simulation results.

## 6 Results FT-Model

Compared to a detailed 1D engine model, the computation speed of the FT-model could be improved by a factor of 350 (real-time factor from 140 to 0.4) at 1000 RPM and by a factor of 133 (real time factor from 252 to 1.9) at 4000 RPM as shown in Fig. 14. The results were determined on a computer with an i7-2.80 GHz processor.

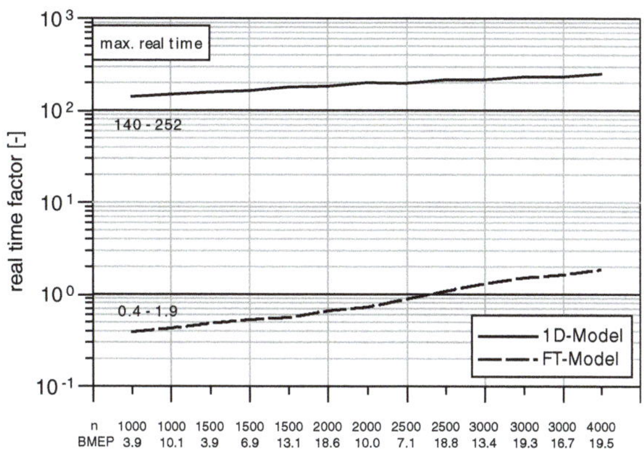


**Fig. 14** Computing speed (Processor: i7-2.80 GHz)

**Fig. 15** Level of model details

| level \ property | 1D geometry optimization | 1D pressure pulsation | flow volumes | simulation speed |
|---|---|---|---|---|
| 1D model | ✓ | ✓ | ✓ | X |
| mean value model | X | X | ✓ | ✓ |
| FT model | X | ✓ | ✓ | ✓ |

In summary, Fig. 15 visualizes the properties of different model levels discussed in this paper and highlights the advantages of an FT-model compared to other model types. The table shows that an FT-model practically forms an optimal combination of a detailed model and a mean value model with respect to the quality of results. However, it remains restricted in predictability regarding geometry optimizations.

## 7 HIL

With around 30 % of vehicle breakdowns caused by ECU-errors, it proves to be highly error-prone advice. At the same time, due to growing requirements, the complexity of its functions keeps growing as well, such that HIL-simulations have emerged to establish high significance for automotive electronics. For HIL-simulation, it is important that engine models are implemented providing a high level of detail. Only then can sensitive functions be tested in a realistic manner and be detected and improved in case of existing errors [23].

In the previous chapters, within the scope of the FT-model development, many steps were taken to speed up the computing time of the model. The following

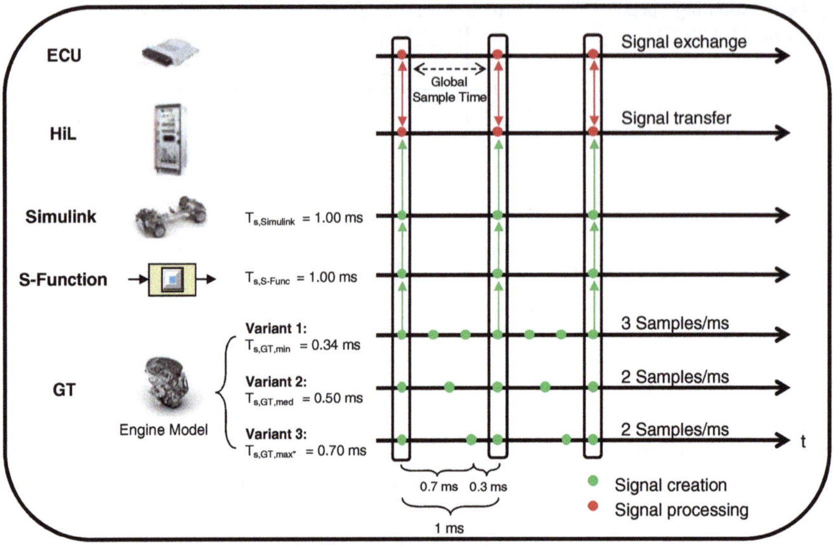


**Fig. 16** Model-based coupling strategies on a HIL-simulator

chapters will show that a strategic model coupling with a HIL-simulator makes it possible to better harness the CPU power and, hence, leads to further valuable speed potentials.

The function design for ECU-control functions can be realized by simulation tools such as MATLAB/Simulink or Stateflow. An overall vehicle model in MATLAB/Simulink, used in this work, enables an implementation of an FT-model designed with MATLAB/Simulink or with GT-POWER. The following presents the second variant.

Figure 16 schematically shows how signal loop is built up, starting from the model until the ECU is reached. The GT-POWER model is implemented by a Simulink-S-Function. For the engine model, a so-called model sample rate (MSR) needs to be predefined, for which three different variants are presented here. According to a global sample rate (GSR) of 1 ms preset by the ECU (see crossbars), the engine model can be synchronized with an evenly distributed MSR (0.34 and 0.5 ms) or a maximum sample of 0.67 ms—which is at the limit of the model's solver stability. The last one forms no common divisor of the GSR, therefore the sampling rate is asynchronous [23].

The results of the coupling for variant 2 are shown in Fig. 17. Apart from a few peaks, the turnaround time (TAT), the time the HIL-simulator needs to perform its computing process, is always below the GSR of 1 ms for engine speeds up to 2500 1/min. The single peaks can be eliminated during the build-process by a proper set of the task-priority control [24].

The result for variant 3 is displayed in Fig. 18. By the asynchronous sampling intervals, the TAT strongly fluctuates between 0.5 ms and 1 ms. This can be explained by the fact that a larger sampling rate of 667 μs first executes a fast

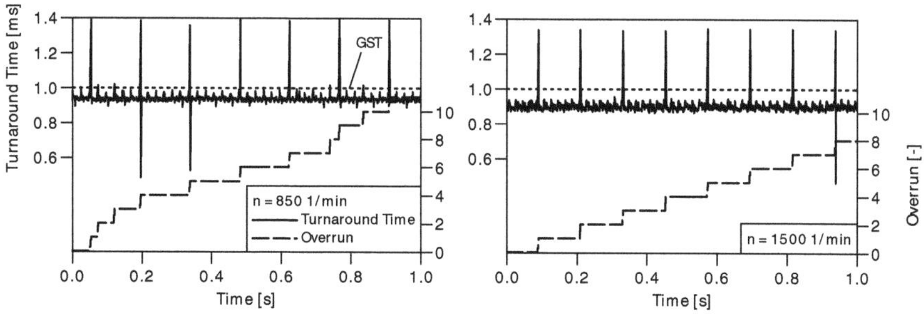


**Fig. 17** Real time capability on HIL-simulator (Var. 2)

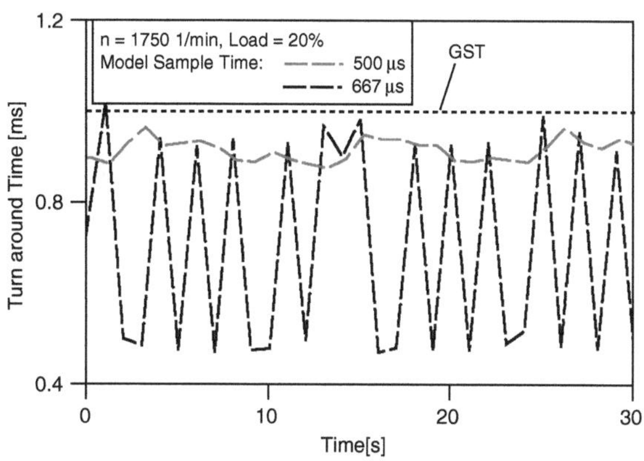


**Fig. 18** Asynchronous sampling rate on HIL-simulator (Var. 2 vs. Var. 3)

computation, followed by a much smaller sampling of 333 µs, which dramatically reduces the speed in the second step. The strong fluctuations render this variant not recommendable.

These results demonstrate that an equally distributed MSR forming a common divisor of the GSR is essential for a stable communication. A further speed boost, therefore, can only be reached if the model is set with a larger common divisor of the GSR. Figure 19 demonstrates this case—the GSR is increased up to a higher integer (2 ms)—thus, a new greatest common divisor (GCD) of 667 µs is obtained.

The result of variant 5 is presented in Fig. 20. As can be seen, by means of this strategy, the computation speed of the overall system can thereby be significantly increased. According to the base configuration (variant 1), this coupling is about 37 % faster on average. This is due to the fact that the model now only calculates 1 sample/ms.

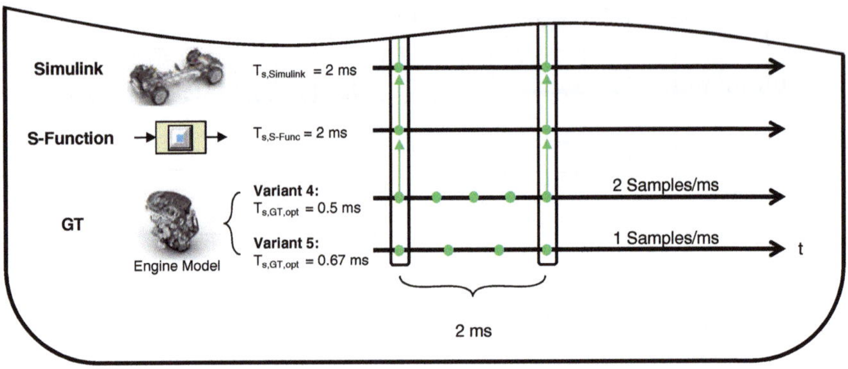


**Fig. 19** Varying the global sample rate (GSR)

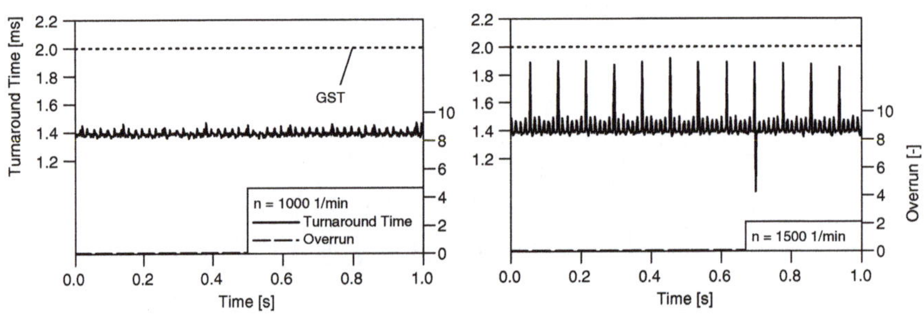


**Fig. 20** Real-time capability on HIL-simulator (Var. 5)

According to the GSR, however, the ECU also sets restrictions especially for subsystems that are triggered within a very short time. For this reason, another variant was investigated (Fig. 21) for which the GSR is brought to a decimal digit. This method allows a much more flexible and individual coupling between the model and the HIL. A MSR of 667 μs from variant 5, for instance, runs at the stability limit of the model solver—only engine speeds of 2000 1/min at maximum

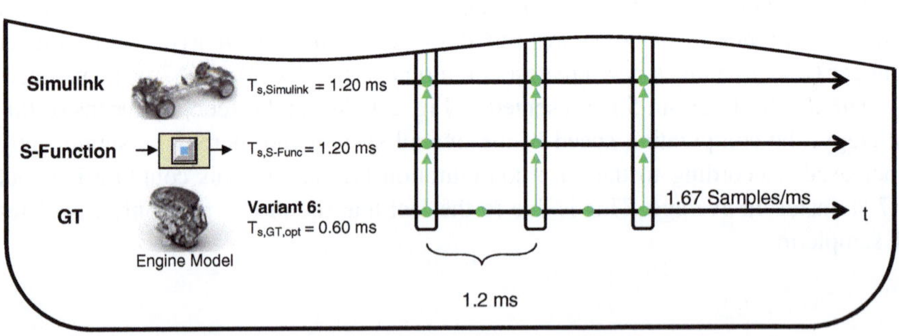


**Fig. 21** Decimal variant of GSR

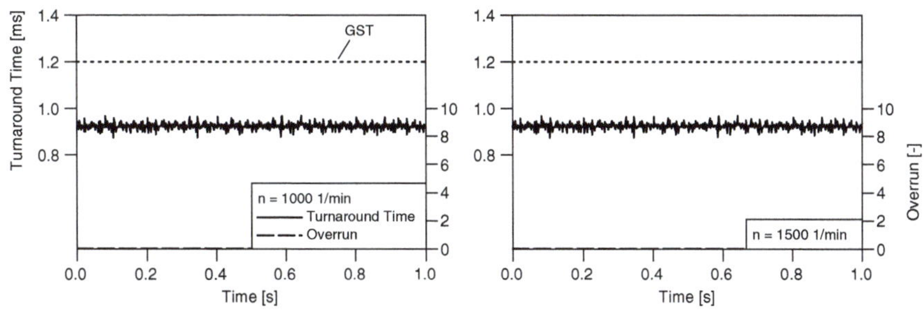


**Fig. 22** Real time capability on HIL-simulator (Var. 6)

can be reached before the solver breaks down. Therefore, a smaller MSR with 600 μs is selected, by this, the model is able to run stably. The GSR is set with a multiple value, namely 1.2 ms.

Figure 22 shows the result of the coupling for variant 6. With an average advantage in speed of 22 %, compared to variant 1, variant 6 presents the best compromise. Both the requirements of the ECU with respect to a restricted GSR and a high model computation speed are brought to their optimum for this variant.

# 8 Results HIL-Coupling

To summarize, all results in computation speed are once more comprised in Fig. 23. Variant 1 sets the reference with standard configuration, where the MSR forms a common divisor of the GSR.

When doubling the GSR, signals are transferred at half rate, which increases the overall speed by 4 %. The resulting chance now in setting a new GCD for the MSR to 667 μs increases the overall speed by further 33 % (V. 3).

**Fig. 23** Potentials of a strategic coupling

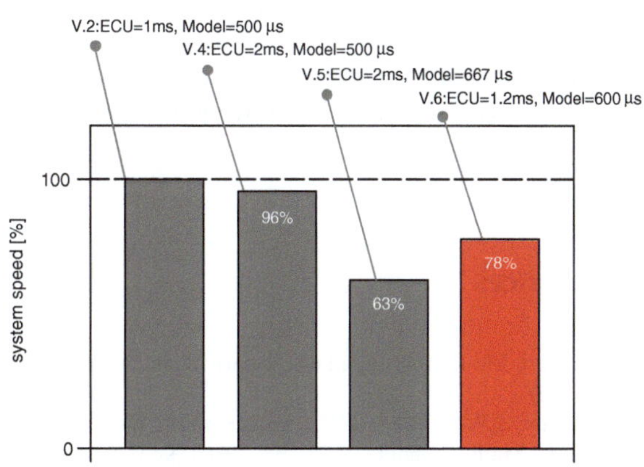

By setting the GSR to a decimal digit, it can be individually fitted to a MSR running at its speed limit (V. 4). Here, a speed potential of still 22 % could be proven. This value, however, is to be considered individually—depending on possible potentials, it can turn out to be considerably greater.

## 9 Summary

In this paper a novel concept for an engine simulation model based on a mean value model was presented. Moreover, it was possible to generate pressure pulsations inside the engine air path by means of the FT-concept. The pressure pulsations were applied using the RDFT method at so-called 'nodes', where a pulsating pressure influences the inner motor behavior. The model still provides high computing speed, such as a mean value model, and represents pressure pulsations of a detailed 1-D engine model without an explicit calculation.

The simulation architecture was built for two conventional simulation tools: GT-POWER and MATLAB/Simulink. For GT-POWER the model needs to be divided into different subsystems—clustering the solvers of each subsystem leads the CPU power to be used more efficiently, which brings an increase of up to 400 % compared to a solution with a single solver.

In a next step of this paper, the FT-model was coupled with a HIL-simulator. By investigating several coupling strategies by varying the model sample rate (MSR) and the global sample rate (GSR), a favored solution could be found.

By setting the GSR as a decimal digit, it provides much more flexibility for the coupling process. This makes it possible stretch the MSR to its maximum to obtain the highest model speed potential. After that, the GSR can be set to the same value of the MSR or to a multiple of it. By this method, an overall speed boost of 22 % could be reached, whereas this value can turn out to be considerably greater if more potential is available.

Besides the possibilities for real-time applications such as HIL, OBD and function development, the FT-model provides many advantages when implementing in longitudinal dynamic models. Replacing commonly used map-based engine models, it renders results of higher quality and predictability.

Especially for highly dynamic certification cycles such as the WLTC etc., it will be of great importance to take effects of the fluid mechanics inside the air path into consideration.

## References

1. Bossert, M.: Einführung in die Nachrichtentechnik. Oldenburg Wissenschaftsverlag, München (2012)
2. Cooley, J.W., Tukey, J.W.: An algorithm for the machine calculation of complex fourier series. Math. Comput. **19**(2), 297–301 (1965)

3. Butzmann, T.: Fourier Transformation für Fußgänger, 7. akt. Aufl., Wiesbaden: Vieweg + Teubner, 2011
4. Pucher, H., Zinner, K.: Aufladung von Verbrennungsmotoren, 4th edn. Springer, Berlin (2012)
5. Mirfendreski, A.: Real-time capable 1-D engine simulation model with the fastfourier transformation (FFT) concept. In: 2nd Conference on Engine Processes, Berlin (2015)
6. Hoffmann, R., Wolff, M.: Intelligente Signalverarbeitung 1, 2nd edn. Springer Vieweg, Wiesbaden (2014)
7. Harris, F.J.: Proceedings of the IEEE, vol. 66, p. 51 (1978)
8. Meffert, B., Hochmuth, O.: Werkzeuge der Signalverarbeitung. Grundlagen, Anwendungsbeispiele, Übungsaufgaben. Pearson Studium, München (2004)
9. Kaiser, G.: A friendly guide to wavelets. Birkhäuser, Boston (1994)
10. Rabiner, L.R., Schafer, R.W.: Digital Processing of Speech Signals. Prentice-Hall, Englewood Cliffs (1978)
11. Terhardt, E.: Akustische Kommunikation. Springer, Berlin (1998)
12. Morlet, J., Arens, G., Fourgeau, I., Giard, D.: Wave propagation and sampling theory. Geophysics **47**, 203–236 (1982)
13. Shensa, M.J.: The discrete wavelet transform. Wedding the À Trous and Mallat Algorithms. IEEE Trans. SP-40 10, S. 2464–2482 (1992)
14. Bäni, W.: Wavelets. Eine Einführung für Ingenieure. Oldenbourg, München/Wien (2002)
15. Bergh, J., Ekstedt, F., Lindberg, M.: Wavelets mit Anwendungen in Signal- und Bildverarbeitung (Übersetzung aus dem Englischen). Springer, Berlin (2007)
16. Blatter, C.: Wavelets. Eine Einführung, 2nd edn. Vieweg, Brauschweig/Wiesbaden (2003)
17. Brigola, R.: Fourieranalysis, Distributionen und Anwendungen. Ein Einstieg für Ingenieure, Naturwissenschaftler und Mathematiker. Vieweg, Brauschweig/Wiesbaden (1997)
18. Zwicker, E.: Psychoakustik. Springer, Berlin (1984)
19. Randall, R.B.: Application of B & K equipment to frequency analysis, 2nd edn. Brüssel & Kjær, Nærum (1977)
20. Beranek, L.L.: Acoustic Measurements. Wiley, New York (1949)
21. Traunmüller, H.: Analytical expressions for the tonotopic sensory scale. JASA 88 (1990), S. 97 – 100. Acustica 10 (1960) 3, S. 185
22. Zwicker, E., Fastl, H.: Psychoacoustics. Springer, Berlin (1990)
23. Lamberg K.: Durchgängiges, automatisiertes Testen bei der Entwicklung von Automobilelektronik, Beitrag zur Tagung Simulation und Test in der Funktions- und Softwareentwicklung für die Automobilelektronik, Berlin (2003)
24. dSPACE, RTI and RTI-MP. Implementation Guide, Release 2013-A

# Part II
# Early Development

# Functional Engineering Platform—A Continuous Approach Towards Function Development

**Christina Stadler and Thomas Gruber**

**Abstract** Over the last years, the developers of vehicle functions faced the challenge of increasing numbers of control units. This led to more and more complex network structures inside a vehicle. To be able to develop vehicle functions under these conditions, all relevant aspects for the function under development have to be considered in the simulation and testing. The methods to satisfy these challenges are Model-in-the-Loop (MiL), Software-in-the-Loop (SiL) and Hardware-in-the-Loop (HiL) simulation. These offer the possibility to develop and test a function in its relevant context inside a vehicle through certain development phases. With the upcoming strong trend towards autonomous driving the current methods of function development are not sufficient anymore. On the way to the ability to drive autonomously, there are many advanced driver assistant systems (ADAS) to be developed. These involve more and more information about the environment of a vehicle. In addition to data from sensors there will be data from other road users and the surrounding infrastructure that influence new vehicle functions via so-called Vehicle-to-X (V2X) communication. Additionally, the vehicle will be able to receive information out of a backend system, like cloud-infrastructures, via wireless local area network (WLAN) or mobile communications. As a result in the future the complexity will not be limited to the networking of control units inside a vehicle, but vehicles will be cross-linked among each other, with their environment and with the world. This ongoing progress makes it inevitable to develop a new approach towards function development for the automotive industry.

**Keywords** Functional engineering · Time sensitive · Safety critical · Distributed simulation · Time synchronization

C. Stadler (✉) · T. Gruber
AUDI AG, I/ET-84, 85045 Ingolstadt, Germany
e-mail: christina1.stadler@audi.de

T. Gruber
e-mail: gruber.thomas@audi.de



C. Gühmann et al. (eds.), *Simulation and Testing for Vehicle Technology*,
DOI 10.1007/978-3-319-32345-9_6

# 1 Introduction

In the future there is not just the increasing number of networking control units within one car, but there will be many cars networking in a traffic scenario. Besides that increasing complexity there is a second challenge the functional engineering aims to overcome. The number of functions that have to be developed and tested will increase rapidly in the near future. That rise will not be manageable by driving tests or domain specific test platforms anymore. Therefore, a generic approach is necessary to reduce the effort to a minimum. That approach is called Functional Engineering Platform (FEP) by Audi. In this paper, the technical approach to the future functional engineering by Audi and a basis for the definition of the time synchronization approach for the FEP will be presented.

Chapter 2 introduces FEP as the answer to the upcoming challenges at Audi. Chapter 3 gives a short introduction into the technical aspects of FEP and defines terms, that are used in the following chapters. On the way to autonomous driving, there will be more and more safety and timing critical vehicle functions to be developed. Functions of that sort require a specific approach on testing. Chapter 4 shows the necessity of a time synchronization mechanism for the engineering platform. The requirements the mechanism has to meet are derived from the use cases of FEP. Afterwards, conceptual questions concerning the mechanism are introduced and possible answers are discussed.

# 2 Audis Answer to the New Requirements

The AUDI AG launched the project "Functional Engineering Platform (FEP)" to face the upcoming complexity in function development resulting out of the increasing networking of vehicle functions. This networking of vehicle functions does not only imply the communication of control units inside one car, but also implies the communication between vehicle functions of various cars, present in a specific scenario. This constellation leads to much more dynamic situations concerning the context of signals a car and its functions are situated in. Thus, it is not enough for the function development to simulate all technical components of a vehicle that interact with the function under development. Additionally, the environment of the vehicle carrying the new function and all other road users interacting with that vehicle have to be simulated. As a result the scope of the simulation and the containing models will have to be significantly extended regarding the function development. The key factor and base for the development of highly networking vehicle functions over the limit of one vehicle is the simulation of the environment. That includes the road topology, the infrastructure, the other road users, even the buildings and trees. Additionally, all information channels a vehicle function uses have to be represented in the simulation. On the one hand there are sensors, on the other hand there is the communication between vehicles and between vehicles and infrastructure. All those sources of information about the environment have to be emulated in simulation models. For example models of vehicle-to-vehicle (V2V) and vehicle-to-infrastructure (V2I) communication based on different technologies have to be available. As an example, this could be a connection of a vehicle to other vehicles or to mobile phones of cyclists or pedestrians. Another possible communication is the one with a server backend system, located anywhere in the world, or with road side units in the surrounding of a vehicle, like traffic lights. That communication cannot be modeled with the common physical methods, because it derives from another domain, the information technology. These models have to represent the following elements the V2X communication consists of:

- Content of the packets sent via the V2X communication
- Protocols defined by standards
- Propagation phenomena of radio waves

The type and the sending frequency of messages sent via a V2X communication are defined by standards. The type, the sending frequency and the content of a message depend on the scenario the sender is located. Additionally, there are standards that define the protocols which realize the communication technology. That protocol stack is defined for different regions in the world. The basis is the standard for wireless local area network (WLAN) communication [1]. In the USA the protocol stack is called wireless access in vehicular environments (WAVE) and is standardized by the Society of Automotive Engineers (SAE) [2]. In Europe there is the intelligent transportation system access layer for the 5-GHz band (ITS G5), standardized by the European Telecommunications Standards Institute (ETSI) [3]. The third aspect, that completes the simulation of a V2X communication, represents the propagation of radio waves. It strongly depends on the environment of a communicating vehicle. For example buildings shadow radio-signals. Depending on the building the shadowing can result in a failure of the communication. Evidentially, there is a locked dependency between these two worlds the computer science, that defines the communication possibilities, and the physical environment, that forms the actual exchange of information. The simulation has to combine those worlds in a realistic way. That requires new modeling methods besides the common physical ones.

Besides the wider scope of the simulation for function development, a second big advantage the FEP will offer is a continuous approach towards function development over the whole V-model and through all collaborating departments. A vehicle function can be influenced by many parameters resulting out of various domains, like mechanical and electrical engineering or computer science and others. There is a variety of specific tools, where each is most suitable for one domain. As a result, if the function under development should be simulated in the context of all its input and output parameters, there will be models designed in many domain specific tools involved in one simulation. Each of these models can be delivered by a domain specific department. In the past, that process made it necessary for the function developer to redesign the received model in his preferred engineering tool. That implies that the developer has knowledge of the foreign simulation tool, its programming language and an in-depth understanding of all models contained in his engineering environment. That means a Hardware-in-the-Loop test engineer needs to have a profound knowledge about around 70 models. Besides that, this approach results in operational overhead and additional costs that produce no added value. The project FEP delivers a technology based on the so called FEP Library, that allows the coupling of various widespread engineering tools. For the function developers this possibility leads to the advantage that they are able to work in their preferred tools and programming languages. In the end, the technology saves unprofitable effort and that means it saves money. The FEP Library delivers a generic communication interface between participants of the simulation network. These participants not always have to be tools and models. Also hardware like a real control unit can be part of a simulation network based on the FEP Library. This technology allows the coupling of models, software and hardware. That approach supports the reuse of models to set up development and test environments through the whole V-Model in the development process.

# 3 Introduction to FEP

## *3.1 Definition of Terms*

To define some FEP specific terms, Fig. 1 shows an exemplary FEP System from the hardware view and the corresponding logical view. The exemplary real system on the left side consist of three computers building a network based on Ethernet. One of these computers is a laptop, on which the engineer develops the vehicle function. One PC runs the simulation of the virtual environment and another one

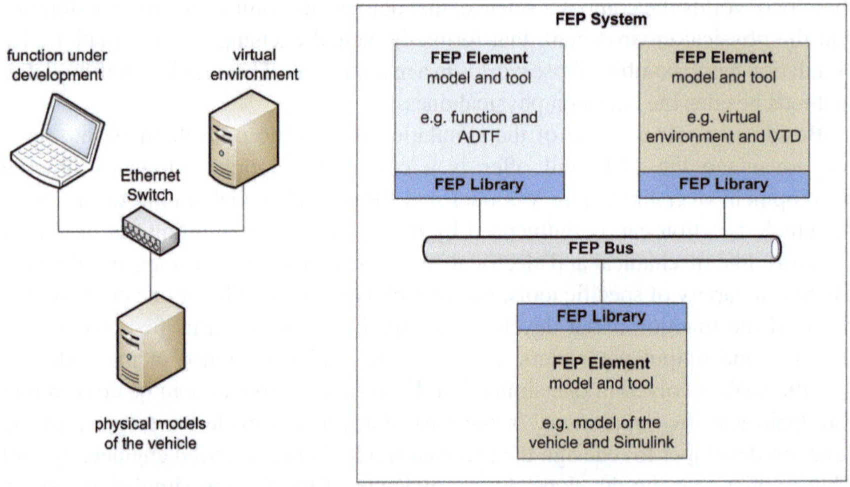


**Fig. 1** Exemplary FEP system: real system (*left*) and logical view (*right*)

runs the simulation of the physical models representing the vehicle that moves in the virtual environment and carries the function under development.

This exemplary FEP System is designed that way that one simulation tool runs on one separate hardware. That does not necessarily occur in a productive system, but it simplifies the definition of terms. The FEP System in Fig. 1 consists of three FEP Elements, which are allocated to separate computers and communicate over the FEP Bus. As already mentioned, the project FEP delivers the FEP Library to couple simulation tools. That library summarizes all features of the FEP and is accessible over a C++ application programming interface (API). To build a FEP Element the FEP Library hast to be integrated into a simulation tool or model. If FEP Elements are allocated to various computers, these are connected via an Ethernet network and exchange data over the logical FEP Bus.

## 3.2 Co-simulation

The FEP Library is not designed as a runtime environment. That means it does not execute any models or compute anything by itself. It enables various engineering tools to communicate with each other in a very autonomous way over a generic interface. A requirement caused by the distributed concept of FEP in combination with the possibility for co-simulation is a synchronization mechanism. Co-simulation means there are at least two simulation tools running at the same time, that require current simulation data of each other. Therefore, the exchange of data between two or more tools requires a coordination. Depending on the use case of a FEP System there are different suitable mechanisms. Two main use cases can be defined:

- Simulation of not timing critical vehicle functions, for example functions concerning the comfort or engine control
- Simulation of timing critical vehicle functions, for example functions concerning the safety or some ADAS functions

The synchronization of a simulation system to develop not timing critical functions can be reached by an event-driven approach. A suitable example for that use case is a mechanism called High Level Architecture (HLA). That synchronization approach was designed by the Defense Modeling and Simulation Office (DMSO) for the U.S. Department of Defense and standardized in the year 2000 [4]. Reid published an evaluation of HLA, that points out the possibilities and limitations of that approach [5]. There are many parallels to the concept of FEP, that make the HLA suitable as an event-driven synchronization approach. HLA is able to synchronize both event- or time-triggered simulators. It is based on the publish and subscribe paradigm, wherein persistent data is handled as objects similar to object-oriented programming languages. Reid proofed that it is possible to implement the HLA mechanism with the programming language C++, but it is not possible to integrate hardware in a system synchronized by HLA. The second disadvantage of HLA is that there is no constantly advancing time. For that reason it is not possible to develop and test timing critical vehicle functions with that approach. The aim is to serve an additional approach on the synchronization of a FEP System with the focus on time sensitive vehicle functions. The mechanism will be based on a constantly advancing time over the whole simulation system. That approach should also be suitable for FEP Systems with real time operating hardware included.

# 4 Synchronization for Time Sensitive Functions

## *4.1 The Necessity of Time Synchronization*

A simulation network based on the FEP Library does not have to, but can consist of more than one computer with simulation tools running on it. That possibility brings a technical issue with it that has to be solved. A personal computer contains a clock, that increases its time by tics given by the frequency of an oscillating quartz. A quartz is a natural product, so there never will be two absolutely identical quartzes available. That fact causes differences in phase and frequency over all computers and results in different behaving clocks [6]. If there are two computers in a FEP System simulating models and exchanging data, they have two different clocks drifting apart because of their quartzes. That drift results in two different internal times of the computers and causes different simulation times. This fact causes problems in the development of time sensitive vehicle functions. If the models have different understandings of the current simulation time, it will result in a wrong reaction or a reaction at the wrong time of the function under development. In the end, this issue leads to the fact that time sensitive functions that have to show a reaction at a specific point of time cannot be developed with that engineering platform. On the way to autonomous driving, more and more time sensitive and concurrently safety-critical functions will be developed. If the functional engineering platform aims to face the future challenges of function development, it has to have a synchronized simulation time over all participants of the simulation.

For example all vehicle functions concerning the phase immediately before a crash happens are time and safety-critical functions. These functions are intended to reduce the consequences of accidents that are not preventable any more. This task results in a very small range of time within that the function has to finish its action. Safety-critical functions cannot be developed and parameterized in real test drives due to the risk of injury. That is one of the use cases the functional engineering platform is intended for. Besides the development of safety-critical functions in the simulation this example includes an additional use case the functional engineering platform aims to. That is the co-simulation of the environment including other road users with the function under development. Additionally, all necessary components of the vehicle that carries the considered function have to be co-simulated. The concept of FEP meets the use cases of the future functional development. As explained, one central precondition for time sensitive functions is a synchronized simulation time over the whole FEP System.

## 4.2 Derivation of the Technical Issues

To be able to develop a concept of time synchronization for FEP, all possible use cases have to be known. Figure 1 shows the derivation of these use cases depending on the number of participating FEP Elements and on the number and type of computers the FEP Elements are allocated to.

The three lower grey boxes (see Fig. 2) contain one FEP Element simulated on one or more computers. The first grey box shows the case where one simulation runs on one computer. That is a classic simulation where no FEP Library is needed to couple FEP Elements. The second grey box shows that a single model runs on more than one computer. That case is meant to reduce the simulation time due to an increased amount of available computing power, but that is not the aim of FEP. A special type of that case would be the one shown in the third grey box. That combination leads to a scenario, where one model runs on more than one computer and at least one of these is a real time operating hardware. That case does not occur in reality. These three types are no use cases of FEP and so have not to be considered for the development of the timing concept. In contrast, the upper blue boxes (see Fig. 2) all represent use cases of the FEP. The co-simulation of more

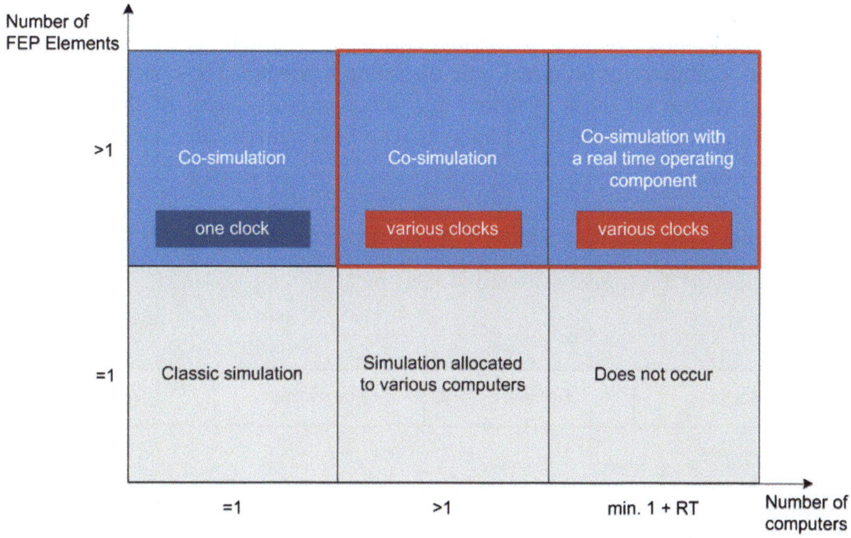


**Fig. 2** Derivation of possible use cases of the FEP (*blue background*)

than one FEP Element allocated on one computer is the simplest use case regarding the timing aspects. That system contains one clock triggering the simulation time. As a result it does not need any mechanism of time synchronization. There are two use cases left that require time synchronization. They are bordered by a red frame in the Fig. 2. On the one hand that is the co-simulation of more than one FEP Element on more than one computer. Various computers lead to various clocks drifting apart. On the other hand that is the co-simulation of more than one FEP Element on more than one computer with the addition that at least one of these computers is a real time operating hardware. In that case the progression of the simulation time regarding the whole FEP System is restricted to the computing time of the real time hardware, that is the wall clock time.

As a result the concept of time synchronization has to enable two types of FEP Systems:

- Co-simulation on various computers
- Co-simulation on various computers including at least one real time operating hardware

Figure 3 shows the ordinary portfolio of simulative development and test environments for vehicle functions. These are classified by the two identified types of time synchronization relevant FEP Systems (see Fig. 2). As a result, if FEP is able to support those two use cases, consequently the common engineering environments for vehicle functions can be implemented based on the technology of FEP.

Model-in-the-Loop systems are realized as a pure co-simulation. Hardware- and Vehicle-in-the-Loop systems contain at least one real time operating component by definition. Software-in-the-Loop systems can be realized as both co-simulation with or without real time operating hardware. That means the real software code of a function can be implemented as a virtual electric control unit (vECU) computed on an ordinary computer or on a real time operating hardware. The concept of FEP is suitable to realize all these systems, that serve various stages of the functional engineering. As a result the FEP Elements with their generic interfaces can be used

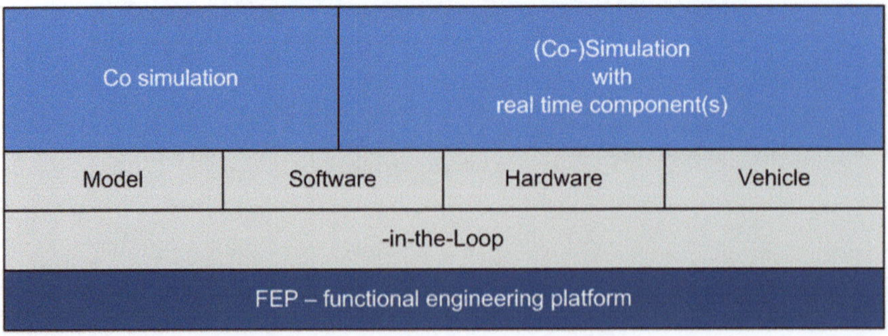


**Fig. 3** Mapping of the ordinary portfolio to the use cases of FEP

through the whole simulation based development cycle of a vehicle function. That fact represents the big advantage of the FEP as a platform for continuous engineering.

## *4.3 Time Synchronization of the FEP*

There are two possible ways to achieve a synchronous communication within a simulation allocated on more than one computer. One way allows a synchronization by events without a constantly advancing base of simulation time. A possible approach of that type has already been discussed in Sect. 3.2. The ability to analyze timing behavior of vehicle functions under development requires a common base of simulation time within a FEP System. In that case, time synchronization over the participating computers is the prerequisite for the synchronization of the communication between various FEP Elements (see Fig. 4). Those mechanisms must not be mixed up. It is not suitable to have one central base of time at the hardware level. There also has to exist one common simulation time over the whole FEP System to reach a communication based on a common understanding of time. Additionally, there has to be a mechanism to synchronize the exchange of data on the basis of a common time inside one FEP System.

**Requirements**
The concept for a time synchronization mechanism has to be developed based on functional and nonfunctional requirements. The use cases of FEP define functional requirements:

- Co-simulation as fast as possible and in wall clock time
- Accuracy
- Robustness
- Minimal load of processing
- Minimal data traffic

The nonfunctional requirements and additional functional ones have to be determined by the developers and users of FEP:

- Minimal costs
- Minimal configuration effort

The two use cases of FEP, that are detected in Fig. 2, make basic functional requirements obvious. The mechanism for time synchronization has to be suitable for variable concepts of the simulation time. That includes the following both use cases. The simulation system runs as fast as possible for a pure co-simulation or it runs in wall clock time for a co-simulation with at least one real time operating hardware participating. As a result the mechanism has to be modular concerning the speed of the simulation time. A further functional requirement is the maximum permitted difference of the clocks at the runtime of a FEP System. That value is

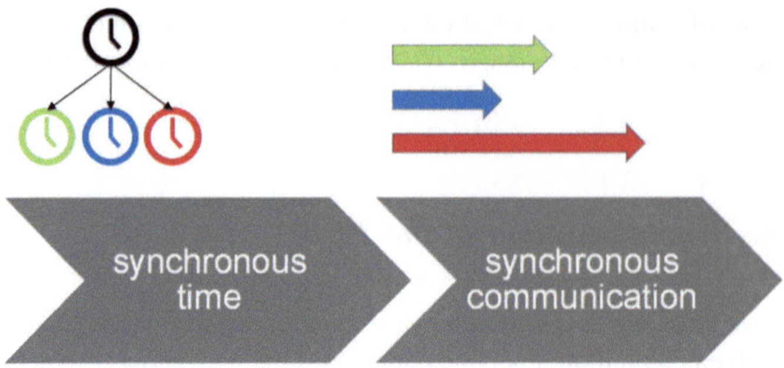


**Fig. 4** Relation of synchronous time and communication

defined by each specific use case depending on the function under development and on the development phase. Additionally, the concept of the time synchronization has to be robust against loss of time source and timing data packets. Two more functional requirements exist regarding the process of the synchronization. During the runtime of a FEP System the load of processing and the data traffic concerning the synchronization have to be as minimal as possible. That is an important requirement to make sure that for every FEP System a high available data throughput can be guaranteed. Nonfunctional requirements are minimal costs to implement the synchronization mechanism on a FEP system and minimal operational effort for the configuration of the mechanism.

**Conceptual aspects**

To define the most suitable time synchronization mechanism for FEP several questions regarding conceptual aspects have to be answered.

1. What is the source of the common time and who takes responsibility for it?
2. How does the distribution of the common time occur?
3. On which layer of the system is the common time located?
4. How does the synchronization happen during runtime?

There are three possible ways to face the question about the source of the common time.

- Coordinated Universal Time (UTC)
- Clock of a computer, that runs FEP Elements
- Clock of a separate time server

The Global Positioning System (GPS) signal delivers a very precise wall clock time, the Coordinated Universal Time (UTC). It is well suited for a FEP System with real time operating hardware involved. A disadvantage of that approach is the necessary hardware to receive the GPS signal. Another possible provider of the common time is the internal clock of a computer inside a FEP System. That computer may be a participant of the system carrying FEP Elements or a separate participant that exclusively provides the common time. The latter option provides a computer, that has no processing load generated by the simulation, but it would be the only source of time. That fact causes a weaker robustness compared with a higher number of possible sources. Concerning that requirement, a general participant that is able to transfer its responsibility to any other participant of the FEP System is the most robust option of these three ones. In the following this approach will be named as the spread source approach.

If there is a source of the common time in the FEP System, there has to be a defined way for the distribution of the reference time. One possible way is that the participant that holds the common time constantly triggers all other participants in the system. That causes high data traffic on the FEP Bus. In contrast a periodic signal inherent in the system could operate as a trigger, but the existence of that signal cannot be guaranteed in all use cases. The third approach is a compromise concerning the accuracy and the data traffic. Every participant of the system carries its own clock that gets synchronized as soon as necessary. The participants have to communicate to compare their current time. With a learning algorithm that communication might be reduced to a minimum load. Additional there lies a high robustness in the system. If that approach gets combined with the spread source of the common time, the simulation keeps on running even if a time delivering participant gets lost for any reason.

The accuracy of a synchronization mechanism depends on its implementation in the system. The sourcing of time can happen at the processor of the network interface card (NIC) or at the central processing unit (CPU) of the computer. The first approach requires the support of a specific hardware, but it enables a higher accuracy of time synchronization. A common NIC collects data and transfers it accumulated. That process consumes time. As a result the FEP Library, which runs on the CPU of a computer, has no knowledge about the processing in the NIC. If the requirements concerning the accuracy of time synchronization are strict enough, the processing time in the NIC is not negligible. That requires a special NIC that supports the synchronization. The same applies to a network switch. They also have to support the synchronization to reach higher accuracy than without special hardware.

A central aim for the operational synchronization mechanism is to cause as less load of processing for the participants and as less data traffic in the FEP System as possible. Therefore a synchronization mechanism has to be chosen, that matches these requirements during the runtime of the simulation.

**Evaluation of time synchronization approaches concerning the functional requirements**

The two most important mechanisms for time synchronization are discussed here, the Network Time Protocol (NTP) and the Precision Time Protocol (PTP). Table 1 shows an overview of the evaluation of these two approaches on time synchronization concerning the functional requirements.

*Characteristics of NTP*

NTP does not focus on local area networks (LAN) it also enables synchronization via the Internet [7]. Hielscher shows an offline synchronization approach with NTP

**Table 1** Evaluation of NTP and PTP concerning the functional requirements

| | NTP | PTP |
|---|---|---|
| As fast as possible and wall clock time | + | + |
| Robustness | + | + |
| Accuracy | – | + |
| Minimum load of processing | + | + |
| Minimum data traffic | + | – |

under the support of a GPS receiver [8]. That approach points out the possibilities of NTP, but it does not meet the requirements concerning the time synchronization of FEP. Because of closed-loops within FEP Systems the time synchronization has to be done during the runtime, that means online. That approach ensures that causal relations within a FEP System are correctly represented. Under optimal conditions, that means in simple networks, NTP enables a precision of the time synchronization of tens of milliseconds over the Internet [9]. Even if these conditions could be guaranteed for every FEP System, tens of milliseconds would not be accurate enough for time sensitive and safety-critical functions. For example functions of autonomous driving require an accuracy in the range below one millisecond. That necessity can be proved by a simple calculation: If the difference of two clocks is 100 ms and the vehicle in the simulation drives 100 km/h, there will be an inaccuracy of 2.78 m in the simulation. That is far too much difference for safety-critical functions. If the difference of two clocks is 1 ms, that will be an inaccuracy of 2.78 cm in the simulation. In the case of a lane keeping assistant 2.78 cm on both sides mean an inaccuracy of 5.56 cm. That example makes evident that an accuracy below one millisecond is crucial to get reliable results out of a simulation. As a result the synchronization by NTP is not precise enough for the FEP. Another characteristic of NTP is the fact, that it needs an external time server. The reference signal might be received via the Internet or via the PPS API [8, 9]. An external time server, like a GPS signal receiver, has to be connected via the API. That approach only will be robust against the loss of the common time, if there are more than one time server available for the FEP System. Under certain conditions NTP is sufficient for long-term tests based on FEP. For the synchronization process only two messages are needed. The client sends a request message to the server, which answers with one response message [8]. That causes a little less data traffic than PTP (see Fig. 5). NTP calculates the offset of clocks by simple mathematical equations. As a result the appropriate processing load is low.

*Characteristics of PTP*

PTP, standardized by the Institute of Electrical and Electronics Engineers (IEEE), provides a protocol for local area networks based on Ethernet with the focus on the precision of time synchronization [10]. It is designed to support multicast communication over a direct connection or switches. These properties meet the characteristics of a FEP System. The implementation of PTP defines the achievable accuracy. As pure software code PTP reaches an accuracy similar to the accuracy of NTP, with the appropriate hardware support it reaches an accuracy in the range of

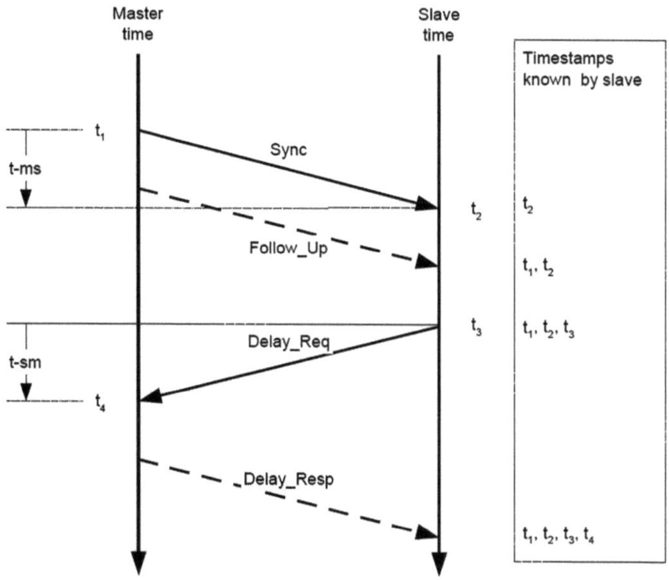


**Fig. 5** Sequence of synchronization messages [10]

microseconds. The latter approach meets the requirement of an accuracy below one millisecond for the development of autonomous vehicle functions. The Best Master Clock Algorithm (BMCA) detects the participant, which is best suited for the role of the time master. That algorithm gets executed every time the topology of the network changes. That meets the requirement concerning the robustness of a time synchronization mechanism. Every participant of a FEP System has the possibility to be the time master. The responsibility changes, whenever a participant exits the network. Four messages are needed to calculate the offset between master and slave. Figure 5 shows the sequence of those messages. In contrast to a continuous triggering signal that approach creates less data traffic. PTP causes low processing load by the calculation of the offset, just like NTP.

*Evaluation of NTP and PTP*

There are three functional requirements in respect of which the characteristics of NTP and PTP differ from each other. That are the robustness, accuracy and the minimum load of data traffic. The robustness is a requirement, that is decisive for long-term tests. Especially on HiL test rigs long-term tests over a few hours or even days occur frequently. In these use cases the robustness of a time synchronization mechanism is crucial. If an error only gets detected after hours, when the test finished, a high amount of time and money will be lost. That example shows, that a robust synchronization mechanism is important for the FEP. Due to the BMCA, PTP provides the required robustness. If a time master gets lost in a FEP System synchronized by PTP, another component will become the time master. That mechanism is available for both use cases concerning the simulation time: As fast as possible and wall clock time. The BMCA does not differentiate whether the common time derives from an external source, like the GPS signal, or form an internal one, like the clock of a computer. NTP also provides mechanisms to ensure robustness against the loss of time servers. Both PTP and NTP meet the requirement for a robust time synchronization.

As mentioned above, NTP does not reach the accuracy of time synchronization required by FEP. The achievable accuracy by PTP depends on the implementation. Therefore, higher demands concerning the accuracy lead to increasing costs caused by the necessity of specific hardware. There is the possibility to synchronize by PTP without hardware support, but many use cases may require an accuracy that can only be reached under the usage of specific hardware.

As mentioned in the characteristics of NTP and PTP, there is a higher load of data traffic caused by PTP than by NTP. The robustness can be ensured by both approaches, but the required accuracy can only be provided by PTP. The requirement concerning the accuracy is more significant than the load of data traffic. Additionally, PTP implies two more degrees of freedom than NTP, so that there are more possible concepts of a FEP System. That degrees of freedom refer to the support of the synchronization by special hardware and the usage of an external time server. Both is possible, but not urgent for PTP. Nevertheless it is advisable to set up a FEP System with NICs and Ethernet switches that support PTP and with an external time server, like the GPS signal, to achieve high accuracy. Under these conditions the advantages of PTP fully benefit the FEP. There are four IEEE standards grouped under the so called Audio Video Bridging (AVB) that define prioritizing and synchronizing mechanisms or streaming of audio and video data for example in LANs. One of these standards, the IEEE 802.1AS, defines a type of PTP as the mechanism for time synchronization in the mentioned use case [11]. Due to that standard NICs that support PTP will gain currency. Therefore, the additional costs for special hardware are expected to decrease. PTP is the solution that meets the previous detected requirements of a time synchronization mechanism for FEP.

**From synchronous time to synchronous communication**

PTP is evaluated to be the approach that will provide the synchronous time in a FEP System. At the beginning of this Sect. 4.3 the difference of a synchronous time and a synchronous communication was mentioned. At that point, the synchronous time is established, but the communication still has to be synchronized in a FEP System. For example, if a HiL is based on FEP, there will be ECUs running parallel to simulations. The time base of ECUs cannot be influenced by the FEP via PTP. To reach a synchronous communication under these conditions, a view at the communication inside a real vehicle is useful. ECUs connected inside a real vehicle are not synchronized altogether. In a FEP System the real constellation should be emulated, to reach a synchronous communication. The simulations that emulate the physical and logical processes inside a vehicle, should run exactly in the real time, the wall clock time, as the correspondent real processes of a vehicle do. That means the common time base is the wall clock time, that is provided by a GPS signal. At that point there is no synchronous communication within the FEP System. Models that simulate the physics of a vehicle require a high amount of processing power. The first challenge is to get them computed fast enough. That condition will just be fulfilled, if the simulation runs faster than wall clock time and delivers the results at the required point of time. That leads to the second challenge. There must be a possibility to slow down the simulation, to reach a synchronous communication on top of the synchronous time.

## 5 Conclusion

This paper showed that the concept of the FEP answers the future questions about the development of vehicle functions on the way to autonomous driving. One issue is the increasing number of time and safety-critical functions that act on a higher level of automation. Another task the FEP is intended for, are functions based on data received from other road users. To meet the challenges that arise on the technical side of that project some further developments are necessary. One of that technical issues is to achieve a common basis of time for a whole simulation system. Therefore the most important approaches on time synchronization are discussed regarding the requirements the FEP use cases imply. It was deduced that PTP fits the best. It contains a degree of freedom concerning the compromise of accuracy and costs, which can be used to tailor the synchronization mechanism for certain use cases. Both PTP and NTP assume the network under synchronization to be symmetric for the calculation of the offset. That fact has to be considered during the conception of a FEP System.

After the time synchronization at the hardware level additional questions have to be answered. How will the common basis of time be available for the FEP System and its FEP Elements to reach one common basis of simulation time? How will a synchronized communication be established based on the synchronous time? How can event triggered messages be integrated in a system based on a constantly advancing time? These questions need to be answered to guarantee a reliable functionality of the FEP for the development of time sensitive vehicle functions.

## References


1. IEEE Computer Society: IEEE standard for information technology—telecommunications and information exchange between systems. In: Local and Metropolitan Area Networks—Specific requirements. Part11: Wireless LAN Medium Access Control (MAC) and Physical Layer (PHY) Specifications (IEEE Std 802.11-2012)
2. Society of Automotive Engineers. http://www.sae.org/
3. European Telecommunications Standards Institute. http://www.etsi.org/
4. IEEE Computer Society: IEEE Standard for Modeling and Simulation (M&S) High Level Architecture (HLA)—Framework and Rules (IEEE Std 1516-2010)

5. Reid, M.: An evaluation of the high level architecture (HLA) as a framework for NASA modeling and simulation. In: The 25th NASA Software Engineering Workshop. Maryland, Nov 2000
6. Latha, C.A., Shashidhara, H.L.: Clock synchronization in distributed systems. In: 5th International Conference on Industrial and Information Systems, ICIIS, India (2010)
7. The Network Time Protocol. http://www.ntp.org/
8. Hielscher, K.-S.: Measurement-based modeling of distributed systems. Dissertation Technische Fakultät der Universität Erlangen-Nürnberg. Erlangen (2008)
9. Mills, D., Thyagarjan, A., Huffman, B.: Internet timekeeping around the globe. In: 29th Annual Precise Time and Time Interval (PTTI) Systems and Applications Meeting. Long Beach, California, Dec 1997
10. IEEE Instrumentation and Measurement Society: IEEE Standard for a Precision Clock Synchronization Protocol for Networked Measurement and Control Systems (IEEE Std 1588-2008)
11. IEEE Computer Society: IEEE Standard for Local and metropolitan area networks—Timing and Synchronization for Time-Sensitive Applications in Bridged Local Area Networks (IEEE Std 802.1AS-2011)

# Part III
# Model-Based Software Test

# Efficiently Testing AUTOSAR Software Based on an Automatically Generated Knowledge Base

**Norbert Englisch, Roland Mittag, Felix Hänchen, Owes Khan, Alejandro Masrur and Wolfram Hardt**

**Abstract** The system architecture AUTOSAR divides an automotive ECU into three main layers—application layer, RTE and basic software layer. Due to clearly specified interfaces, each of these layers and its software modules can be developed partly or completely by different companies with different tool chains. This promotes competition between companies and enables the hardware independence of applications, but it also leads to challenges during the integration phase of an ECU, the mapping towards different ECUs, and the reuse of functions. To ensure a high quality of AUTOSAR developments, we provide an approach that checks automatically the source code and configuration files of a project before compilation, i.e., performing static tests. Our approach is independent of the tool chain used and of the AUTOSAR version. The consistency and AUTOSAR compliance tests are based on a knowledge base which is filled semi-automatically by our parsers. The knowledge base contains information about the structure of an AUTOSAR ECU, i.e., basic software modules, layers, function names and parameters, hardware information, etc. We create the information by parsing once AUTOSAR meta-models, base projects and configuration files of the tools automatically. These inputs provide most of the information, necessary for our tests. Our static tests

N. Englisch (✉) · R. Mittag · F. Hänchen · O. Khan · A. Masrur · W. Hardt
Department of Computer Science, TU, Chemnitz, Germany
e-mail: enn@cs.tu-chemnitz.de

R. Mittag
e-mail: mitro@cs.tu-chemnitz.de

F. Hänchen
e-mail: haenf@cs.tu-chemnitz.de

O. Khan
e-mail: owes@cs.tu-chemnitz.de

A. Masrur
e-mail: a.masrur@cs.tu-chemnitz.de

W. Hardt
e-mail: hardt@cs.tu-chemnitz.de

 
C. Gühmann et al. (eds.), *Simulation and Testing for Vehicle Technology*,
DOI 10.1007/978-3-319-32345-9_7

check the consistency between configuration files and source files, validate the configuration, and parse source files for non-compliant code within the given AUTOSAR project. Besides a test report, we visualize test results in a graphical user interface (GUI) which shows the layers and modules of the AUTOSAR project. Data for visualization is further provided by our knowledge base. All problems like inconsistencies or missing modules are visualized by flags on the corresponding elements to support testers and developers finding solutions. Our approach increases the quality of AUTOSAR projects by detecting inconsistencies, errors, and non-compliant code with AUTOSAR. In addition, by finding inconsistencies/errors in an early phase, our semi-automated test approach provides an optimal basis for the dynamic test of an AUTOSAR ECU supporting and complementing acceptance tests. Finally, all known AUTOSAR tool chains with readable configuration files and source files can be integrated with our approach.

**Keywords** AUTOSAR · Knowledge base · Static test

## 1 Introduction

Traditionally, applications in the automotive domain were developed for a specific platform, defined by operating system and hardware (HW). In contrast to that, the system architecture AUTOSAR [1] defines a function-oriented development approach. Therefore, applications are developed independent from the hardware. AUTOSAR (AUTomotive Open System ARchitecture) defines three horizontal layers—the application layer, the runtime environment (RTE) and basic software (BSW). The RTE realizes the communication between applications and the connection to the BSW. After mapping the applications to concrete electronic control units (ECU), configurations containing parameters for hardware drivers, operating system and tasks of the application for the target platform need to be created. If the application or a part of the application is migrated to another ECU or another platform, only the configurations need to be adapted but not the code of the application itself. Hence, the AUTOSAR approach saves time and resources in a project, starting with the second development cycle after introducing the system architecture. Besides the advantages, AUTOSAR creates new challenges for the test of ECUs. Localizing sources of error in AUTOSAR becomes a problem due to the large number of modules involved. Moreover, due to the heterogeneous tool environment a wrong or non-optimal configuration may occur.

The approach in this paper provides a set of static test techniques [2] for all layers of an AUTOSAR ECU, as displayed in Fig. 1. Static tests do not execute the test object, but they analyze it. The test cases are independent from the used tool provider and can be defined for each AUTOSAR version separately. The test cases are dependent on a database, which stores abstract AUTOSAR knowledge. This database can be filled automatically by our project parsers for the most parts. Manual changes in the suggested AUTOSAR database can be done additionally.

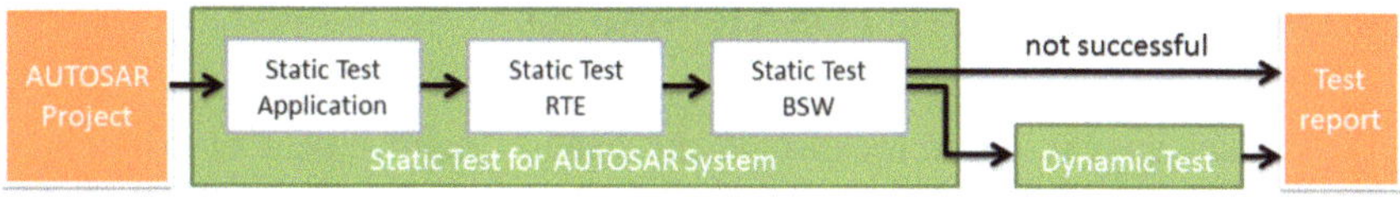


**Fig. 1** Abstract workflow for test of AUTOSAR systems

We have integrated our approach in a tool for static and dynamic test of AUTOSAR systems. The workflow for generation and execution of dynamic test sequences of AUTOSAR systems is explained in [3]. This tool was used successfully for different developments of our Automotive Demonstrators.

## 2 State of the Art

Independent from the system architecture, an efficient combination of static and dynamic tests is important to enhance the quality of the development. Static test techniques will mainly support the developer during the development process. Dynamic tests will check the functionality in a later development phase, when the test object is executed and compared to the specification. Especially commercial tools are providing static tests for configuration of different AUTOSAR layers. But these test cases are mostly company-specific and cannot be used, if the developer team changes the provider in the tool chain, like shown in Fig. 2. Realistic changes could be the basic software, the RTE generator and the application architecture tool. This effect is increased when parts of the application are integrated into projects using other tools or when an application is reused after many years. Moreover, due to tools which are used for programming or configuration of one AUTOSAR layer, errors may occur after integrating the layers to one system. Therefore, static tests which consider more than one layer are needed for quality enhancement. Other approaches, like in [4], an abstract test pattern is introduced for testing where aspects are classified which serve as base for the test pattern. Here, the interfaces of the SWCs are statically tested. Besides the application and their SWCs no other modules are statically analyzed and only some basic software modules are tested in a simulation model. [5] describes some methods for testing AUTOSAR software components. The approach concentrates only on the application layer. Some static test methods are also described. In general they check the code against the MISRA-C rules in their proposed approach. But BSW modules are still not checked at all. A common approach for static test of a complete AUTOSAR system is again required.

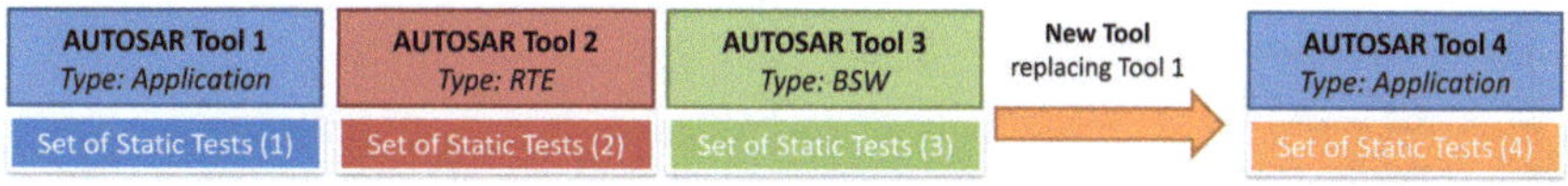


**Fig. 2** Exchange of an AUTOSAR tool results in a new set of static tests

## 3 Static Testing of AUTOSAR Systems

The approach in this paper focuses on the quality improvement of AUTOSAR systems by static testing. These static tests are independent from the used tool provider and target platform. Moreover, our test cases can be defined for different AUTOSAR versions. Therefore our approach contains an AUTOSAR knowledge base which consists of architectural know-how, created mostly automatically by parsers. Our static tests analyze source code and configuration files for consistency and compatibility.

Especially the static tests which contain checks for configuration and source files across different layers are improving the quality of software. Considering that mostly different tool providers are used for the separated layers, it is obvious that these checks are necessary. Moreover, changing a tool provider in projects should not decrease the analysis quality of configuration and source code.

In the following chapters we explain selected examples of the light control of our automotive demonstrator. This application is implemented in AUTOSAR version 3.1, running on a STM SPC560 [6]. The basic software is an Elektrobit tresos product [7]. The application development is done with dSpace SystemDesk [8]. The light control is responsible for the interior and exterior illumination of the demonstrator. There are several lights which can be accessed by the light control. While the exterior light control actuates all lights attached to the outside of the car (e.g., low beam, daytime running light, and indicators) the interior light control handles the ambient light of the demonstrator. Besides the illumination, some logical checks are executed. E.g., the indicator switch position is checked. The indicator switch cannot be in two positions at the same time. Moreover some sensors are used to realize functions like an automatic light switch which is based on the environment brightness.

### *3.1 Static Test for Runnables*

As shown in Fig. 3, the runnables in the application layer can be checked. The implementation of software components (SWCs) is realized by runnables.

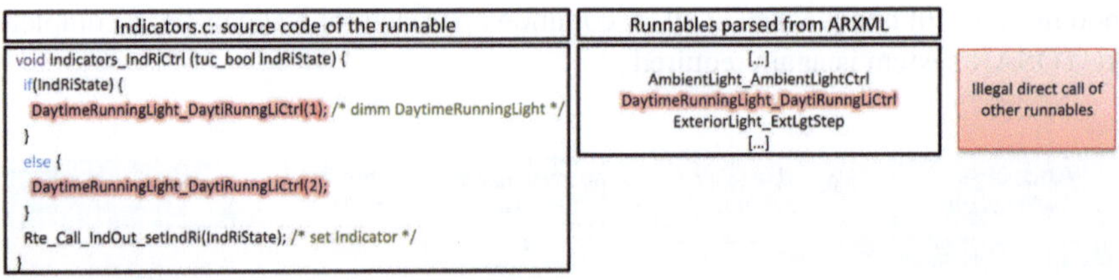


**Fig. 3** Checking runnables in the application layer

Runnables are functions which are started by the RTE and can be mapped to OS tasks. Each SWC consists of a set of runnables which is schedulable. Therefore we analyze them and check if there are illegal direct calls of other runnables. We do that by parsing the ARXML file and saving all available runnable names to our knowledge base. Then we parse the corresponding source files and check if some of the used function calls are corresponding to known runnable names. Furthermore illegal calls of basic software functions can be detected by verifying all found functions with the module interfaces stored in the knowledge base.

### *3.2 Static Test for Task Definition*

When considering the RTE, an important part is correct task mapping of runnables. This can be a problem if the RTE is changed without adapting the runnable mapping and hence this part has to be tested. In the first step (1), as seen in Fig. 4, runnables that are mapped to a specific task are analyzed. This is done by parsing the RTE calls of a runnable and depending on the RTE generator, specific functions must also be parsed. In this case, the function call "ActivateTask" specifies the task that should execute this RTE call. In the present case this is only Task14. The next step is to check if this task exists in the RTE. This can be verified in the extract of the "Rte.c" file (2). Here it is checked if the task which was found in the previous step is bound to a macro. In this case this check would be successful. The last step (3) would be to check the configuration of the operating system if this specific task is declared. Therefore, all "DeclareTask" functions are parsed and checked for all found tasks.

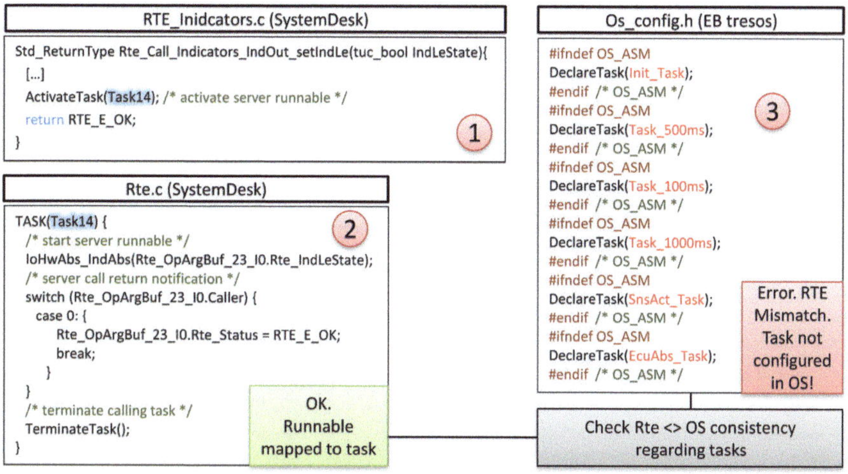


**Fig. 4** Checking task definitions in basic software and RTE

The test case would fail because no such task is declared. This can happen when the user replaces the RTE files without adjusting the configuration of the basic software. The knowledge base stores the distinctive function calls for specific RTE generators.

## *3.3 Static Test of Application Data Types*

One aspect that is tested in the application layer is the consistency of the used datatypes. We parse the ARXML file of the system, like displayed in Fig. 5, and save important information, like datatype definitions, datatypes used by interfaces etc. In the shown case, the port "IntrLiGreenIn" is checked and from its definition, a reference to the used interface can be found. The next step involves in checking of interface definition of the used datatype. Next we check the runnables and determine if the datatypes that are used match the target. Therefore, RTE function calls that access ports of a SWC and also the variables that are used in these calls are parsed. In this case it is recognized that the variable "green" is of type "tuc_bool" but the corresponding RTE call will return "tuc_uint8". If any deviations are encountered, we raise an error message to inform the user. Depending on the compiler and its configuration, the compilation process can result in a warning. Thereby, unexpected results can occur during runtime. Detecting or localizing this kind of error is mostly hard during runtime in the later dynamic test process. Additionally the datatypes are checked against values from the knowledge base where compliant datatypes can be found for each AUTOSAR version.

## *3.4 Static Test of Module Versions in Basic Software*

As shown in in Fig. 6, the software version of basic software modules can be checked. This kind of analysis is part of the static tests for the basic software

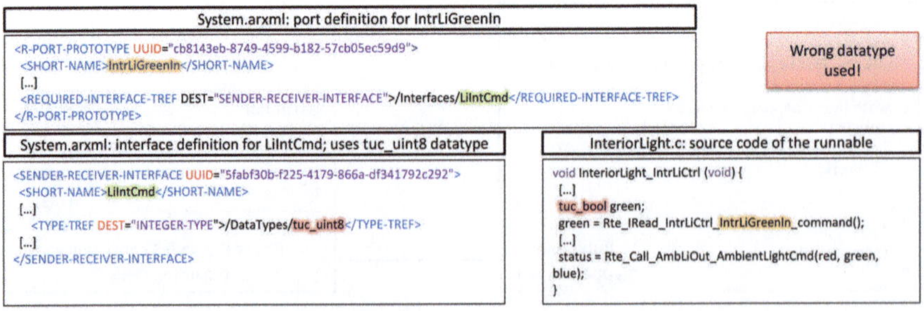


**Fig. 5** Checking datatype conformity between software components

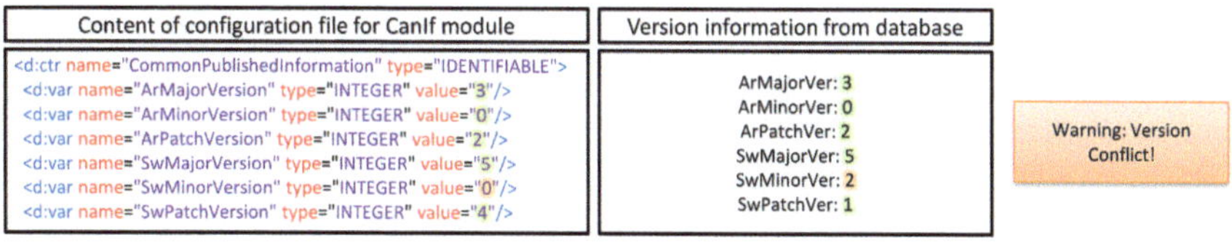


**Fig. 6** Checking basic software module version in configuration file

(see Fig. 1). This approach checks the module consistency among the basic software modules. One part of this check is to analyze version information of the modules that are part of a project and compare them with our reference basic software for a specific platform. The reference is stored in our database. If the analysis encounters deviations we warn the user. This shall prevent unexpected errors or results. Another aspect checked is the integrity of the basic software configuration. With the help of the knowledge all necessary modules are verified for their presence. For example, if it is found that the "Can"-Module is present, but not the "CanIf"-Module then a warning is raised.

# 4 AUTOSAR Knowledge Base

The AUTOSAR knowledge base contains all the details of the BSW. Modules created by different vendors that are particular to a hardware platform or AUTOSAR version can be included in this knowledge base. Each software layer of the system architecture, its properties, and contained modules are represented. For every software module, the dependencies to other modules, respective interfaces and version details are stored in a SQLite database. We use this knowledge base for efficient and automated static and dynamic test techniques of our automotive developments.

## *4.1 Population of Knowledge Base*

As depicted in Fig. 7, the AUTOSAR knowledge base can be populated by different sources. Firstly, the AUTOSAR project is parsed. Since the contents of project configuration files are standardized, it forms a good data foundation. To complete this foundation, additional vendor specific file formats can be used as data source to be parsed for reusable information. In addition to this, a manual data administration is also featured. Furthermore, extraction of module interface is done by parsing BSW module source codes. The sequential process of data base population can be seen in Fig. 7. Table 1 represents information that is extracted from AUTOSAR project data, more specifically from BSW modules configuration files. The files contain general information about the version of a module and its vendor.

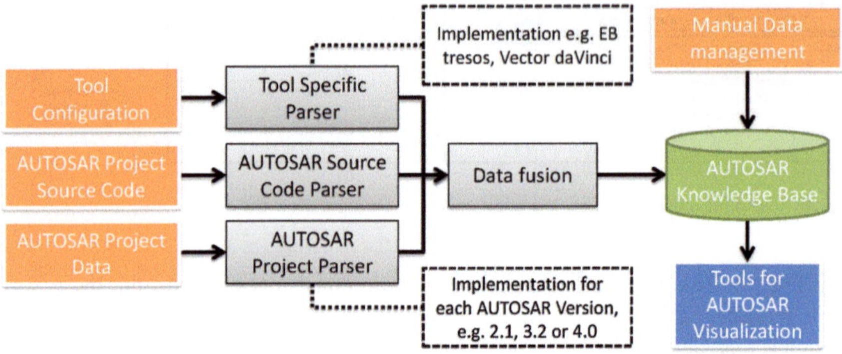


**Fig. 7** Population sequence of AUTOSAR knowledge base by parsing different sources

**Table 1** Details of the AUTOSAR project files for AUTOSAR knowledge base

| XML element | Description |
|---|---|
| ArMajorVersion | AUTOSAR version of the module |
| ArMinorVersion | |
| ArPatchVersion | |
| SwMajorVersion | Software versions of the module |
| SwMinorVersion | |
| SwPatchVersion | |
| ModuleID | AUTOSAR ID of the module |
| VendorID | AUTOSAR ID of the BSW vendor |

Table also contains the BSW modules' vendor specific implementation version and as well as the AUTOSAR specification version. An extract of the tool specific configuration file of Elektrobits tresos studio is shown in Table 2. These files contain module related detailed information, like its name and if this module is

**Table 2** Descriptions of tool specific configuration for the AUTOSAR knowledge base

| XML element | Description |
|---|---|
| label | Name of the module |
| mandatory | Obligatory module |
| allowMultiple | Multiple instances of the module are allowed |
| description | Short of description of the module |
| copyright | Module vendor |
| relVersionMajor | Parent AUTOSAR versions of the complete BSW |
| relVersionMinor | |
| relVersionPatch | |
| target | Target platform |
| derivate | Version of target platform |

obligatory, short description and allowance of multiple instances. It also specifies to which parent AUTOSAR version a particular module belongs to and on which platform this module is intended to run.

In the first phase of parse operation, the corresponding installation folder of a BSW configuration tool is analyzed to check for AUTOSAR configuration templates, tool-specific configuration files and module source codes. Folders with irrelevant data are ignored through an exclusion criteria provided in a list. After analysis and parsing of these files, the obtained data is processed. In order to merge information from various sources, the distinguishing feature of the module name is used as it occurs throughout. Before the input of data into the database, the BSW vendor is checked to avoid redundancy. Multiple versions of the same vendor can exist and are not considered as redundant information. Similarly it is elicited for HW platforms. After data comparisons, the inputs for modules and their interfaces are created.

## *4.2 Schema of Database*

Based on the formalization, the knowledge of our knowledge base can easily be used for visualization of AUTOSAR knowledge. Our visualization is used checking the data and training for students and developers. As shown in Fig. 8, the database schema for the basic software is divided into roughly three parts. The green tables contain information to uniquely identify the BSW modules. The important details that are required for interfacing of modules in the visualization tool are stored in the yellow tables. And the third part is the blue tables where detailed information of the individual modules is populated.

To identify each table uniquely the following tables are necessary: AutosarVersion, Manufacturer, Platform and Vendor where the first table provides all the existing versions of AUTOSAR in the database. As the name suggests, the table Vendor specifies the provider of individual BSW modules. Table Manufacturer enlists all existing HW manufacturers and similarly Platform incorporates their respective HW platform. Information of each BSW module is incorporated in the following tables: AutosarStack, AutosarBSWModule, AutosarBSWModuleVersion, BSWLayer, BSWModuleDescription, CHeader, HeaderOfModule, InterfaceBlob, InterfaceParameter, ModuleInterface, SpecificationBSWLayer, SpecificationBSWStack, TestType. Basic data of a module like name and version are stored in AutosarBSWModule and AutosarBSWModuleVersion. The version information not only includes the software version, but also the AUTOSAR specification based on which a particular BSW module has been implemented. Textual descriptions of each module are encapsulated in BSWModuleDescription. The table CHeader contains names of all header files, whereas the table HeaderOfModule maintains a list of all the headers that belong to a particular module. Interfaces and

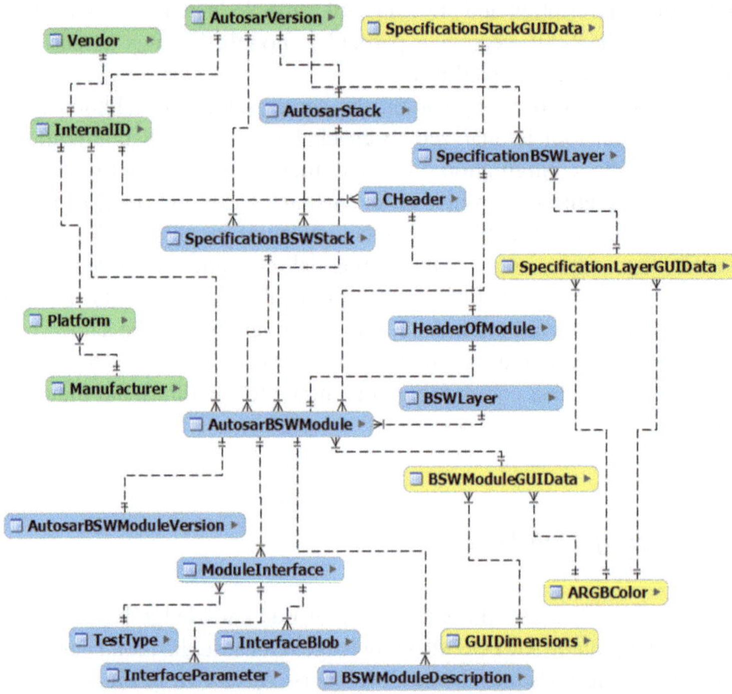


**Fig. 8** Database schema for AUTOSAR knowledge base

their parameters can be found in ModuleInterface and InterfaceParameter respectively. The table TestType specifies which interfaces need to be tested and how these interfaces should be tested.

## 5 Summary

The presented approach provides comprehensive static tests for AUTOSAR systems, independent from the used development tools. By using this approach, the quality of configuration and implementation of an AUTOSAR project can be increased. Thereby, our technique provides an optimal basis for the following dynamic tests. The mentioned static test technique supports the development process of AUTOSAR systems by analyzing source code and configuration files. To this end, we proposed creating a knowledge base consisting of interface names and their corresponding parameters, module names, etc. This knowledge base is created by parsing existing project folders and manual data management and constitutes the main contribution of this work.

## References

1. AUTOSAR, AUTOSAR Webpage: www.autosar.org. Accessed 26 Feb 2016
2. Spillner, A., Linz, T., Schaefer, H.: Software Testing Foundations, 4th edn. Rocky Nook, March 2014. ISBN: 978-1-937538-42-2
3. Englisch, N., Hänchen, F., Ullmann, F., Masrur, A., Hardt, W.: Application-driven evaluation of AUTOSAR basic software on modern ECUs. In: Proceedings of the 13th IEEE/IFIP International Conference on Embedded and Ubiquitous Computing (EUC). IEEE Computer Society, 2015. ISBN: 978-1-4673-8299-1
4. Michailidis, A., Spieth, U., Ringler, T., Hedenetz, B., Kowalewski, S.: Test front loading in early stages of automotive software development based on AUTOSAR. In: Proceedings of the Design, Automation and Test in Europe Conference and Exhibition (DATE), 2010
5. Park, G., Ku, D., Lee, S., Won, W., Jung, W.: Test methods of the AUTOSAR application software components. In: Proceedings of the ICCAS-SICE, 2009. ISBN: 978-4-907764-34-0
6. ST Microelectronics Documentation—SPC560P50L5: http://www.st.com/web/en/catalog/tools/FM116/SC959/SS1675/PF259567. Accessed 26 Feb 2016
7. dSpace SystemDesk Product Page: http://www.dspace.com/en/inc/home/products/sw/system_architecture_software/systemdesk.cfm. Accessed 26 Feb 2016
8. Elektrobit tresos Studio Product Page: https://www.elektrobit.com/products/ecu/eb-tresos/studio . Accessed 26 Feb 2016

## References


1. AUTOSAR: AUTOSAR Webpage. www.autosar.org. Accessed [illegible]
2. [illegible]
3. [illegible]

# A Method for an Integrated Software-Based Virtualisation of System Inputs

**Andreas Kurtz, Bernhard Bauer and Marcel Koeberl**

**Abstract** The influence of the consumer electronic industry to the automotive industry leads to shortened development times for the automotive domain. To keep up with the competitors and to enable the customer maximum connectivity demands new ways of testing. On a model based testing approach this paper shows a software-based virtualisation approach integrated in the distributed system itself for testing the system. The goal is the virtualisation of the environment as near as possible to the customers usage behaviour. The approach uses the deployment information for generating a mapping of the abstract interfaces to the specific implemented software. The shown approach is described regarding to an AUTOSAR architecture and has been implemented in an automotive electronic control unit for validation. A software-based testing approach enables to keep up with the fast changing development tasks.

**Keywords** Automation · AUTOSAR · Model based testing · System model · Test model · Virtualisation

## 1 Introduction

Vehicle development has become a software-centric approach, because software is infiltrating every component of the automobile. This change in the automotive industry is the enabler for innovations. Automotive manufactures have to become

A. Kurtz (✉) · M. Koeberl
BMW Group, 'Integration Electric/Electronics, Software', Munich, Germany
e-mail: andreas.kurtz@bmw.de

M. Koeberl
e-mail: marcel.koeberl@bmw.de

B. Bauer
Institute for Computer Science, Software Methodologies for Distributed Systems, University of Augsburg, Augsburg, Germany
e-mail: bauer@informatik.uni-augsburg.de

 
C. Gühmann et al. (eds.), *Simulation and Testing for Vehicle Technology*,
DOI 10.1007/978-3-319-32345-9_8

software developers but still stand at the beginning of the development. To keep up with the development speed of the consumer electronics industry the software cycles in the automotive domain have to become shorter. The main question is: How to ensure software quality when the complexity rises and the development time decreases?

Often the developer try to reduce development time by using tools or automation of tools. They use tools instead of methods. What is the main statement out of this: There is a need to develop methods in combination with tools. Tools are useful for executing the methods or to support using methods. In the following, the presented approach shows a new automation method via virtualisation of the user input.

## *1.1 Model Based Testing*

"Testing is a process in which we create mental models of the environment, the program, human nature and the tests themselves. [...] The art of testing consists of creating, selecting, exploring, and revising models" [1]. Based on this definition, testing is always based on models. Often these models are available but only as mental models. Model based testing is the key for automation because test cases can be generated out of the models [2] (Fig. 1).

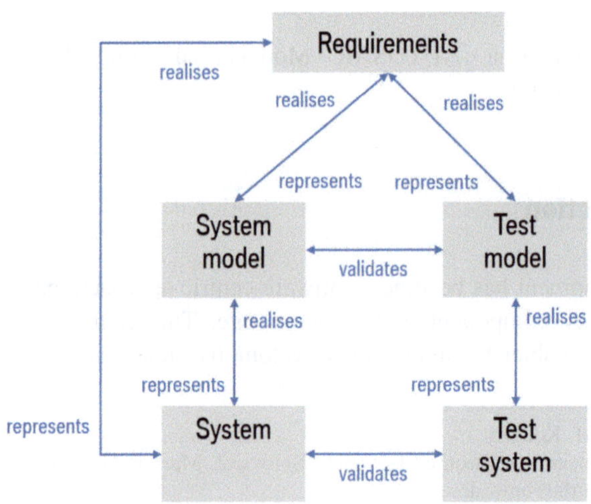


**Fig. 1** Model based testing [2]

### 1.2 Test Automation

Test automation is the enforcement of manual activities by machines. This contains all activities for testing software quality in the development process, in the different development periods and testing levels [3]. Test automation for software test improves the quality of testing, because of increasing the reproducibility.

## 2 The Problem Statement

The development raises several challenges to test automation, for this approach three were identified as main challenges.

### 2.1 The System Architecture

The system architectures of actual vehicles is a distributed system. A distributed system is the merger of multiple embedded systems, respectively electronic control units (ECU). Referring to the Introduction the car itself is not the product. It is the networking of car with the environment to enable more functionality and more comfort. With a system boundary 'vehicle', we speak about a distributed system in 1st order. If we add e.g., a mobile device like a mobile phone which is a distributed system itself to the system 'vehicle', and extend the system boundary, these two become a distributed system 2nd order (Fig. 2).

In the following, the term distributed system includes systems any order. This trend expands the customer functions beyond the classical system boundaries and leads to the need of new testing methods.

### 2.2 Sensors

Another aspect challenging the test methods is the change of sensors. In todays vehicles, classic switches have almost become obsolete. Displays with touch control

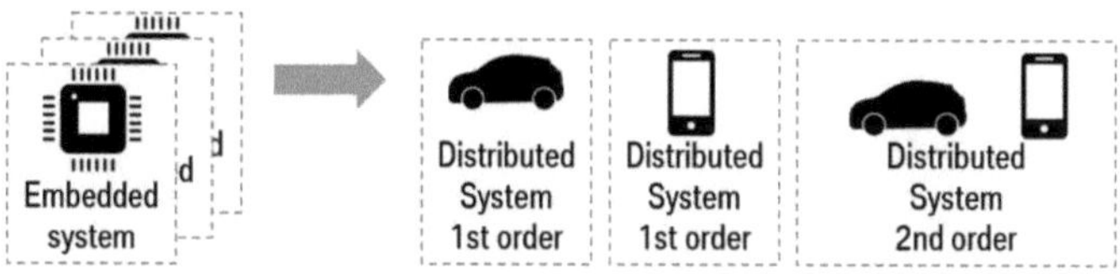


**Fig. 2** Embedded and distributed system 1st and 2nd order

or sensor surfaces dominate the interior of actual vehicles. This kind of sensors variate in their technology, surface and size, which complicates the test automation.

### 2.3 Development Speed

New operating concepts are pushing from the consumer industry to the automotive industry. The consumer industry has a product update cycle at about less than 1 year and forces the automotive industry to keep up. Loosing the connection to the new networked technology does not mean to be competitive.

## 3 Related Work

An approach with focus on "automation, modularization and compatibility of all equipment to do measurement, calibration and diagnosis" [4] is the CAN Calibration Protocol (CCP). The Protocol is used for calibration and data acquisition. It is realised as a driver with access to the internal ECU memory. This part of the protocol has an effect on the central processing unit (CPU) load. During a session using CCP a "continuous logical connection" [4] is established to transfer data from the ECU to the master device (off board test automation).

Figure 3 shows a simplified communication. The command processor allows the ECU to receive commands from the measurement system. The Data Acquisition (DAQ) processor is responsible for sending the data list at the appropriate time. The CCP has the main goal of data acquisition in contrast to data simulation, with

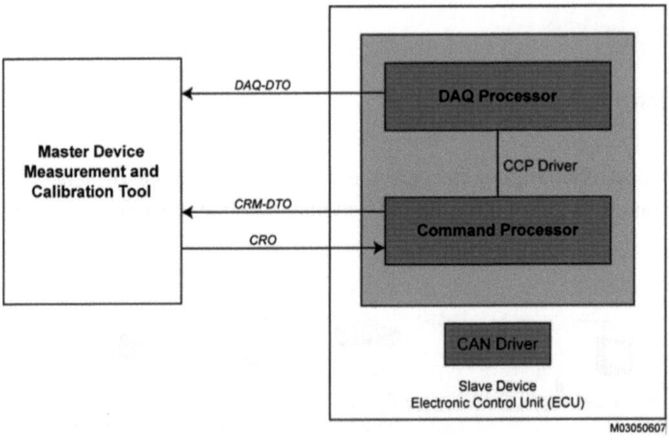


**Fig. 3** CCP-Communication [4]

an open data connection to the ECU. This allows getting access to the driver layer during runtime but this has an effect on the communication behaviour of the ECU.

Through further development, additional bus systems like FlexRay or Ethernet got in use in the automotive domain. The limitation of the CCP demands the enhancement to a new protocol, called Universal Measurement and Calibration Protocol (XCP) [5]. The XCP is separated in the protocol and transport layer. In detail, the communication of the XCP has two objects, the "Command Transfer Object" (CTO) and the "Data Transfer Object" (DTO). The master (off board) sends a CTO via the communication paths to the ECU. This CTO is acknowledged by the ECU. The DTOs are used for the data transfer and provide the event based reading of data.

# 4 The Approach

The new approach is to virtualise the sensor signals with a software-based solution for a decoupling of sensors at the hardware interface. The goal is to virtualise the environmental interaction with the system in shape of software signals. The approach was developed on a Automotive Open System Architecture (AUTOSAR) [6] specific architecture (Fig. 4) but can be projected to different architectures. The

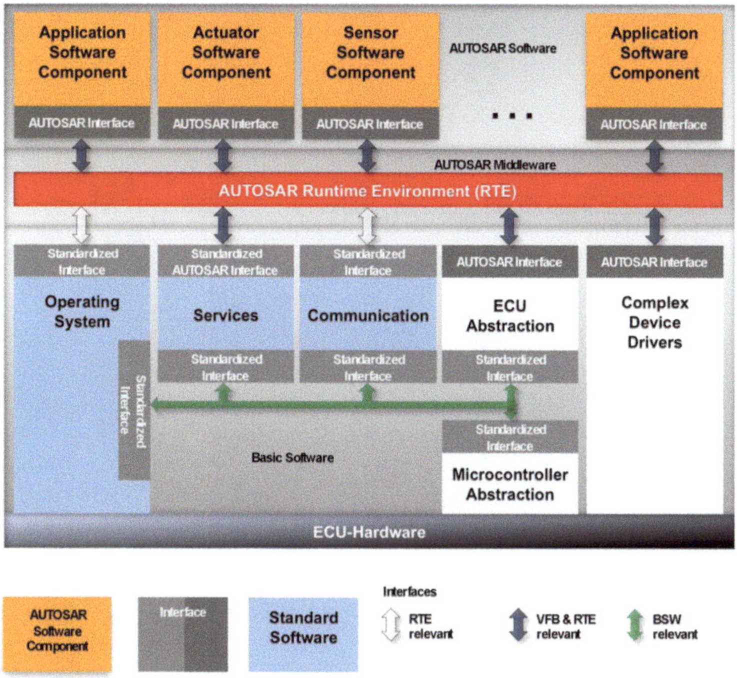


**Fig. 4** AUTOSAR architecture [6]

approach enables the test automation of any kind of embedded or distributed system.

Referring to the model based testing the requirements are the major input for the system- and test model. The following explains the approach guided at the model based testing approach.

## *4.1 System Model to Specific Software*

The system model describes the software functionality and is modelled in software components (SWC). The deployment process in Fig. 5 shows the SWCs connected with a virtual functional bus (VFB). The VFB does only specify the communication between the SWCs but not the used technique. At the deployment, the specific system and the ECU description determine which SWC is deployed on which ECU. This abstract modelling enables the reuse of SWCs respectively functions and the optimization for each derived derivate. This information of the ECU and System description is essential for the automation approach.

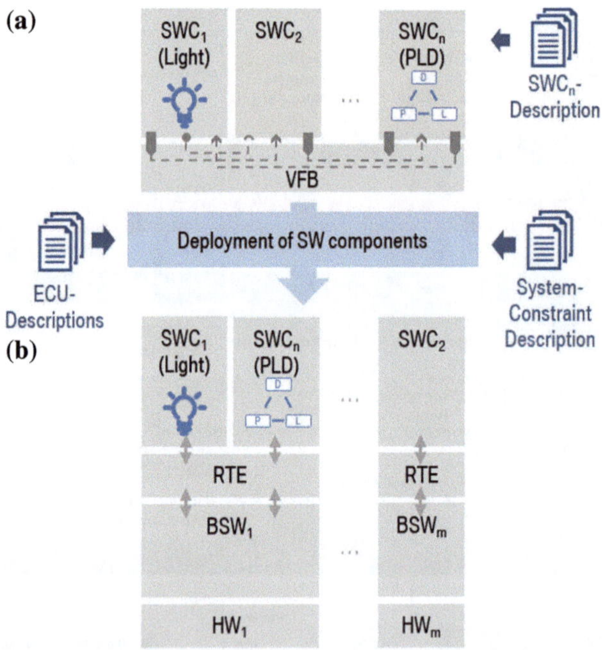


**Fig. 5** AUTOSAR deployment

At the end of the deployment, every $SWC_{1 \ldots n}$ is mapped to one $ECU_{1 \ldots m}$ ($HW_{1 \ldots m}$). The information, which SWC is mapped to which ECU as well as the mapping of function inputs and outputs, is stored in the ECU description and System constraint description.

## 4.2 Test Model to Abstract Test Cases

The test model describes the expected behaviour of the system in the designated environment. The environment also includes the user of the system. The detailing of the model depends on the testing level. This can be e.g., a use case model. The system behaviour has to be described abstract to enable a reuse of the test model and enables the generation of test cases within the abstract information.

Figure 6 shows and test model example in shape of a state chart. The test model design follows rules:

- Transitions (representing system input events) contain the abstract interface information and the system input data
- States (representing system reactions) include the abstract interface and the system output data.

In summary, with the information of the deployment of the SWCs to the ECUs and the abstract test cases it is possible to map the abstract interfaces to the specific implementation of SWC. Using an approach like this has an essential advantage. Once the test cases are generated and the deployment changes, there is no need in changing the test cases.

## 4.3 The Integration of Software-Based Virtualisation

The novelty is an extension of the existing AUTOSAR driver module to get access at the AUTOSAR basic software (BSW) level. The driver is located in the

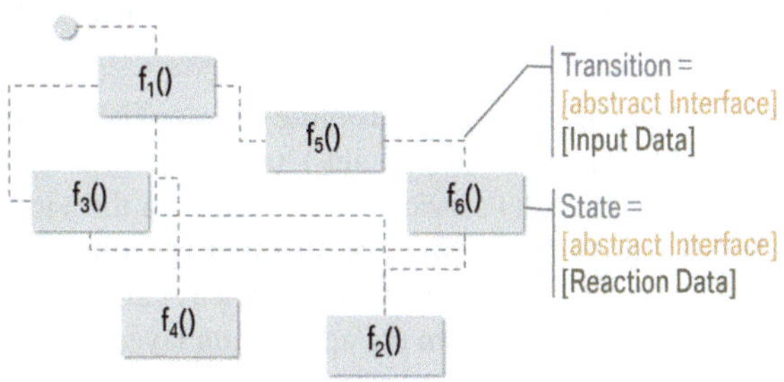


**Fig. 6** Abstract test model example

microcontroller abstraction layer (MCAL), Fig. 4. With the help of this additional interface, a software-based virtualisation at this layer is possible.

The interaction at a low layer allows to reduce the signal types for simulation and all layers above the driver will be tested as well. This exceeds the chain of reaction through the system to a maximum length.

With the help of a simple generic SWC called SimAgent, test cases, or better said signals, can be virtualised via a software solution. This module is integrated at the abstract system model and will be deployed to all ECUs of the distributed system. The test case data is temporary stored in the ECU memory and being executed via the SimAgent. The SimAgent has only a few functionalities to keep it simple and small. The only mechanisms implemented are stopping mechanisms. In the case of an error or failure of the system, the execution of all SimAgents can be stopped immediately. Unlike CCP the approach, no data connection is open during tests and therefore reduces the influence of additional communication or CPU load.

With the SimAgent as a standardised software component and a gateway it is possible to simulate (software-based) signals at the driver layer.

## 5 The Detailed Approach

In the following some details of the implementation and the simulation process are shown. The Sim module is a merger of three components. Figure 7 shows the architecture of the Sim module with its components:

- *SimAgent* is the deployed application SWC executing the test cases.
- *SimGW* is a gateway module for the connection to the driver layer. This is necessary to be in compliance with standards.
- *DioSim* is the additional driver interface to enable virtualisation at the driver layer.

This module is integrated in the existing AUTOSAR architecture (Fig. 4) and gets thereby access to the BSW.

### *5.1 The Implementation of the Additional Driver Interface*

The implementation of the driver interface is quite simple. The AUTOSAR driver is extended with an additional readSimChannel() service. This is specified like the e.g., original DIO_ReadChannel() service [7] but refers to the virtualised memory data. Other AUTOSAR driver types do have a similar specification like the digital input output (DIO) driver [8]. A Gateway (SimGateway) allocated in the complex device driver is used for the access from the application layer to the driver layer [9].

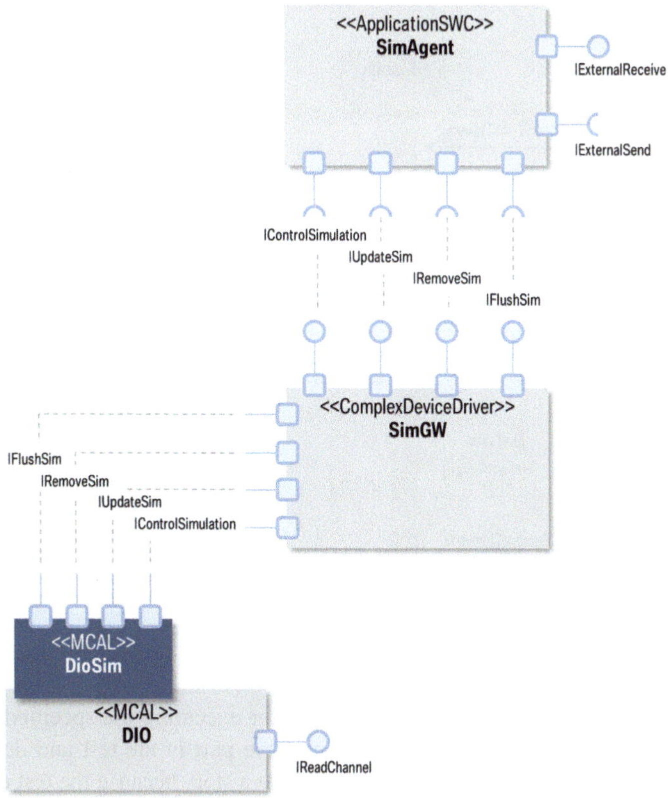


**Fig. 7** Simulation module architecture

## *5.2 The Implementation of the SimAgent*

The SimAgent is designed as simple as possible and operates like a reactive agent. The behaviour of the SimAgent SWC is leaned on agent behaviour, it receives the test tasks and executes them without questioning.

## *5.3 The Simulation Process*

The simulation process is shown in Fig. 8. The process describes how the off board test automation system (SimMaster) interacts with SimAgent of the system model. It shows also how the mentioned deployment data and test model data flows in.

At the beginning the *user* (Fig. 8) executes a test case and the following simulation process steps are:

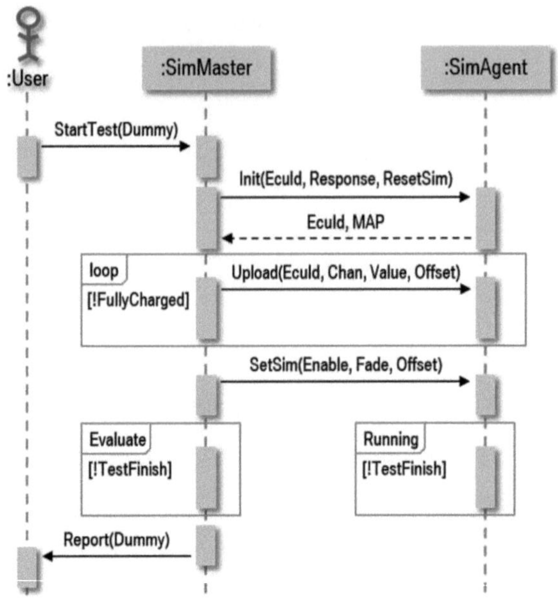


**Fig. 8** The simulation process

- *Init*-The SimMaster checks the parameters for executing the specified test case. It sends an ident to all ECUs taking an active part in the test and demands an answer of the SimAgents. No answer leads to a stop, because the test case could not be executed.
- *Upload*-The test case is send via the available communication path via broadcast to the system model. The data includes the receiver identifier that the assigned SimAgent can upload the test case into the memory.
- *SetSim*-This enables the test case execution at a specified time. This is needed to synchronise the SimAgents. There is only one timestamp for all SimAgents because of safety. This helps to avoid uncoordinated late starts of Agents, which can lead to unexpected system behaviour.

The *SimAgent* executes the simulation sequences as specified, starting at the defined timestamp. During this step, the off board system is waiting until the end of the simulation.

- *Evaluate*-The evaluation of the system reaction is done with proven techniques already in use. This is for example automated with a state machine aided analysis of communication data.
- *Report*-The SimMaster generates a report for the user.

The automation of the test case execution can also be implemented. This demands on the off board test automation system.

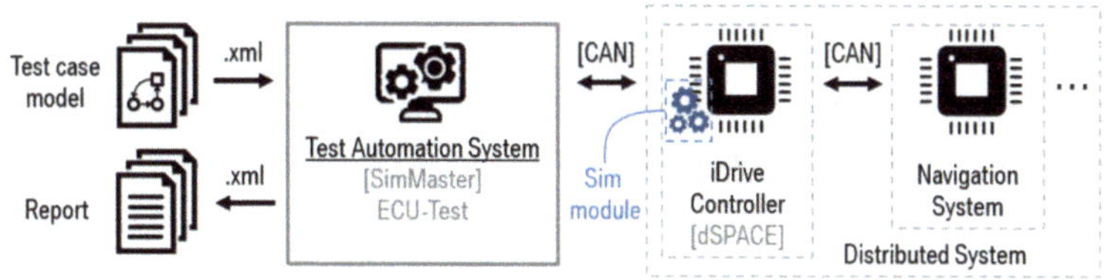


**Fig. 9** The test automation structure

## 6 Validation

For the validation of virtualised customer input the Sim module was implemented on a ECU and integrated in a distributed system. For this purpose, a dSPACE system was used and the corresponding series SWC of the central control element (iDrive-controller) was recreated. Further, the AUTOSAR driver was extended with the described interface and connected with the gateway. Figure 9 shows first validation setup to enable functional, performance and quality analysis. The aim is the testing of the navigation system in the distributed system with automation on the iDrive controller.

ECU-Test 6.3 (TraceTronic GmbH) is used as off board test automation tool. This tool supports the import of the test model and automated report generation. Moreover, a integration of the described process into the tool is realised. The internal accuracy of the sim module is less than 100 ms resulting in a highly reproducible system behaviour. The software-based customer virtualisation approach shows a system behaviour close to the customer use without additional hardware.

## 7 Conclusion

This paper shows an approach for virtualisation of the environment in the system under test itself. This approach improves the testing of the system model functional logic. One main advantage is the generating of the mapping out of the deployment process and merging it with the abstract test cases. This method enables flexibility and reduces test case generating. With the help of the additional software module and the interface, the specific test sequences can be easily injected in the system model. This new approach enables the testing of the system model comparative to the normal customer use.

The simulation itself is realised as a distributed system integrated in the distributed system. The modules are synchronised via the system time. The Modules have to upload the test sequence into the local ECU memory that there is no additional traffic compared to the normal customer use.

The approach has been realised for evaluation and shows promising results. The method does not depend on a special tooling or a system architecture.

## References


1. Beizer, B.: Software testing techniques, 2 edn. International Thomson Computer Press, New York (op 1990)
2. Roßner, T.: Basiswissen modellbasierter Test. dpunkt.verl., Heidelberg, 1 edn. (2010)
3. Seidl, R., Baumgartner, M., Bucsics, T.: Praxiswissen Testautomatisierung, 1 edn. dpunkt, Heidelberg and Neckar (2011)
4. Kleinknecht, H., Krüger, A., Kunz, H.G., Maier, R., Schröter, H., Zaiser, R.: Can calibration protocol—version 2.1 (1999)
5. Association for Standardisation of Automation and Measuring Systems (ASAM): ASAM MCD-1 (XCP): Universal Measurement and Calibration Protocol (2015)
6. AUTOSAR Partnership: AUTOSAR Components. http://www.autosar.org/fileadmin/images/media_pictures/AUTOSAR-components-and-inte.jpg (2014)
7. AUTOSAR Partnership: Specification of DIO Driver. http://www.autosar.org/fileadmin/files/releases/4-1/software-architecture/peripherals/standard/AUTOSAR_SWS_DIODriver.pdf (2013)
8. AUTOSAR Partnership: AUTomotive Open System ARchitecture: Enabling Innovation. http://www.autosar.org/ (2014)
9. AUTOSAR Partnership: AUTOSAR Layered Software Architecture. http://www.autosar.org/fileadmin/files/releases/4-2/software-architecture/general/auxiliary/AUTOSAR_EXP_LayeredSoftwareArchitecture.pdf (2014)
10. Köberl, M.: Integration softwarebasierter Automatisierungsmethoden in eine Test-ECU. Master thesis, University of Augsburg, Augsburg (2015)
11. Myers, G.J., Sandler, C., Badgett, T.: The Art of Software Testing, 3 edn. Wiley, Hoboken and N.J. (©2012)
12. Pezzè, M., Young, M.: Software Testing and Analysis: Process, Principles, and Techniques. Wiley, Hoboken and N.J., (©2008)

# Part IV
# Test Data Generation

# Usage Models for the Systematic Generation of Test Cases with MODICA

**Peter M. Kruse and Jörg Reiner**

**Abstract** The choice and determination of suitable tests for modern integrated hardware software system represent a challenging task. In this work we presented a new approach using a formal description of the system behavior for automatic test generation. We rely on usage models based on Statecharts for the automatic generation of tests. MODICA allows the creation of usage models and thereof the test generation for complex systems. MODICA helps to reduce expenses in test creation and helps to decrease system error rate.

**Keywords** Test generation · Usage models

## 1 Introduction

The choice and determination of suitable tests for a modern system consisting of software and hardware is a challenging task. This is especially true for complex networked system functions, where the test determination can be time consuming and error prone. In model-based testing, the modeling of the system to be tested is an essential task after the definition of system boundaries.

The approach taken in this work consists of a usage model, which describes the system from an outside view and its usage (for example, by user and/or neighboring systems). The test tool MODICA implements the introduced approach.

P.M. Kruse (✉)
Berner & Mattner Systemtechnik GmbH, Gutenbergstr. 15, Berlin, Germany
e-mail: peter.kruse@berner-mattner.com

J. Reiner
Berner & Mattner Systemtechnik GmbH, E.-v.-Kreibig-Str. 3, Munich, Germany
e-mail: joerg.reiner@berner-mattner.com
URL: http://www.berner-mattner.com

 
C. Gühmann et al. (eds.), *Simulation and Testing for Vehicle Technology*,
DOI 10.1007/978-3-319-32345-9_9

The usage model serves two purposes:

- It provides a compact understandable description of the system (the documentation of system behavior).
- It forms the basis for the test case generation.

Typically, test cases are a sequence of events, i.e. the system to be tested is stimulated in a defined manner and subsequently the reaction(s) of the system are compared with the expected values. In typical test environments, each test case is a concrete scenario and thus accurately describes a sequence of interactions with the system under test. This means that the creator of test cases considers the relevant procedures and describes them in the form of test cases.

The more complex the system under test is and the more different interactions with the system are possible, the more test cases are needed. In real systems this results in set of all possible conceivable test cases being greater than the number of test cases that can be written and executed. The aim is to find a minimum set of tests to be performed, which is suitable to confirm the maturity of the system under test (SUT).

Rather than creating specific test cases manually in a conventional manner, the approach of MODICA is to describe the behavior of the system in terms of its externally visible properties in so-called usage models. The creator of these models does initially not need to deal with the operational procedures of a particular test, but describes the system behavior as experienced by the user. A major advantage is that the model creator can easily be assisted by domain experts. These usage models will then be used for the generation of test cases [1]. The built-in MODICA algorithm provides for that the test objectives (e.g. covering the associated requirements) fully achieved or coverage gaps are identified and thereby optimizes the amount of test cases. The aim is to achieve a full coverage of the test targets with the smallest possible number of test steps in the test cases.

### *1.1 Structure*

Section 2 shows, how usage models are created in MODICA (structural and methodical). In Sect. 3 the usage models are used for automated test case generation. Section 4 contains a short evaluation.

## 2 Modeling

Generally models describe the allowed or expected system behavior. The description consists of *states* (nodes) and state *transitions* (edges) (see Fig. 1).

The system is at any time in exactly one state. States usually have a name (label). As special conditions there is an anonymous start state in which the system initially

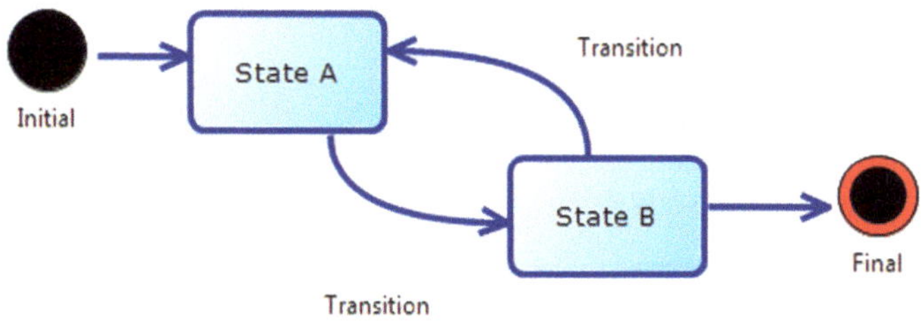


**Fig. 1** Abstract usage model

resides (e.g. immediately after powering). In addition, a final state (*Terminal Node*) may be defined. When the system reaches this, the execution is completed.

States can be hierarchical, so they contain more sub-states.

State transitions determine the allowable sequence of states of the system. They can be conditionally restricted (i.e. *Guards* included), then they can be traversed only if certain conditions are met. State transitions can be *triggered* by external *events*.

States (and sub-states) are used to describe states of a system which are different from other states of the system in a perceptible way (e.g. different behavior or reaction on conditions), and to group complex sequence of events belonging together in sub-states. Only very simple systems can be modeled in a usage model with just a single layer. Whenever a system has complex functionality, refinements in terms of sub-states must have to be modeled.

Test oracles can also be modeled as part of refined usage models.

## 2.1 Example

The modeling with MODICA will be illustrated using the example of the *Adaptive Cruise Control (ACC)*. The ACC in a vehicle is a speed control system which considers the distance to the vehicle driving ahead.

Figure 2 gives the usage model on the up most level. The ACC is initially turned of, the system is in state *Turned off*. In the further course the ACC can either be turned on (Transition *turn on*) and will then be in state *Turned on* or the *Final State* is reached and the run ends.

The state *Turned on* is a refined state. The sub-states are given in Fig. 3.

The ACC is initially inactive after turning on, the system resides in inner state *Inactive* of the outer state *Turned on*. The ACC can now be activated (Transition *activate*) and will then be in state *Active* (it is then active). In state *active* there are several actions and events, which can be taken or happen. State *active* also is a refined state itself.

When the state *Active* is reached its refinements are always traversed before further transitions. In our example, we include a simple test evaluation here (Fig. 4).

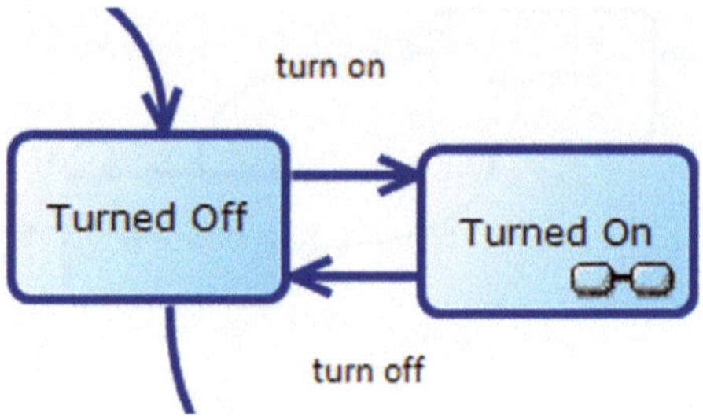


**Fig. 2** Usage model of *ACC* with states *off* and *on*

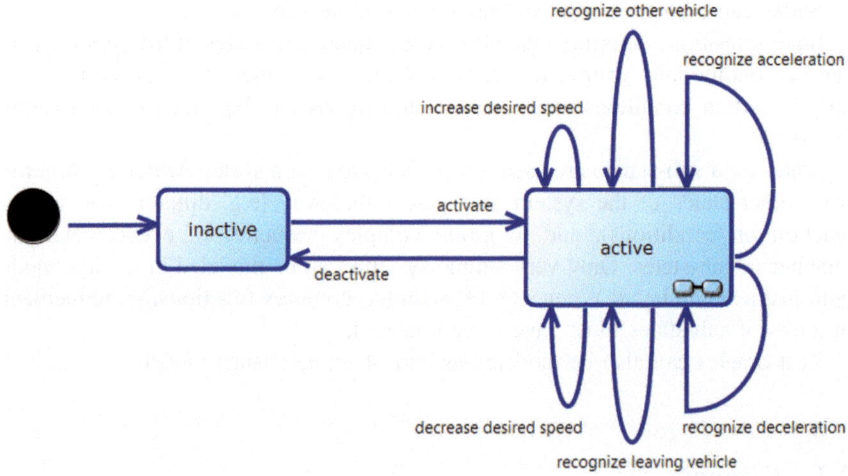


**Fig. 3** Refined usage model for state *turned on*

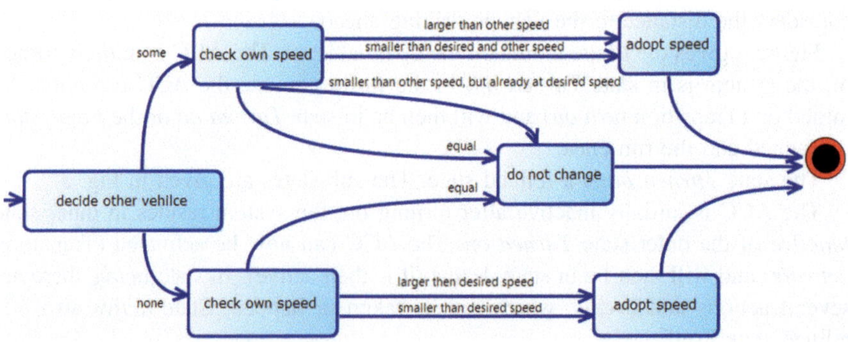


**Fig. 4** Test oracle

In test evaluation, there is an initial decision (state *decide other vehicle*) whether there is a vehicle driving ahead (possible outcomes are states *none* and *some*). In each case, the current vehicle speed is checked (states *check own speed*) with possible next states *smaller*, *equal*, *larger*. The current vehicle speed is then adjusted accordingly (states *adopt speed*, *do not change*). Then the final state is reached. This gives the sequence of: Check other vehicle, check current speed, adjust speed.

## 3 Test Generation

MODICA uses the test generation technology by Conformiq [1]. The following coverage criteria [2] are currently supported:

- Selection of requirements and requirements coverage,
- Selection of states and state coverage,
- Selection of transition and transition coverage
- Path coverage.

During generation, the courses through the model are basically selected by the generator so that the cover criteria are met. If the coverage criteria can not be met, so not covered by the amount of generated test cases, these are presented separately to the user.

**Requirements**: The usage model can contain annotated *requirements*. These can be used in test generation to specifically generate test cases covering all requirements (Fig. 5).

The test generator tries to generate a minimum set of test cases covering all requirements. It favors many short sequences over few long ones.

| | Testing Goals | 18 | 19 | 20 | 21 | 22 |
|---|---|---|---|---|---|---|
| 100% 16/16 | ◢ Requirements | | | | | |
| ✔ | ACC can be activiated. | X | X | X | X | X |
| ✔ | ACC can be deactivated. | | | | | |
| ✔ | ACC can be turned off (PowerOff). | X | X | X | X | X |
| ✔ | ACC can be turned on (PowerOn). | X | X | X | X | X |
| ✔ | Desired speed can be decreased by 1 kmh pressing th | | | | | |
| ✔ | Desired speed can be increased by 1 kmh pressing th | | | | | |
| ✔ | Do not change speed. | | | | | X |
| ✔ | Recognized vehicle decreases speed (brakes). | X | | X | | |
| ✔ | Recognized vehicle increases speed (accelerates). | | | | X | X |

**Fig. 5** Met requirements coverage

| | Testing Goals | 18 | 19 | 20 | 21 | 22 |
|---|---|---|---|---|---|---|
| 100% 32/32 | ◢ Transitions | | | | | |
| ✓ | InitialState -> Turned Off | X | X | X | X | X |
| ✓ | Turned Off -> FinalState | X | X | X | X | X |
| 100% 22/22 | ◢ Turned On | | | | | |
| ✓ | activate (inactive -> active) | X | X | X | X | X |
| 100% 13/13 | ◢ active | | | | | |
| ✓ | adopt speed -> final state | X | X | X | X | X |
| ✓ | adopt speed alone -> final state | | X | | | |
| ✓ | do not change -> final state | X | X | X | X | X |
| ✓ | eqaul (speed check alone -> do n( | X | X | X | X | X |

**Fig. 6** Transition coverage reached

**Structure Elements**: In the group *Statechart* elements from the model can be selected to be covered as part of test generation. The supported model elements are *States*, *Transitions* and *Transition Pairs*. Figure 6 shows the coverage of transitions.

**Sequence Rules** are used to influence the test generator. They allow to direct the traversal of the model to special test sequences (e.g. likely to be error-prone), otherwise hard to reach in the model, and to speed up the generation for very complex models. Their use requires expert knowledge of the system under test.

For the ACC example, it could be desirable to have only test sequences where another vehicle is detected at least once before adjusting the current speed.

### *3.1 Test Implementation and Execution*

Usage models in MODICA can be enriched with EXAM function calls [3]. All generated test sequences can be exported to EXAM and are ready for execution there. For upcoming versions, other execution environments are currently considered.

## 4 Experiences

MODICA is currently used successfully for testing embedded functions in the automotive domain. In addition to the example of an adaptive cruise control system shown here, there are first usage models for stationary air conditioning.

In the trial MODICA has proved itself a valuable support for the testers. In particular, the automatic generation of test implementations is an essential work relief.

## 5 Related Work

Heimdahl et al. briefly surveys a number of approaches in which test sequences are generated using model checking techniques [4]. The common idea is to use the counter-example generation feature of model checkers to produce relevant test sequences.

Model checking aims to prove certain properties of program execution by completely analyzing its finite state model algorithmically [5, 6]. Provided that the mathematically defined properties apply to all possible states of the model, then it is proven that the model satisfies the properties. However, when a property is violated somewhere, the model checker tries to provide a *counter-example*. Being the sequence of states the counter-example leads to the situation which violates the property. A big problem with model checking is the *state explosion problem*: The number of states may grow very quickly when the program becomes more complex, increasing the total number of possible interactions and values. Therefore, an important part of research on model checking is *state space reduction*, to minimize the time required to traverse the entire state space.

The *Partial-Order Reduction* (POR) method is regarded as a successful method for reducing this state space [6]. Other methods in use are *symbolic model checking*, where construction of a very large state space is avoided by use of equivalent formulas in propositional logic, and *bounded model checking*, where construction of the state space is limited to a fixed number of steps.

Two temporal logics are compared and debated extensively [7], *Linear temporal logic* (LTL) and *Computation Tree Logic* (CTL). Temporal logics describe model properties and can reduce the number of valid paths trough the model. For the verification of classification tree [8] test sequences [9], Krupp and Müller introduce an interesting application of CTL [10]. Using a real-time model checker, the test sequences and their transitions are verified by combining I/O interval descriptions and CTL expressions.

Wimmel et al. [11] propose a method of generating test sequences using propositional logic.

Ural [12] describes four formal methods for generating test sequences based on a *Finite-State Machine* (FSM) description. The question to be answered by these test sequences is whether or not a given system implementation conforms to the FSM model of this system. Test sequences consisting of inputs and their expected outputs are derived from the FSM model of the system, after which the inputs can be fed to the real system implementation. Finally, the outputs of the model and the implementation are compared.

Bernard et al. [13] have done an extensive case study on test case generation using a formal specification language called *B*. Using this machine modeling language, a partial model of the GSM 11-11 specification has been built. After a system of equivalent constraints was derived from this specification, a constraint solver is used to calculate boundary states and test cases.

Binder [14] lists a number of different oracle patterns that can be used for software testing, including the simulation oracle pattern. The simulation oracle pattern is used to simulate a system using only a simplified version of the system implementation. Results of the simulation are then compared to the results of the real system. We can regard the formal model of the system as the simulation of the system from which expected results are derived.

## 6 Conclusion

With MODICA and modeling usage models a powerful approach to test generation is available for the tester. Unlike other tools, the user has a full generation of executable tests. In particular, the aspect of test completeness is much better addressed than in manual test creation.

## References


1. Huima, A.: Implementing conformiq qtronic. In: Testing of Software and Communicating Systems, pp. 1–12. Springer (2007)
2. Friske, M., Schlingloff, B.-H.: Abdeckungskriterien in der modellbasierten Testfallgenerierung: Stand der Technik und Perspektiven. In: Dagstuhl-Workshop MBEES: Modellbasierte Entwicklung eingebetteter Systeme II, pp. 27–34 (2006)
3. Thiel, S., Zitterell, D.: EXtended Automation Method (EXAM) zur automatisierten Funktionserprobung von Steuergeräten in der Automobilindustrie. In: GI Jahrestagung (2), pp. 625–630 (2008)
4. Heimdahl, M.P.E., Rayadurgam, S., Visser, W., Devaraj, G., Gao, J.: Auto-generating test sequences using model checkers: a case study. In: 3rd International Workshop on Formal Approaches to Testing of Software (FATES 2003) (2003)
5. Bošnački, D., Edelkamp, S.: Model checking software: on some new waves and some evergreens. Int. J. Softw. Tools Technol. Transf. **12**, 89–95 (2010)
6. Jhala, R., Majumdar, R.: Software model checking. ACM Comput. Surv. 41:21:1–21:54 (2009)
7. Vardi, M.Y.: Branching vs. linear time: final showdown. In: Proceedings of the 7th International Conference on Tools and Algorithms for the Construction and Analysis of Systems, TACAS 2001, pp. 1–22, London, UK, 2001. Springer-Verlag (2001)
8. Grochtmann, M., Grimm, K.: Classification trees for partition testing. Softw. Test. Verif. Reliab. 3(2):63–82 (1993)
9. Kruse, P.M., Wegener, J.: Test sequence generation from classification trees. In: 2012 IEEE Fifth International Conference on Software Testing, Verification and Validation (ICST), pp. 539–548. IEEE (2012)
10. Krupp, A., Müller, W.: Modelchecking von Klassifikationsbaum-Testsequenzen, 1 April 2005. GI/ITG/GMM Workshop "Methoden und Beschreibungssprachen zur Modellierung und Verifikation von Schaltungen und Systemen", München
11. Wimmel, G., Loetzbeyer, H., Pretschner, A., Slotosch, O.: Specification based test sequence generation with propositional logic (2000)
12. Ural, H.: Formal methods for test sequence generation. Comput. Commun. **15**, 311–325 (1992)

13. Bernard, E., Legeard, B., Luck, X., Peureux, F.: Generation of test sequences from formal specifications: Gsm 11-11 standard case study. Softw. Pract. Exper. **34**, 915–948 (2004)
14. Binder, R.V.: Testing object-oriented systems: models, patterns, and tools. Addison-Wesley Longman Publishing Co., Inc, Boston, MA, USA (1999)

# TASMO: Automated Test Data Generation for Simulink Model Coverage

**Benjamin Wilmes**

**Abstract** Model-based development of embedded software with MATLAB/Simulink is a well-established practice in the automotive industry. Simulink models are usually the first executable artifacts in the development process. Testing these models is therefore particularly relevant for an efficient quality assurance process. Considering the expanding role of software in modern automobiles, going hand in hand with rising testing costs, the automation of testing activities is highly desirable. One technique which has shown its capabilities in automating software verification is search-based testing. The tool TASMO utilizes search-based algorithms as well as static analysis techniques to generate input data (test data) for structural coverage of Simulink models. TASMO has been the subject of research for more than 7 years and is now part of the testing tool TPT. This paper introduces TASMO and delivers insight into how the search for model-covering test data is automated.

**Keywords** Model testing · Simulink · Structural testing · Test case generation

## 1 Introduction

The development of embedded systems and embedded software relies heavily on systematic testing activities to assure functional correctness and other relevant quality criteria. Following the most commonly used process model, the V-model, each testing activity, or phase, usually targets particular artifacts or interconnected artifacts of the system or software. Subject to the condition that the development is modular and model-based, the functional models, software code, interconnected software units, electronic control units, interconnected hardware units, and the entire system are among these artifacts. The testing techniques applied in practice vary depending on artifact type and environment. The main goal, however, is

B. Wilmes (✉)
PikeTec GmbH, Waldenserstraße 2-4, 10551 Berlin, Germany
e-mail: benjamin.wilmes@piketec.com

 
C. Gühmann et al. (eds.), *Simulation and Testing for Vehicle Technology*,
DOI 10.1007/978-3-319-32345-9_10

essentially the same: proving functional correctness of an artifact against its functional specification. Since 'proving' in a mathematical sense is most often only a theoretical ideal and practically unachievable, testing is a challenging task. Performing testing of system or software exhaustively, i.e. for any possible scenario, is hardly ever feasible, let alone cost-efficient. Testing sketchily and missing out relevant scenarios, however, bears a high risk—especially in the automotive domain where many developed systems are safety-relevant. Picking the right strategies, methods, and tools, out of a pool of well-tried and innovative solutions, is usually the key to mitigate this risk.

One such strategy is front-loading of verification activities, i.e. to test the product for compliance with its functional requirements as early as possible in the development process. Functional models of a software unit, also called implementation models, are usually the first artifacts produced that are executable, and therefore testable, in an automated way. In order to find any functional incorrectness, or other flaw, as early and as close to its source as possible, testing these models is particularly important. For the most part, testing a functional model is focused on achieving functional coverage, i.e. requirements coverage. However, structural coverage, i.e. covering certain logical states or guaranteeing the execution of certain model fragments, should not be disregarded. As daily testing practice shows, full requirements coverage rarely implies full structural coverage, and missing structural coverage effects that potentially faulty or unwanted functionality becomes part of artifacts in the following development phases. Ultimately, problems might be carried much further and be revealed in a later phase, or in the worst case, become part of the final product.

Unfortunately, achieving high structural coverage often requires considerable manual effort. In the automotive domain, functional models tend to be very complex and implemented time-dependent behavior complicates this task further. Manually finding a scenario that would trigger the model to reach certain internal states may take some time. Having an automation of this task available not only saves time and costs but also allows a tester to focus more on testing functional model aspects.

The tool TASMO (*T*esting via *A*utomated *S*earch for *Mo*dels) addresses itself to this task. Tailored to functional models designed with MATLAB/Simulink [1] or TargetLink [2], TASMO can automatically generate input data resulting in high structural model coverage. The novelty of TASMO is in the way it generates the input data. First of all, it is possible for the user to put restrictions on the input data to be generated, e.g. requiring it to match certain signal characteristics. Secondly, TASMO utilizes search-based algorithms, also known as evolutionary algorithms, to automatically search for desired input data. So far, such techniques predominantly have been subject to research only [3]. Thirdly, TASMO combines these search-based algorithms with special static analysis techniques to improve overall performance. The tool is the outcome of several research works [4, 5] and has recently been integrated into the testing tool TPT [6].

In the following, the use-case and scope of TASMO is introduced in more detail. The paper presents the tool's workflow and its integration in TPT. In addition, a rough insight into the underlying techniques of the tool is given.

## 2 Model Testing

Code of embedded software is usually written in the programming language C. In the vast majority of cases the code is not written by hand. In the automotive industry, the tool MATLAB/Simulink is predominantly used to graphically design and implement the functionality of a software module or unit. From the resulting models, C code is then automatically generated using tools like the Simulink Coder [7] or TargetLink.

For the most part, Simulink is a graphical data-flow language that is geared towards describing the behavior of dynamic systems. Such systems are time-dependent, i.e. input data, output data and internal data are essentially value sequences (signals). Syntactically, a Simulink model consists of functional blocks and lines connecting them, while most of the blocks are equipped with one or more input ports as well as output ports. The semantics of such a model result from the composed functionalities of the involved block types, e.g. sum blocks, relational blocks or delay functions. In principle, the relationships between a model's blocks form a mathematical system of equations that is being solved as the model is executed. A solver executes the blocks in a certain order repeatedly over time, each block at certain time steps. In case of a fixed-step solver for example, these time steps are generally the same ones for every block. Further concepts like conditionally executed model fragments introduce exceptional behavior to this general mechanism. In addition to its model-and-line-based design principle, Simulink allows the realization of event-driven or state-based functionalities via Statechart-like automata. For this purpose, so-called Stateflow [8] blocks may be integrated in a Simulink model just as any other block. The block's behavior is designed graphically by the user.

The execution of a Simulink model is usually called a simulation since the model calculations might differ from those in the generated code. The main cause for such semantic differences is the application of scaling to variables during code generation for code efficiency reasons and to keep the precision of values under control. Nonetheless, Simulink models are major subjects to testing activities in industrial practice. As pointed out before, testing such a model primarily focuses on checking whether it complies with the (usually textual) functional requirements. For this purpose, correspondent test cases are defined, for the most part manually by a tester. Each test case contains test data, i.e. input data for the model, and comparisons or check routines for evaluating the model's behavior on the basis of its resulting output data. The functional testing approach treats the model generally as a 'black box'. From this point of view one can imagine that it's rather unpredictable whether the internal structures of the model all have been executed/covered or not.

**Fig. 1** Code example

```
1: if (a > b)
2:     x = 0;
3: else
4:     x = 1;
5: c = x + 2;
6: if (a == c || b > c)
7:     x = x – 1;
8: return x;
```

However, industrial standards like ISO 26262 demand the consideration of coverage metrics when performing low-level tests—this means on either or both, model and code level.

In contrast to functional testing, structural testing is commonly aimed at creating test data based on the internal structure of the test object. Traditionally, structural testing is exercised on the code level. For example, one could consider creating a set of test data which executes all statements of a code function, or all paths in the code's control flow graph. In the coding world, various coverage criteria exist, such as statement coverage, branch/decision coverage and condition coverage. An example: Fig. 1 shows a piece of code and Fig. 2 shows its control flow graph. Branch coverage for example, requires that at nodes 1 and 6, both the *true* and *false* case (i.e. both edges leaving the nodes in the graph) are taken at least once when the code is executed. For condition coverage, each atomic sub-condition of nodes 1 and 6 (like $a == c$) would have to be executed at least once *true* and *false*.

For structural testing of Simulink models, very similar coverage criteria exist. However, coverage criteria like decision coverage or condition coverage address the model's data flow rather than its control flow. The reason is that Simulink models contain only very little control flow since all blocks are normally executed

**Fig. 2** Control flow graph for the code example

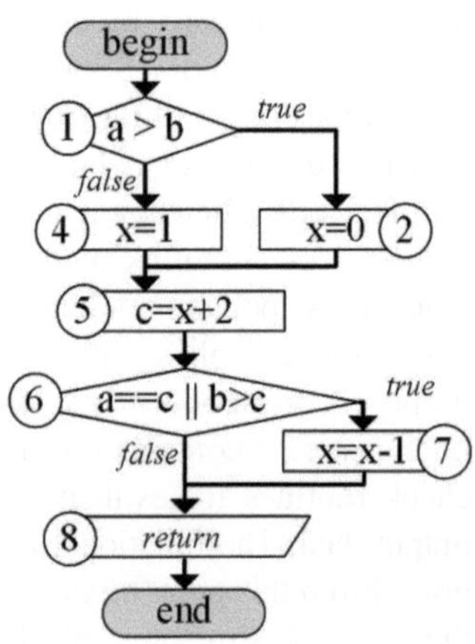

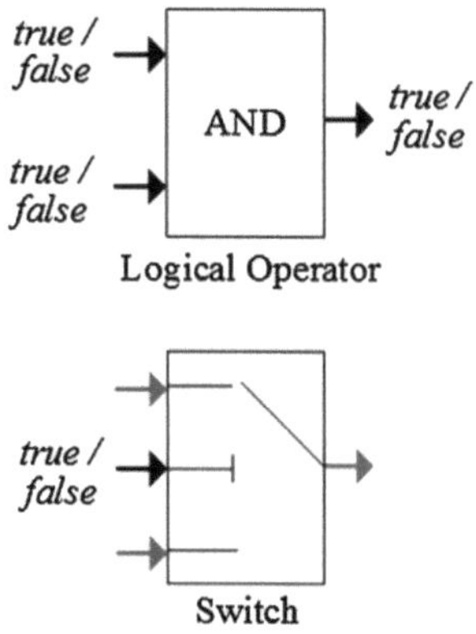


**Fig. 3** Examples of coverage-relevant model blocks

at every execution cycle. Concepts like conditional subsystems however, result in control-flow which is also addressed by the coverage criteria. Figure 3 includes Simulink's logical operator block as an example for a coverage-relevant block. Decision coverage for Simulink models requires that the output of this block is at least once *true* and *false*. Condition coverage, in contrast, requires that each block input is at least once *true* and *false* during an execution. Considering that this block might be positioned somewhere deep in the hierarchy of a very complex model, the search for model input data that triggers such *true* or *false* states can become a very time-consuming task when done manually.

In industrial practice, Simulink's 'Verification and Validation' toolbox [9] is often used to measure the model coverage while executing (requirements-based) test cases. Extending or modifying a test suite in order to increase the total model coverage is most commonly still a manual task though.

## 3 Automated Test Data Generation

The automation of test case design and implementation on a requirements-based level has always been and still is a difficult undertaking. In industrial practice, requirements are hardly ever specified in a sufficiently formal way and the possibilities of automated textual analysis are still limited. Structural testing however, is much more amenable to automation since control flow graphs or Simulink models, for the most part, are formal enough and easier to analyze in an automated way.

Automated test data generation for structural testing has been a research topic for over 40 years. While traditionally focused on code coverage, the most prominent techniques are symbolic execution [10], random testing [11], concolic testing [12], special usage of model checking [13], and search-based testing [3]. Generally, these techniques are subdivided into either static or dynamic ones. Static techniques usually do not execute the test object—they solely analyze a formal and possibly abstract representation of the test object. Dynamic ones such as random testing, on the contrary, execute the test object. While static techniques tend to scale poorly with increasing test object complexity and are helpless in case of unknown internal structures (an external function call for example), dynamic techniques most often lack efficiency.

When attempting to automatically generate test data for structural coverage of Simulink models, these issues emerge even more—especially in case of comparatively large models. Various research works tried applying the above-named techniques on the model level [14–17]. However, these studies rarely paid attention to the complexity of industrial-size models. Tools for the task of generating model-covering test data exist, such as Simulink Design Verifier [18]. However, even these tools struggle with the complexity of models from industrial practice. The tool presented in this paper therefore is based on an approach that none of the currently available tools utilize: search-based testing. The general idea of the search-based approach is pretty simple: the test data finding problem is transformed into an optimization problem by defining a cost function, called fitness function. This function rates any test data generated by the deployed search algorithm based on information gained from executing the test object with it. An iteratively working search algorithm uses the fitness ratings to distinguish good test data from bad, and based on this, generates new test data in each iterative cycle. This fully automated procedure continues until desired test data is found, or until the search gives up. Evolutionary algorithms and genetic algorithms are suitable choices of search algorithms in this instance. Furthermore, the required fitness functions can be automatically derived from the search goals in question, i.e. the desired internal states in the model and their surrounding model structures.

As mentioned, such a dynamic test data generation approach normally lacks efficiency since lots of intermediate test data might need to be generated before the desired test data is found. Accordingly, the model needs to be executed many times. In order to speed up the automatic search, recent research [19] extended the search-based test data generation approach in application to Simulink models by incorporating custom-built static techniques, such as static analysis and symbolic execution. This way, for example, unreachable model states and such ones that are difficult to tackle with an automatic search can be identified. The resulting 'hybrid' test data generation technique has been evaluated in case studies involving relatively large Simulink models from automotive software development projects. The results show that these additional techniques can reduce the run-time of the automated test data generation by over 90 %—compared to using a purely search-based approach [19].

## 4 TASMO Test Data Generator

TASMO was initially developed as a prototypic tool for evaluating the search-based test data generation approach and its extending techniques, as described in the previous section. The groundwork on the prototype [5] had been carried out at the

Berlin Institute of Technology, in collaboration with Daimler. In the recent past, the prototype has been enhanced gradually and, most importantly, became part of the testing tool TPT [6].

TPT is developed and distributed by PikeTec. It is aimed at testing embedded software, or more generally, embedded systems. It supports various types of test objects, such as C code, MATLAB/Simulink (or TargetLink) models, ASCET models, and HiL (hardware-in-the-loop) platforms. Apart from TASMO's capabilities, TPT already supports certain structural testing activities, e.g. measuring code coverage or model coverage during test execution. However, TPT is primarily a functional testing tool. Test cases are created systematically and graphically by the user, via mechanisms like automata or very intuitive step lists. The behavior expected from the tested system can be defined either directly within these constructs or, if more complex checks need to be done, using assessment scripts for instance. For the definition of both test case behavior and expected test object behavior, TPT is purpose-built for handling time-dependent processes. In addition, test execution, assessment and documentation are completely automated.

For testing a particular test object with TPT, the user connects it to the tool by configuring a type-specific platform. For testing Simulink models for example, TPT provides a special platform. Using this platform, the model's interface and parameter set can be automatically analyzed and imported. TPT then automatically generates a so-called test frame—a Simulink model which establishes the communication between TPT and the model under test. In this test frame, both model under test and TPT are represented by blocks. They interact with each other both ways so that model-in-the-loop testing becomes possible. This means that the test case behavior may depend on the test object's outputs.

In the following, the integration of TASMO in TPT and the workflow of using TASMO is described. Note that Fig. 4 visually summarizes this description. In general, TASMO is an extension to TPT's platform for Simulink models. As long as the platform is sufficiently configured to run functional test cases, it is also ready to be used with TASMO. Before the tool starts its automatic search for model-covering test data, the user is asked to provide details about the nature of the model's inputs (see Fig. 5). For instance, the user can specify that a certain input signal must match certain value bounds or that the signal may only contain linear or sine wave-form characteristics. Such user-defined restrictions effect that the generated test data is 'realistic' and better readable. The performance of the search algorithm might also be influenced positively since restrictions on the input data narrow the space in which the tool has to search. After specifying the test data to be generated, the user must select one or more coverage criteria. Currently, TASMO supports decision coverage and condition coverage—also for Stateflow diagrams. After this step, the test data generation finally starts (see Fig. 5). As indicated in the previous section, the tool then generates test data after test data—by utilizing the intelligence of a search-based algorithm—and executes the test frame model with it. Note that TASMO works with a copy of the original test frame for these executions since the model is automatically modified and instrumented for measuring the achieved coverage.

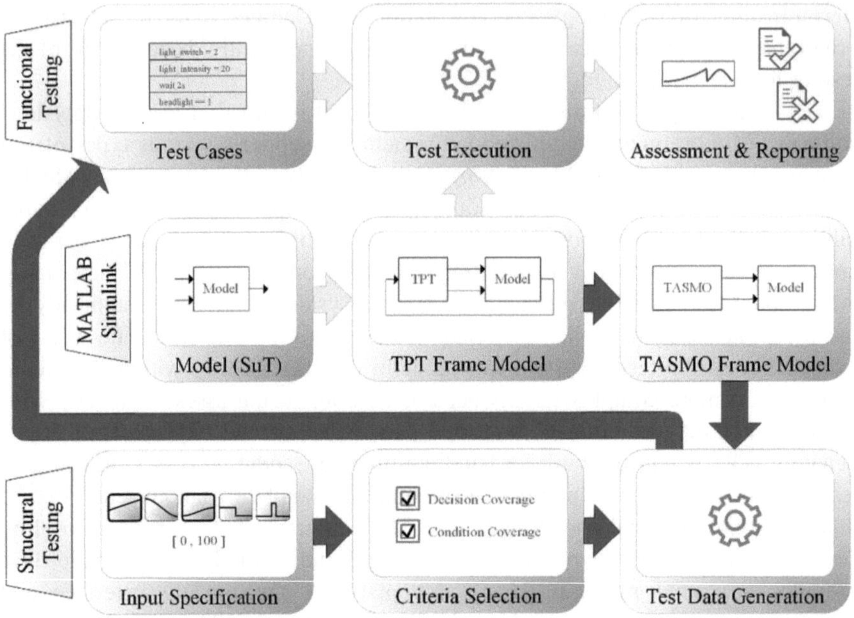


**Fig. 4** Workflow using TPT for functional testing and its TASMO tool for structural testing of a MATLAB/Simulink model

When TASMO is done with its test data generation, the test data is ready to be added to the project's test suite. TASMO directly creates common TPT test cases—in the format of the aforementioned step lists. Such step lists also contain annotations with details about which part of the test case definition is responsible for reaching the covered model states. Note that only a very small portion of the collectively generated test data is added. Minimization techniques are used to keep the number of test data entries that lead to the achieved model coverage as low as possible.

From a functional testing point of view, the user might want to evaluate the test object's functional behavior for the generated test data next. The test cases generated by TASMO do not contain any specific checks or assessments related to functional requirements. However, if any assessments for existing requirements-based test cases have been specified in a test-case-independent way (forming a global test oracle), they can be directly used for functionally evaluating the auto-generated test cases as well.

An automated test data generation as provided by TASMO is particularly useful (but not limited) to the following testing tasks:

1. extending a collection of requirements-based test cases with automatically generated test cases to increase the total degree of model-coverage
2. creating a collection of test cases for performing back-to-back tests.

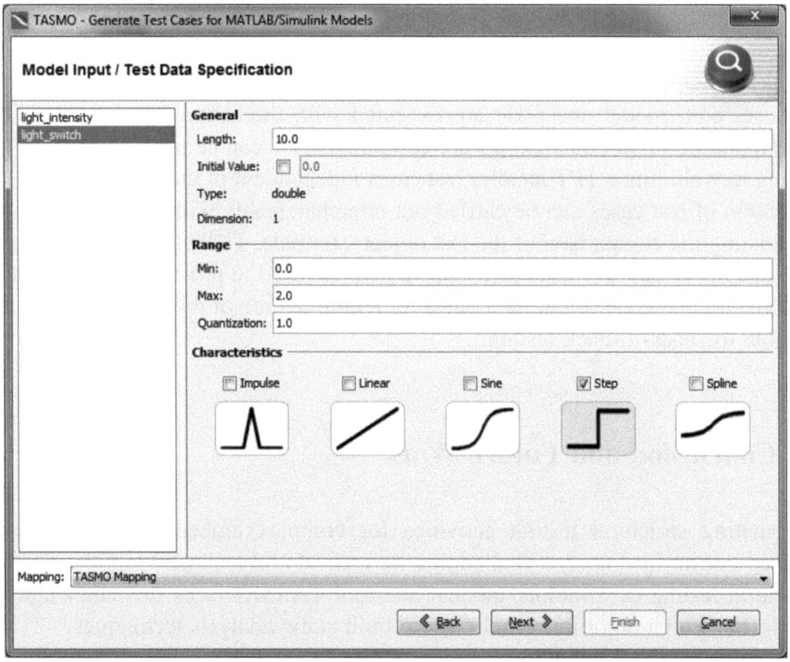


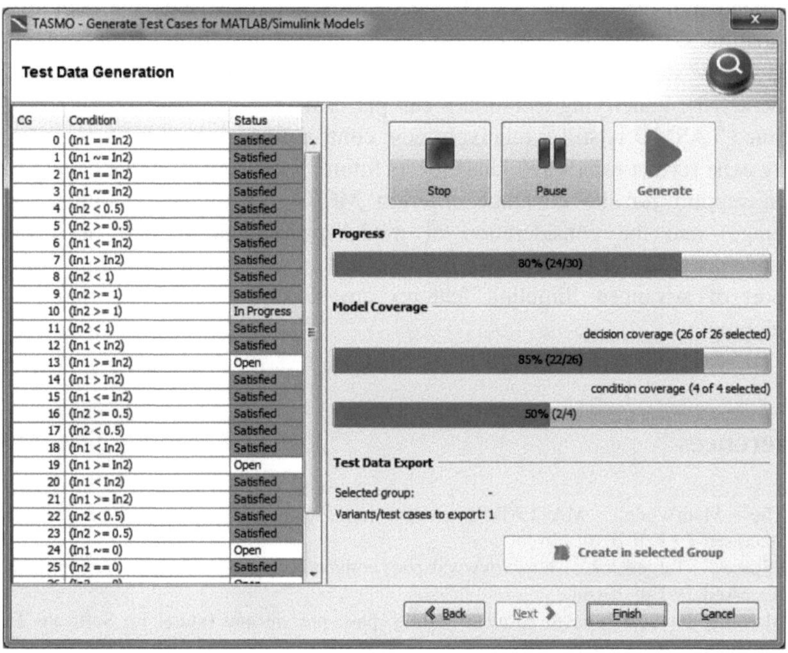


**Fig. 5** Insight into TASMO: influence the nature of the generated test data by specifying each model input in advance (*top*) and follow the test data generation progress (*bottom*)

The first use case is the one primarily discussed in this paper. The second one addresses a form of testing in which the behavior of the (usually generated) code is compared with the behavior of the model that the code is based on. For this purpose, both model and code are executed with the same input data, and then corresponding output data entries are compared. TPT can be used for back-to-back testing as well. Since TPT handles test cases independent of the test object type, the execution of test cases can be carried out interchangeably with model or code. For automating the comparison of the test object's outputs, TPT offers a special type of assessment. Within this use case, TASMO can be used to provide the input data for both model and code. Input data effecting a high degree of coverage is particularly suitable for back-to-back testing.

## 5 Conclusion and Future Work

Automating structural testing activities for complex embedded software is, as pointed out in this paper, a challenging theoretical and technical topic. Tailored to structural testing of Simulink models, the tool TASMO faces this challenge with intelligent search-algorithms and custom-built static analysis techniques.

As presented in this paper, the tool provides an extensive automation to this task. Integrated into the testing tool TPT and thusly combined with the manifold test automation skills of TPT, it contributes to improving efficiency of unit/module testing. Case studies carried out with a prototype of the tool have demonstrated how efficiently the underlying techniques can perform.

Since TASMO is still a relatively new contribution to the TPT tool, there naturally exist further topics and ideas for its future development. Among these topics are a support for the coverage criterion MC/DC (modified condition/decision coverage) and the consideration of model parameters as part of test data. Improvements to the performance of the test data generation algorithm and a full support of advanced Simulink features are also subject to the ongoing tool development.

## References

1. The MathWorks: MATLAB/Simulink. http://www.mathworks.com/products/simulink. Accessed 29 Feb 2016
2. dSpace: TargetLink. https://www.dspace.com/en/ltd/home/products/sw/pcgs/targetli.cfm. Accessed 29 Feb 2016
3. McMinn, P.: Search-based software testing: past, present and future. In: Software Testing Verification and Validation Workshops, ICSTW '11, pp. 153–163 (2011)
4. Windisch, A.: Suchbasierter Strukturtest für Simulink Modelle. PhD thesis, Berlin Institute of Technology (2011)

5. Wilmes, B.: Hybrides Testverfahren für Simulink/TargetLink-Modelle. PhD thesis, Berlin Institute of Technology (2015)
6. PikeTec: TPT. http://www.piketec.com. Accessed 29 Feb 2016
7. The MathWorks: Simulink coder. http://www.mathworks.com/products/simulink-coder. Accessed 29 Feb 2016
8. The MathWorks: Stateflow. http://www.mathworks.com/products/stateflow. Accessed 29 Feb 2016
9. The MathWorks: Simulink verification and validation. http://www.mathworks.com/products/simverification. Accessed 29 Feb 2016
10. King, J.C.: A new approach to program testing. In: SIGPLAN Not., vol. 10, Issue 6 (1975)
11. Renfer, G.F.: Automatic program testing. In: Proceedings of 3rd Conference of the Computing and Data Processing Society of Canada, University of Toronto Press (1962)
12. Sen, K., Marinov, D., Agha, G.: CUTE: a concolic unit testing engine for C. In: Proceedings of the 10th European Software Engineering Conference, pp. 263–272. ACM (2005)
13. Fraser, G., Wotawa, F.: Improving model-checkers for software testing. In: International Conference on Quality Software. IEEE Computer Society, Los Alamitos, CA, USA, pp. 25–31 (2007)
14. Hamon, G.: Simulink design verifier—applying automated formal methods to simulink and stateflow. In: AFM'08: Third Workshop on Automated Formal Methods (2008)
15. Gadkari, A.A., Mohalik, S., Shashidhar, K., Yeolekar, A., Suresh, J., Ramesh, S.: Automatic generation of test-cases using model checking for SL/SF models. In: Proceedings of the 4th model-driven engineering, verification and validation workshop, pp. 33–46 (2007)
16. Roy, P., Shankar, N.: SimCheck: an expressive type system for simulink. In: Proceedings of the 2nd NASA Formal Methods Symposium (NFM 2010), pp. 149–160 (2010)
17. Satpathy, M., Yeolekar, A., Ramesh S.: Randomized directed testing (REDIRECT) for Simulink/Stateflow models. In: Proceedings of the 8th ACM International Conference on Embedded Software, pp. 217–226 (2008)
18. The MathWorks: simulink design verifier. http://www.mathworks.com/products/sldesignverifier. Accessed 29 Feb 2016
19. Wilmes, B.: Static preprocessing for automated structural testing of simulink models. Int. J. Adv. Syst. Measur. 6(3/4), 310–323 (2013)

# Part V
# Gear

# Simulation of Gear Microgeometry for Gear Whine Noise Reduction in Passenger Car

**Dnyaneshwar Shelke and Koji Sakurada**

**Abstract** In Indian automobile market, customers are become more sensitive about cabin comfort in all carsegments. Cabin noise is one of the important factor which plays vital role in overall passenger comfort inside vehicle. Gear whine noise is the critical factor towards customer's overall NVH perception in cabin. Gear whine is the undesirable tonal noise experienced in cabin, originated in the transmission. Reduction of gear whine noise becomes an important subject for transmission engineer in this competitive market. Transmission error acts as an excitation source during gear mesh and generates vibration which passes through gears, shafts and bearings to the housing which vibrates to produce noise. Gear whine can be minimized by t/m error optimization through optimizing gear tooth profile geometry. Traditional method of gear tooth optimization is lengthy and time consuming. This paper explains a simulation based approach to predict the gear whine at early stage of design. Targeted transmission model is prepared in Romax software by accurate modeling technique. This model is analyzed with different load conditions to get gear profile optimized for centralized gear tooth contact pattern and lower transmission error. For confirmation of software results, gear proto parts are manufactured with desired Microgeometry values. Transmission with these gears is tested on bench NVH test set up. Very good correlation is observed in gear tooth contact pattern of software and actual bench set up. This paper will present experience of simulation of Transmission model in Romax software for optimizinggear geometry and correlation with testing Results.

D. Shelke (✉)
Transmissions Design & Development Department,
Maruti Suzuki India Limited, Gurgaon, India
e-mail: Dnyaneshwar.Shelke@maruti.co.in

K. Sakurada
Automobile Transmission Design Department,
Suzuki Motor Corporation, Hamamatsu, Japan
e-mail: sakurada@hhq.suzuki.co.jp



C. Gühmann et al. (eds.), *Simulation and Testing for Vehicle Technology*,
DOI 10.1007/978-3-319-32345-9_11

# 1 Introduction

With the advancement in automotive technology, gear whine has become a prominent contributor for cabin noise as the masking has been decreased. Whine is not the loudest one, but it is high tonal noise which often perceived as highly unpleasant. It is uniquely audible in the passenger cabin due to its particular frequency. With the significant improvement that has been made in reducing the engine and other background noise level, gear whine is becoming more apparent and perceptible to passenger.

It has been recognized that source of gear whine is at the gear tooth mesh. The gearbox is a self-exciting system. Small errors at the gear tooth mesh, caused by gear tooth geometry and misalignment in the gearbox, result in a constant harmonic displacement excitation called transmission error. Transmission error acts as an excitation source and these vibration pass through gears, shaft and bearings to the housing which vibrates to produce noise in surrounding air and structure borne noise via the mounts on the casing. The harmonic excitation is transferred via structural and airborne routes to the driver's ear location, resulting in gear whine [1].

When there is requirement to minimize the gear whine in a vehicle several action can be taken; reduce or remove the excitation, reduce the amplification through the system, isolate or cover the source of the noise by applying damping. Among these options, noise reduction by reducing the excitation result in improvement at no extra cost and does not require design change in other than gear parts [2].

As product development time is decreasing day by day, traditional method of gear tooth profile optimization, which requires repetitive prototyping, will not continue as an effective tool. Hence software orientated approach is used to simulate gear tooth profile to predict transmission error for whine noise reduction in design stage only. Tooth optimization is focused on the reduction of excitation force as a countermeasure. Therefore transmission noise issues can be handled before any prototyping is done.

A CAE method is developed to calculate and optimize the TE at the gear mesh and ultimately reduce the gear whine into the passenger car application. The method utilizes a full system model including all components which contribute to misalignment and unique nonlinear bearing representation. The method uses optimization techniques through software which is validated with real test data.

# 2 Simulation of Gear Microgeometry

## 2.1 *Need of Simulation Approach*

Gear design has two fundamental aspects: Durability and NVH. Current gear design approach is necessarily biased towards durability. As a result, NVH tends to come up second. A typical traditional approach of gear deign would be as follows [3]:

- Design gear macrogeometry for maximum application load (possibly with some consideration for NVH, but often not)
- Determine gear durability based on maximum load and duration
- Apply some default microgeometry to avoid tooth contact durability problems
- Hope it is not too noisy
- Refine microgeometry by experience/experiment and repeated prototyping.

Improving microgeometry which is basically being carried out by examining gear tooth contact pattern in actual. Prototype gears are painted with orange color and assemble in transmission which is mounted on bench for different torque similar to vehicle. Position and variation of tooth hitting occurs due to the engagement of gears and it removes color at engagement location. Tooth profile optimization is done by deciding whether tooth hitting is at center position. But this process has some disadvantages [4];

- Manufacturing variation can be not considered
- Manpower, cost and time is required for prototyping and testing
- Decision is based on judgment of person-in charge.

To overcome these disadvantages, transmission designer requires a CAE methodology to take into account gear whine noise evaluation quickly during the design process, a tool that can predict gear whine noise performance purely by simulation to ensure that the transmission is quiet. Therefore, the transmission NVH issues are optimized before any prototyping is done, significantly reducing development time and cost.

It is understood that, considering a gear in isolation is not sufficient, since the generated noise is greatly affected by its mesh misalignment, which in turn depends on the influence of housing and bearing stiffness. Thus, it is necessary to model the entire transmission system, including the needle roller bearings under synchronized gears, shaft supporting bearings, differential and gearbox housing.

## 2.2 *Preparation of Transmission Model*

A software tool, which has the ability to simulate transmission system and predicts TE, is utilized to resolve gear whine issue. An image of the model used for this study is shown in Fig. 1. This represents five speeds, front-wheel-drive manual transaxle.

This is fully detailed model comprises of following primary components [5–7]

- Simple shafts—modeled as 1-dimensional Timoshenko beams
- Complex shafts—modeled as 3-dimensional FE components, e.g. differential.
- Rolling element bearings—modeled as detailed 6-degree of-freedom non-linear components.
- Gears—modeled including all detailed macro and microgeometry parameters.

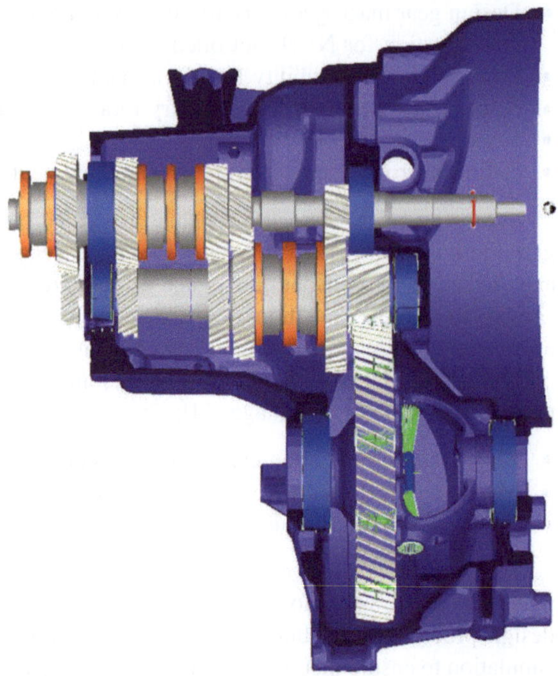

Fig. 1 3D model of five speed manual transaxle

- Clearances—radial and axial clearances between components modeled using non-linear contact elements.
- Housing—modeled as a 3-dimensional FE component.

Flexible elements, such as the transmission housing, differential were imported from finite element models, extracting the mass and stiffness information using various component mode synthesis techniques. Radial clearance at bushes and internal clearance of bearings were calculated from design tolerance. The bearings are connected to FEM housing via RBE3 elements.

During analysis setup, input torque is applied at the input shaft and output torque at differential center. During analysis run, CAE Software takes this torque as input and solves iteratively to determine the deflections of the entire system. From these deflection values, misalignment of the gear mesh is calculated.

Three load cases are prepared with three different torque values for both drive and coast condition for gear set to be optimized.

## 2.3 Optimization Technique

There are several methods for the solution of noise and vibration problem [2],

- Move the modes of vibration out of the frequency range
- Reduce or remove the excitation

- Increase damping to the structure to reduce the excitation amplitude
- Change or remove vibration transfer path.

Several of these options are unfeasible, as it is virtually impossible to move all modes of vibration out of the frequency range of gear whine, addition of damping and extra material can be difficult to model accurately and costly, as is changing or removing the transfer path of the vibration. Therefore reduction of the excitation should be considered as a best possible option.

The aim of microgeometry optimization is to ensure uniform distribution of load across the tooth face while simultaneously keeping the transmission error (TE) low in the operating range as much as possible. TE is the source of gear whine. It is caused by the non-conjugacy of motion in the gear pair due to a combination of misalignment and the deflection of gear teeth under load [3].

Target parameters to be considered to start optimization for the gear [8];

- Transmission Error—Try to minimize TE 1st harmonic
- Gear tooth contact—Try to centralize gear tooth contact at working facewidth.

The software includes optimization tool which assist designer to provide weightage to microgeometry targets as per specific application. For example TE under higher load is given slightly lower weightage, as gear whine noise is often perceived to be less of a problem at high load.

The microgeometry variables considered during optimization includes;

- Involute slope
- Barreling
- Lead slope
- Crowning.

The optimization is not performed in terms of TE only; additionally it is performed in terms of the gear edge loading and gear stress values. For a design that is robust and insensitive to manufacturing tolerances, a sensitivity study is performed at each candidate within the manufacturing tolerances. This allows the selection of gear tooth profile to be based upon a combination of TE and insensitivity to manufacturing tolerances. Therefore the best design is not necessarily the one with the lowest TE levels.

## 2.4 Optimization Results

It is assumed that example of second gear pair of five speed manual transaxle in acceleration (drive) condition. In 2nd gear pair, input gear which is fixed on input shaft is considered as fixed and output gear which is mounted on counter shaft through needle bearing is made target gear. Both gears are assigned with some default microgeometry variables. After optimization changes in variables of 2nd output gear are shown in Table 1.

**Table 1** Microgeometry variables before and after

| Variables (μm) | Before optimization | After optimization |
|---|---|---|
| Involute slope | −3 | −4 |
| Lead slope | 0 | 6 |
| Crowning | 4 | 3 |

All three variables, involute slope, lead slope and crowning are modified after optimization. Left side tooth edge was loaded with more loads compared to right side, which was optimized so that both edge receives balanced load as shown in Fig. 2. Peak load value which is at center of effective facewidth is reduced by 5 %.

The contact load distributions show that the optimizer has avoided edge contact by introducing end relief, and the load is distributed over a wide area of the tooth—thus reducing overall peak load. Maximum contact load is close to the center. Contact load distribution shows improvement over the previous design in that contact is more centralized across the load range, and peak load is reduced at the highest torques [8].

Transmission error is reduced at all load cases. Difference in before and after optimization values is less because originally TE across the torque range was less as shown in Fig. 3.

In brief, with below outcomes of optimization, it is concluded that NVH of 2nd gear pair is improved through simulation.

- Uniform contact area
- TE is decreased.

The optimization process described requires thousands of gear contact analysis. For this reason, the analysis algorithm must be very fast to make this a practical tool for the designer.

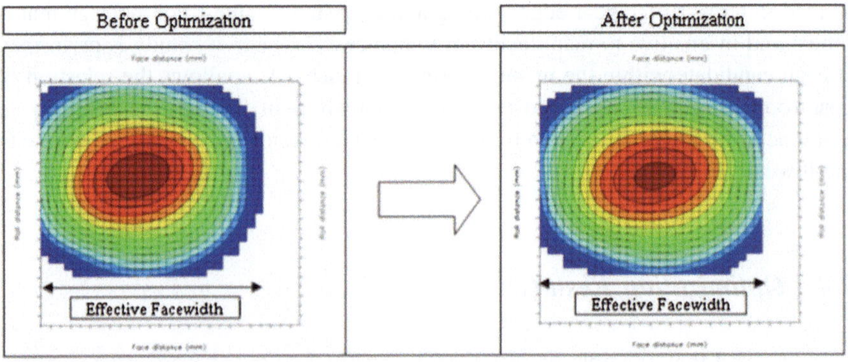


**Fig. 2** Peak load and contact pattern before and after

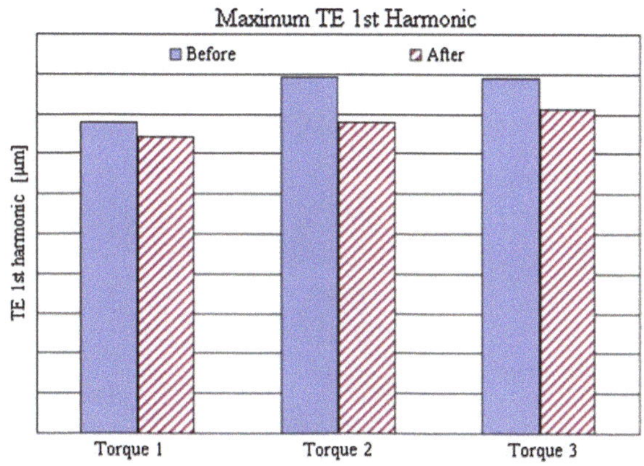


**Fig. 3** Maximum TE—1st harmonic comparison in tolerance

# 3 Correlation of Simulation Results

The optimization process produces a design, which is nominally the best, but it takes no account of the potential robustness of the design. A design with very good performance, when made to exact optimization specifications, may give poor performance in experimental conditions. It is important to correlate the simulation results with actual experimental results of a gear pair to confirm accuracy of transmission model and analysis condition during CAE [9].

## *3.1 Gear Proto Manufacturing*

In current study the gears are manufactured considering microgeometry as per simulation results exactly. To ensure that the manufacturing process should not significantly affect the desired test samples microgeometry the gears were finished by grinding process. Microgeometry values of proto gear sets were inspected.

Tooth contact location on the tooth flank is dependent on the relative difference in values of involute slope and lead slope. Considering that tooth contact of gear pair with simulation results locate at the center of the tooth. Two types of tests are planned for proto gear sets;

- Gear tooth contact pattern check
- Noise and vibration data measurement.

## *3.2 Gear Tooth Contact Check*

Selected gear sets are painted with orange color and assembled in transmission which is mounted on bench set up. Contact pattern checked at torque values considered during CAE and some revolution of input shaft on test bench.

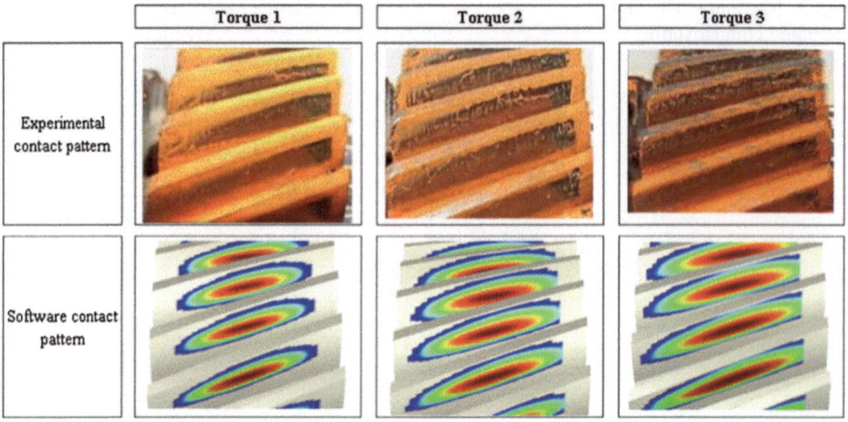


**Fig. 4** Gear tooth contact pattern comparison (test vs. simulation)

Transmission dismantled to check tooth contact pattern which occurred due to gear engagement. Example images of actual contact pattern & contact pattern taken through software by inputting microgeometry values from inspection report of proto gears are shown in Fig. 4.

Figure 4 shows an example of contact pattern comparison for location at tooth center position. Actual test contact pattern and simulation contact pattern shows correlation which is judge by considering below points;

- Contact location is same as prediction
- Shifting of contact position with respect to torque
- Area of contact zone.

Same exercise is repeated to capture contact pattern at other location on gear tooth.

## 3.3 *Noise and Vibration Measurement*

For acquiring gear noise and transmission vibration data the same gearbox was used. Gear noise data was acquired by using microphones on gear noise tester. Noise and Vibration data was acquired in acceleration and deceleration condition with all three torque values applied at input shaft of gearbox (considering max engine torque). Noise and vibration data is measured along tooth flank at similar location where tooth contact checked [9].

Example graph for 2nd gear in deceleration noise is shown in Fig. 5.

Graph displays noise level improvement in dB for 2nd gear set at optimized microgeometry values. It was concluded that simulation results of gear profile modification are confirmed by gear tooth contact correlation and noise measurement on bench set up.

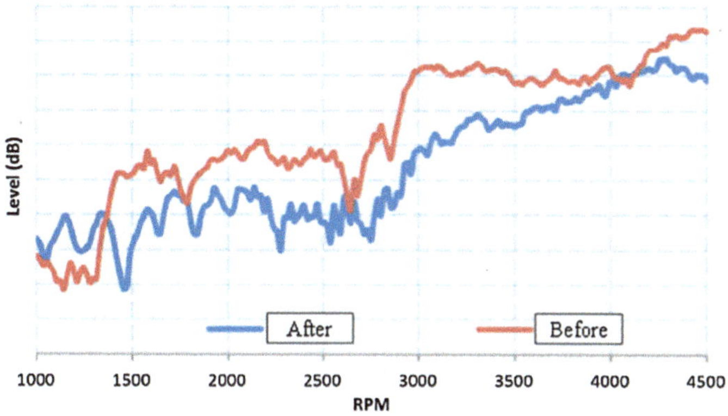


**Fig. 5** Gear noise data in 2nd gear deceleration

# 4 Conclusion

In the transmission development process lead time and cost are critical parameters; designer has to deliver the best design within timeline and with less prototyping. This implies that durability, efficiency and NVH must be confirmed as early as possible during product development. The approach used here is to reduce gear whine by minimizing the transmission error. In this paper transmission error was predicted for gear whine noise by system level simulation of transmission. This model was constructed with accurate details such as clearances, measured micro geometry and housing stiffness. The mathematical model that is created can be used to optimize the micro-geometry on the teeth to ensure that the transmission error is minimum without compromising the tooth load contact pattern, within the existing manufacturing tolerances. The comparison of the simulation result to physical test result displayed good correlation. By using this optimization technique, traditional method of experimentation with repeated prototyping can be avoided.

In summary it proves a new analytical method to reduce vehicle NVH by minimizing gear excitation induced by TE. The optimization technique greatly improves design process efficiency and design quality by comparing current trial and error design.

**Acknowledgments** The authors would like to thanks Suzuki Motor Corporation, Japan for sharing their knowledge for this methodology preparation. Authors also would like to thanks Maruti Suzuki India ltd. for permission to publish this paper.

# References

1. Smits, J.D.: Gear Noise and Vibration. Marcel Dekker, INC
2. Curtis, S., et al.: An Analytical Method to Reduce Gear Whine Noise, Including Validation with Test Data. SAE Paper No. 2005-01-1819

3. Platten, M., et al.: Optimization of gear microgeometry for durability and NVH. Sound Vibr (2009)
4. Arai, M.: Technology to reduce gear whine noise by gear micro geometry optimization. Published in Suzuki Technical Review, vol. 38, March 2012
5. Mane, Y., et al.: Manual Gearbox Gear Whine Noise Prediction and Importance of Parametric Sensitivity in NVH. SAE Paper No. 2013-26-0069
6. James, B., et al.: Development of a Gear Whine Model for the Complete Transmission System. SAE Paper No. 2002-01-0700
7. Korde, A., et al.: Driveline Noise Reduction by Prediction Mesh Transmission Error and Microgeometry Optimization. SAE Paper No. 2011-01-1553
8. Romax Technology Ltd.: Helical Gear Microgeometry Optimization-Usage Notes (2005)
9. Singh, P.K., et al.: Study of Effect of Variation in Microgeometry of Gear Pair on Noise Level at Transmission. SAE Paper No. 2015-26-0130

# Wear Modelling for Remaining Lifetime Prediction of Dry Clutches

**Daniel Strommenger, Clemens Gühmann, René Knoblich and Jörg Beilharz**

**Abstract** Modern passenger cars are equipped with effective systems to increase vehicle dynamic performance, comfort and reduction of fuel consumption. Therefore it is crucial to sustain a robust powertrain performance for increasing requirements. The clutch is an important component of the powertrain system with wear-based behavior, which substantially influences comfort. Therefore knowledge of clutch wear is essential to estimate influences on the powertrain system behavior. The system robustness can be increased and comfort can be sustained by measures, which are based on estimation of abrasion. Furthermore the remaining lifetime can be predicted from material loss and friction behavior. That implies that future maintenance can be forecasted with high accuracy. On the assumption that the clutch wear is known, the remaining lifetime of the system can be increased by an optimization of the clutch control. The contribution summarizes and compares existing methods for modelling wear and remaining lifetime of clutches. Subsequently a data-based approach for dry clutch wear modelling will be presented. The approach is capable to estimate online the current abrasion of the clutch and predict future wear. Identification and verification of the wear model will be performed by experiments on a test bench.

**Keywords** Dry friction clutch · Wear prediction · Remaining lifetime · Condition monitoring

D. Strommenger (✉) · C. Gühmann
Institute of Energy and Automation Technology, TU Berlin, Berlin, Germany
e-mail: daniel.strommenger@tu-berlin.de

C. Gühmann
e-mail: clemens.guehmann@tu-berlin.de

R. Knoblich · J. Beilharz
Department of Transmissions and Hybrid Systems, IAV GmbH, Berlin, Germany
e-mail: rene.knoblich@iav.de

J. Beilharz
e-mail: joerg.beilharz@iav.de



C. Gühmann et al. (eds.), *Simulation and Testing for Vehicle Technology*,
DOI 10.1007/978-3-319-32345-9_12

## 1 Introduction

Requirements for modern passenger cars are challenging in respect to fuel consumption, comfort and performance. Therefore the improvement of powertrain components is still in focus of the automotive industry. The clutch is one of the most important driveline components, which have a significant effect on launch and shift process quality. Every mechanical system, as well as the clutch, shows a wear dependent behavior. Sustain robust performance for an aging system is a demanding challenge for system control. This task can be facilitated by estimation of clutch wear. The detailed knowledge of clutch fatigue offers opportunities to improve control robustness and to enhance lifetime of the clutch system. Thus estimation and prediction of wear by an appropriate model is desired. The identification of wear models is complex, because on the one hand causes and effects of clutch fatigue are not fully understood yet. On the other hand gather measuring quantities, which describe the amount of wear, is challenging. The clutch rotates in the compact transmission case. Hence, placing sensors for data acquisition is a complex task.

Different approaches for the identification of clutch wear models are introduced in literature. Most of them consider reduction of friction lining volume as wear. A general statistic approach for predicting friction lining volume was proposed by Watson et al. [1]. Watson presents a generic wear-modelling approach, which is based on the law of Archard. According to the law of Archard the degrading volume is proportional to the work done by friction forces. The proportionality between volume and work is described by the wear coefficient. Watson's model estimates the wear coefficient and the clutch temperature from measurement signals. However, Watson did not explain the model structure in detail.

More detailed approaches for estimating the wear volume of wet clutches are explained by Saito et al. [2], Li et al. [3], Yang et al. [4] and Rao [5]. Saito und Li presented a two phase model for wear volume prediction. Unfortunately, the validation of both models by experiments was not presented. Yang used a simple approach based on the Archard law to estimate wear volume for a wet clutch. Rao introduced a detailed model for estimating friction and wear volume of wet clutches. The validation of this model shows good accuracy for predicting steady-state wear.

Botonakis et al. [6] developed an approach to estimate wear volume for a double dry clutch. In this approach the wear volume is calculated from the surface roughness of the friction linings. Botonakis model was able to plausibly describe the run-in process by using the surface roughness.

Estimation of wet clutch lifetime by the friction coefficient instead of wear volume was done by Ompusunggu et al. [7]. Ompusunggu proved the correlation between slipping time and friction coefficient. The slipping time was calculated from the speed sensor signals and the clutch actuator control signal. Due to the dependency between friction coefficient and lifetime he was able to predict the remaining useful lifetime of a wet clutch system.

Fundamentally, all methods base on estimation of clutch lifetime through one parameter, namely volume or friction coefficient. Furthermore, these methods explain not fully all known clutch effects. Especially temporary dynamic effects for high loads are not described.

This paper presents an approach for wear estimation for dry clutches, which combines wear volume and friction coefficient as parameters for wear prediction. The model shall be able to describe dynamic wear effects and predict the clutch lifetime.

First, the definition of wear and the wear effects for dry clutches are presented in Sect. 2. Afterwards follows a presentation of the wear model in Sect. 3. In Sect. 4 results are discussed and next steps are named. Finally in Sect. 5 a conclusion is given.

## 2 Dry Clutch Wear

Clutches are used to control the torque flow between combustion engine and gearbox. This is achieved by separating and coupling the crank shaft and drive shaft. The clutch can establish a force-fitted connection between both shafts (Fig. 1).

By applying normal force $F_N$ the clutch starts to close. The clutch locks as soon as the resulting friction force $F_R$ is sufficient to transfer the engine torque. The relation between normal force and friction force is described by the friction coefficient μ.

$$F_R = \mu F_N \tag{1}$$

The formula of the resulting clutch torque $T_c$ is generally known and can be calculated from the friction force (1), the number of friction plates $z$ and the friction radius $r$.

$$T_c = z r \mu F_N \tag{2}$$

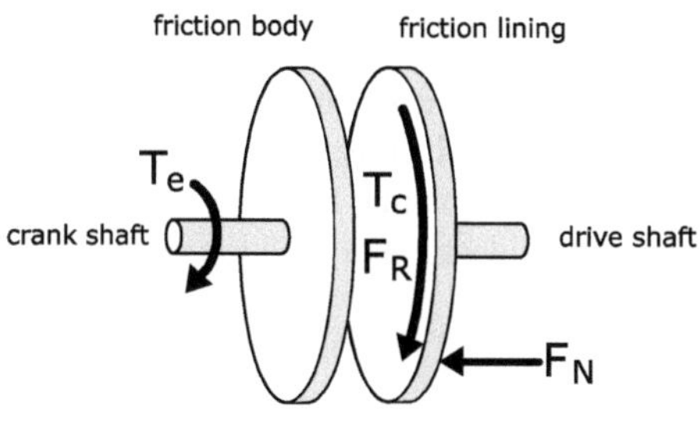


**Fig. 1** Principle of torque transfer for dry clutch

## 2.1 Characteristic Wear Values

The dry clutch wear results from the relative movement of two solid surfaces in contact, also called dry friction. The clutch contains different contact surfaces, which are stressed by friction. Nevertheless, wear of the friction lining has the highest influence on clutch lifetime. Therefore, other components, like the release bearing, will not be taken into consideration for the wear model.

As a first step for modelling dry clutch wear, the definition of wear and its influence parameters need to be clarified. Describing wear by measurable quantities is desired. The essential performance of the clutch can be used to define such quantities. The clutch shall be able to transmit torque from the engine to the vehicle without unwanted slipping. This will be guaranteed, if the clutch capacity $T_c$ is greater than the maximum engine torque $T_{e,max}$ [8].

$$T_c > T_{e,max} \tag{3}$$

The abstract parameter w describes the wear of the clutch. Clutch wear w will be assumed as 0, if the clutch guarantees the designed clutch capacity. This pertains for new assembled clutches. Clutch wear w will be assumed as 1, if Eq. (3) is not satisfied. According to Eq. (3) it is obvious to find influence parameters of the wear by the related clutch capacity. Therefore Eq. (1) can be extended as follow.

$$T_c = z\, r\, \mu(F_N, \Delta\omega, \vartheta_{fl})\, F_N(s_c, s_w, \vartheta_c) \tag{4}$$

The number of friction plates $z$ and the friction radius $r$ will be assumed as constant values. Hence, the clutch capacity depends on the friction coefficient $\mu$ and the normal force $F_N$. Influence parameters of the friction coefficient $\mu$ are normal force $F_N$, clutch slipping speed $\Delta\omega$ and temperature of friction linings $\vartheta_{fl}$. The normal force depends on clutch position $s_c$, total reduction of the friction lining height $s_w$ and clutch housing temperature $\vartheta_c$.

Some of the influence parameters like slipping speed and clutch position can be measured and others need to be estimated. The temperatures are directly dependent to the friction energy E, which occur during the slipping time $t_{slip}$.

$$\vartheta_c = f_{\vartheta_c}(E) \text{ and } \vartheta_{fl} = f_{\vartheta_{fl}}(E) \text{ with } E = \int_0^{t_{slip}} T_c \Delta\omega_c dt \tag{5}$$

The above named influence parameters are dependent to the actual operating point and are independent to previous load. Only the friction lining height reduction $s_w$ and the friction coefficient $\mu$ show a dynamic behavior, which depends on the previous load. Therefore wear $w$ is considered as a function of these characteristic values.

$$w = f_w(\mu, s_w) \tag{6}$$

The influence parameters of the friction coefficient $\mu$ are already defined in Eq. (4). For the friction lining height reduction $s_w$ the influence parameters were not considered yet. According to the research of Archard and Fleischer, the friction lining height reduction is proportional to friction energy and to the wear coefficient [5]. The wear coefficient depends on temperature of friction linings. Therefore friction lining height reduction $s_w$ depends on friction energy $E$ and temperature of the friction linings $\vartheta_{fl}$.

$$s_w = f_s(E, \vartheta_{fl}) \tag{7}$$

## 2.2 Wear Effects

The general tribological mechanism for dry clutches is abrasive wear, because of different hardness of friction lining and friction bodies [9]. Abrasion decreases the total clutch friction lining height in every load cycle. The loss of friction lining height per cycle is defined as $\Delta s_w$. Hence, the total reduction of friction lining height $s_w$ can be calculated by summing all losses.

$$s_w(i) = \sum_{j=0}^{j=i} \Delta s_w(j) \tag{8}$$

Normally, the friction coefficient decreases slowly over the clutch lifetime. This results from two facts. First $\mu(F_N, \Delta\omega, \vartheta_{fl})$ is correlated to $s_w$ through normal force $F_N$ and second the surface roughness is changing slightly.

Furthermore, temporary wear effects for dry clutches were observed in different experimental studies. As first effect the run-in process is considered. This effect takes place, when a new assembled clutch starts to operate. During clutch operation a friction layer of particles is formed between friction body and friction lining. The friction coefficient value will stabilize after the friction layer was formed. At the beginning of the run-in process, the friction lining height will decrease faster than in normal operation due to surface roughness [6] (Fig. 2).

Secondly, the friction coefficient decreases during operation with high thermal load (above 200 °C). If the high load is interrupted after a short time, the friction coefficient starts to regenerate. These effects are known as fading and regeneration. Damage of the friction layer occurs during operation with high load, which causes a decrease of the friction coefficient. For subsequent operations with normal load, the friction layer starts to rebuild like in the run-in-process [10] (Fig. 3).

If operation with high load (above 400 °C) is not interrupted, the friction lining and the surface of the pressure plate and flywheel will be damaged irreversible. The friction coefficient will drop and the clutch will start to slip unwantedly. This effect is called thermal destruction.

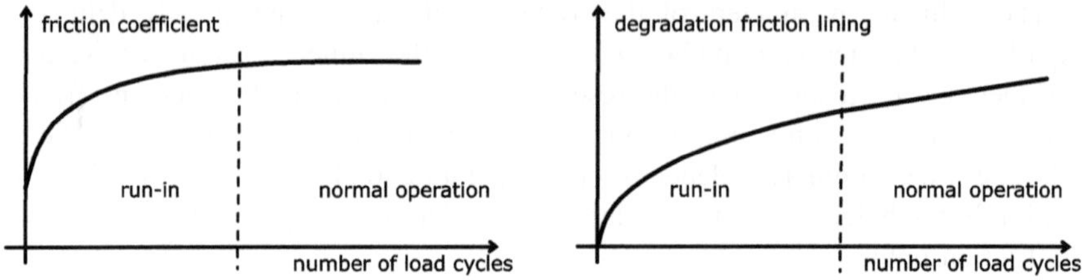


**Fig. 2** Schematic run-in-process for both characteristic wear values

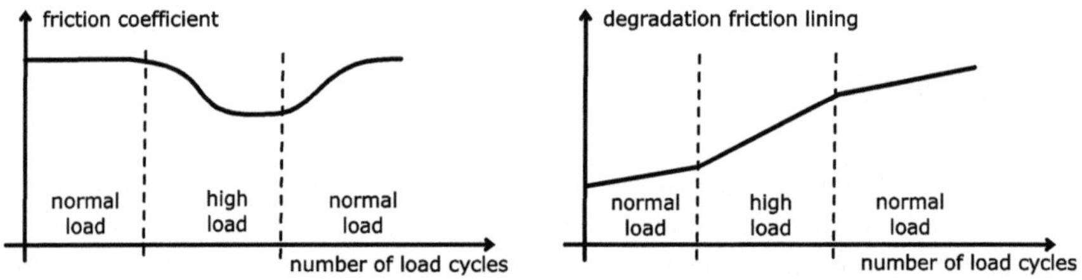


**Fig. 3** Schematic of fading and regeneration for both characteristic wear values

In summary the mentioned wear effects can be distinguished in two groups by their dynamic behavior and their cause.

- Slow dynamics: decreasing of friction lining height
- Fast dynamics: changes of friction coefficient (run-in process, fading and regeneration, thermal destruction)

## 3 Wear Model

The wear model shall be able to describe the named effects of Sect. 2.2. This can be achieved by predicting wear using the friction coefficient and the friction lining height. Estimating the characteristic values is done by a data-based model, which is shown in Fig. 4.

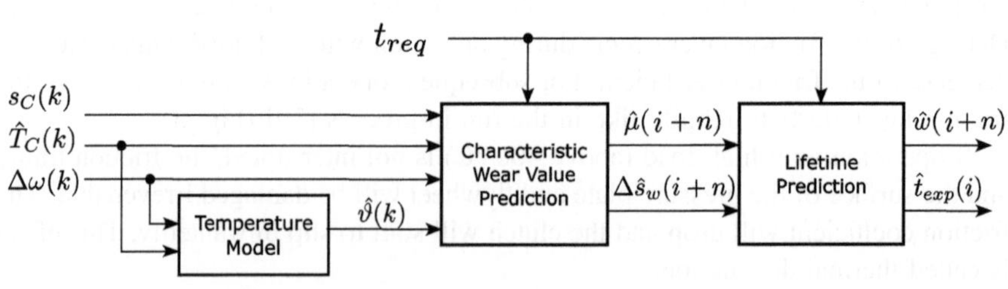


**Fig. 4** Block scheme of the wear model

The model inputs are clutch position $s_c$, slipping speed $\Delta\omega$ and estimated clutch torque $\hat{T}_c$ and the outputs are the wear characteristic values. It should be noted that the inputs are discrete-time data with $k$ as current sample and the outputs are discrete-event data with $i$ as current load cycle. The model calculates the friction coefficient $\hat{\mu}$ and the friction lining height $\Delta\hat{s}_w$ per each load cycle. Wear can be predicted by assuming a constant load for the future cycle $n$. The constant load can be estimated from the mean previous load.

The wear model in Fig. 4 consists of a temperature model, a characteristic wear value model and a lifetime prediction. For temperature estimation a physical model like in [11] is used. By dividing the clutch components in thermal resistors and thermal capacities, the temperature of the clutch can be calculated from friction energy and environmental temperatures. The thermal resistors and thermal capacities of this model need to be identified by experiments.

The main part of the characteristic wear value model can be simulated by simple approximations. These approximations are illustrated in Fig. 5 and will be discussed subsequently in detail. In contrast to Fig. 4, the normal force $\hat{F}_N$ and the friction energy $\hat{E}$ are illustrated in Fig. 5 as inputs. The normal force is estimated from clutch position $s_c$ by characteristic curve and the friction energy is calculated from clutch torque $\hat{T}_c$ and slipping speed $\Delta\omega$ (see Eq. 5). The conversion of the input parameters is not shown in Fig. 5, in order to avoid an unclear illustration.

First, the static friction coefficient $\hat{\mu}_s$ is calculated from a characteristic map. The characteristic map is based on [12].

$$\hat{\mu}_s\left(\hat{F}_N, \Delta\omega, \hat{\vartheta}_{fl}\right) = \alpha + \beta\frac{\hat{F}_N}{F_{N_0}} + f_\mu\left(\hat{\vartheta}_{fl}\right) + \mu_\Delta\left(\tanh\left(\frac{\Delta\omega}{\Delta\omega_0}\right) - 1\right) \tag{9}$$

According to Eq. (9), the normal force is linear proportional to the friction coefficient. The hyperbolic tangent function is used to describe the slipping speed influence. A polynomial function is used to describe the temperature influence.

The dynamic friction coefficient is calculated by first order lag elements from the static friction coefficient.

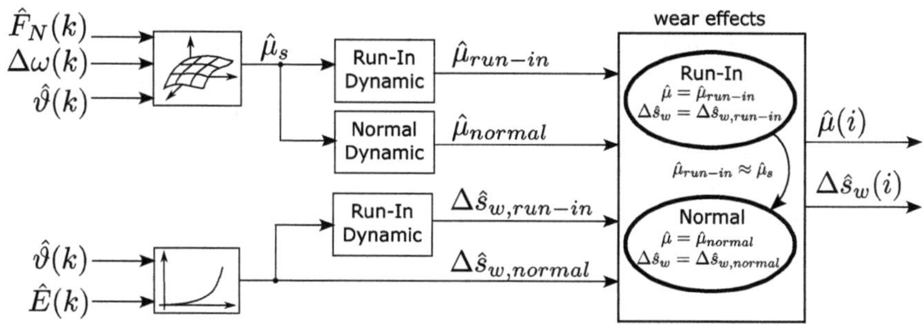


**Fig. 5** Block scheme of characteristic wear value simulation

$$\hat{\mu}_{\text{run-in}}(\text{i}) = \text{a}\,\hat{\mu}_s(i) + \text{b}\,\hat{\mu}_{\text{run-in}}(\text{i}-1) + \text{c}\,\hat{\mu}_s(0) \tag{10}$$

$$\hat{\mu}_{\text{normal}}(\text{i}) = \text{d}\,\hat{\mu}_s(i) + \text{e}\,\hat{\mu}_{\text{normal}}(\text{i}-1) + \hat{\mu}_s(0) \tag{11}$$

Two first order lag elements with different time constants (a, b, d and e) are used to describe the run-in process $\hat{\mu}_{\text{run-in}}$ and the normal dynamic $\hat{\mu}_{\text{normal}}$ with fading and regeneration. The parameter $c$ is the initial value for the run-in process. The model output $\hat{\mu}$ is dependent on the wear effect. During the run-in-process $\hat{\mu}_{\text{run-in}}$ is set as model output. If the dynamic coefficient reaches the range of the static value $\hat{\mu}_s$, the run-in process will be assumed as finished and $\hat{\mu}_{\text{normal}}$ is set as model output.

In normal condition the reduction of friction lining height $\Delta\hat{s}_w$ is calculated through a look-up-table from temperature and friction energy. The look-up-table is based on data from manufacturer datasheets. During the run-in-process the reduction of friction lining height decreases exponentially. This results from the fact that the wear volume per cycle is constant, but the material ratio is not constant during the run-in-process due to surface roughness [6]. The parameters f, g and h in Eq. (12) need to be identified by surface analysis in the run-in-process.

$$\Delta\hat{s}_{w,run-in}(\text{i}) = \text{f}\,\Delta\hat{s}_w(i) + \text{g}\,\Delta\hat{s}_{w,run-in}(\text{i}-1) + \text{h}\,\Delta\hat{s}_w(0) \tag{12}$$

The irreversible wear behavior for thermal destruction is implemented by temperature and friction energy thresholds. If the temperature threshold $\vartheta_{\text{max}}$ or friction energy threshold $\text{E}_{\text{max}}$ is exceeded, the minimal wear coefficient $\mu_{\text{d,wear}}$ is set permanently as model output.

The presented model was implemented in Simulink. Besides this model a vehicle model was used to generate input signals for the wear model. The vehicle model contains an automated manual transmission with a dry clutch. By variation of the input signals different wear effects can be simulated.

In Fig. 6 the model response is presented for one engagement during vehicle launch. Slipping speed, clutch torque and friction power were generated by the vehicle model. The friction coefficient and reduction of friction lining height are output values of the wear model.

In Fig. 7 several engagement and disengagement cycles are illustrated. Instead of slipping speed and clutch torque, the friction power is depicted as model input. During the simulation the load was changed, hence it was possible to simulate all named wear effects. In Fig. 7 the run-in process lasts 800 s. Afterwards, the model switches to the normal wear effect state. Fading was simulated from 1500 until 2000 s. Regeneration starts after 2000 s, then normal load was applied. Thermal destruction is not illustrated.

Lifetime prediction in Fig. 4 is considered as the expected number of cycles when the wear limit $s_{w,max}$ will be reached. The expected lifetime $t_{\text{exp}}$ can easily be calculated by normalized friction height $s_w/s_{w,max}$ and current cycle number $i$.

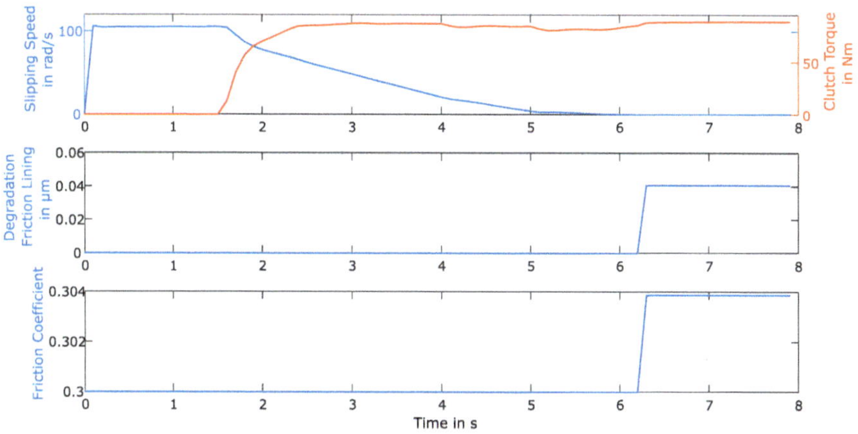


**Fig. 6** Simulated wear behavior for one cycle (vehicle launch)

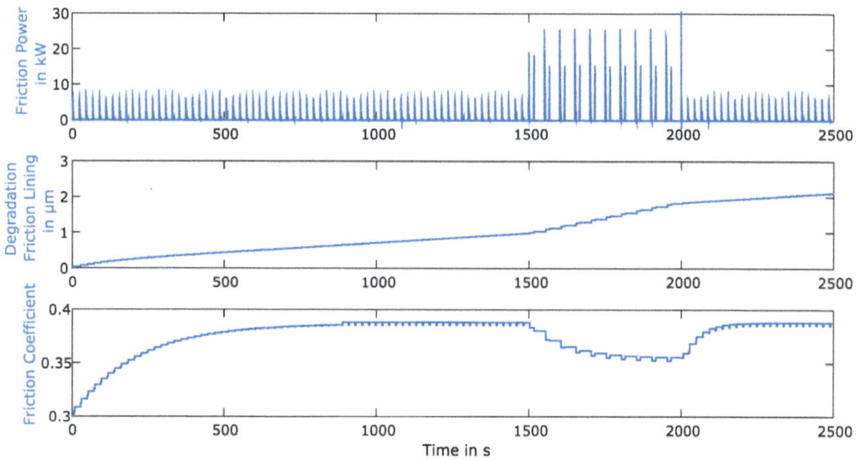


**Fig. 7** Simulated wear behavior for several cycles (run-in, fading and regeneration)

$$t_{\exp} = i\frac{s_w}{s_{w,max}} \tag{13}$$

## 4 Discussion and Further Procedure

The presented approach offers advantages in comparison to existing models. By the combination of friction coefficient and reduction of friction lining height the wear model will be able to describe all known wear effects for dry clutches. The combination of two characteristic values is expected to improve the prediction accuracy. Moreover, the knowledge of characteristic wear values offers different opportunities to integrate the predicted wear into the clutch control taking system behavior changes over lifetime into account. These changes can be compensated by an adaptive control strategy. Shortening of remaining useful lifetime can be prevented by adjusting the control targets in case of high wear. Hence, the expected lifetime of the clutch can be guaranteed.

In Sect. 3 a simulation model was proposed. The simulation model will be used to generate data for wear model identification. This data will be used to prove feasibility for different data-based model structures, as for example dynamic Gaussian-process model, recurrent neural network or Kalman filter. The simulation results are reasonable, as the friction coefficient remains within the expected range. The time response of the friction coefficient during run-in is similar to [6, 10]. During fading and subsequent regeneration is the simulated time response similar to [10]. The degraded friction lining height is about 2 μm after 200 cycles, which corresponds to 0.1 % of the maximum friction lining degradation. The ratio of the total friction energy and the resulting wear, which is equivalent to the wear coefficient, is in agreement with literature values of [10]. The run-in behavior of the friction lining height reduction is too small compared to [6]. However, the assumptions for the applied friction energy and the friction area in [6] are valid for a different clutch system. On the basis of measurement data it can be verified, whether the run-in behavior of the friction lining height reduction is within the proper range.

As a next step the presented model will be identified by experimental data. Data acquisition will be done on a test bench [13]. Different loads can be simulated on this test bench by adjusting different parameters like starting value of engine speed before slipping, clutch torque, load torque and vehicle inertia. The clutch temperature will be measured by wireless electronics, which is integrated in the flywheel [14]. The friction coefficient will be calculated from clutch torque and normal force. The reduction of friction lining height will be measured by weighting the clutch disk after a defined number of cycles.

The wear model in Sect. 3 is based on characteristic values, which cannot be measured in an operating vehicle. Identification of the presented model needs to be done for each clutch type during prototype development through test bench results. The model validation in vehicle can be achieved by using other sensor signals. In the vehicle the friction coefficient can be estimated by an observer. Hence, this model output can be validated through the observer value. Determining the friction lining height reduction in a driving vehicle can be estimated by displacement variation of the clutch touch point. The touch point is commonly determined by clutch position measurement and clutch torque estimation.

Transferability to wet clutch systems is under investigation. Wet clutches show different degradation and friction behaviors than dry clutches. For wet clutches the friction coefficient can decrease over lifetime [7], however it stays nearly stable for dry clutches after run-in process with normal load [10]. Further investigations and experiments are necessary to clarify the transferability.

## 5 Conclusion

Existing approaches for clutch wear modelling are using the wear volume or the friction coefficient to estimate the remaining useful lifetime. Some models were able to describe the run-in-process as well. For existing wear models the accuracy is limited, because the estimation is based on constant loads and temporary effects like fading are not modelled. To guarantee a constant lifetime for the clutch system, it is necessary to estimate wear for all known effects and different loads properly.

It was shown in this contribution, that wear can be defined by significant influence parameters for a dry clutch. Known wear effects were summarized and illustrated qualitatively. An approach for characteristic wear value prediction was presented. Different wear effects were simulated by simple approximations, like characteristic maps or first order lag elements. By prediction of two characteristic wear values, the model can describe all known wear effects. Especially the characteristic wear values can be determined more accurately for high loads by the presented model.

Further research will focus on the identification of the temperature model, the characteristic wear value model and the integration of the characteristic wear values in the clutch control. The aim is to integrate the wear model in the control of clutch in order to guarantee a required lifetime.

## References


1. Watson, M., Byington, C., Edwards, D., Amin, S.: Dynamic modelling and wear-based remaining useful life prediction of high power clutch systems. In: ASME/STLE International Joint Tribology Conference (48), pp. 208–217 (2005)
2. Saito, T., Kotegawa, T., Matsuura, Y., Tanaka, S., Ohtsuki, K.: Study of durability prediction with focus on wear properties for multiple plate clutches. In: SAE World Congress and Exhibition (2007). doi:10.4271/2007-01-0240
3. Li, M., Khonsari, M.M., McCarthy, D., Lundin, J.: On the wear prediction of the paper-based friction material in a wet clutch. Wear **334–335**, 56–66 (2015). doi:10.1016/j.wear.2015.04.005
4. Yang, Y., Twaddell, P.S., Chen, Y.-F., Lam, R.C.: Theoretical and experimental studies on the thermal degradation of wet friction materials. In: SAE International Congress and Exposition (1997). doi: 10.4271/970978
5. Rao, G.: Modellierung und Simulation des Systemverhaltens nasslaufender Lamellenkupplungen. Dissertation, Technische Universität Dresden (2011)
6. Botonakis, I., Schwarze, H., Adamis, P., Deters, L.: Modellierung von Reibung und Verschleiss bei trocken laufenden Kupplungsreibbelägen. Tribol. Schmierungstech. **57**, 5–10 (2010)
7. Ompusunggu, A.P., Papy, J.-M., Vandenplas, S., Sas, P., van Brussle, H.: Condition monitoring method for automatic transmission clutches. Int. J. Prognostics Health Manag (2012)
8. Tirović, M.: Dry clutch. In: Crolla, D., Foster, D.E., Kobayashi, T., Vaughan, N.D. (eds.) Encyclopedia of Automotive Engineering, pp. 1–30. Wiley (2014)
9. Popov, V.L.: Kontaktmechanik und Reibung. Springer-Verlag (2010)

10. Gauger, D.: Wirkmechanismen und Belastungsgrenzen von Reibpaarungen trockenlaufender Kupplungen. Dissertation, Technische Universität Berlin (1998)
11. Beitler, H.: Untersuchung zum Temperatur- und Wärmeabgabeverhalten einer Einscheiben-Trockenkupplung. Dissertation, Universität Karlsruhe (2008)
12. D'Agostino, V., Senatore, A., Pisaturo, M.: Improving the engagement performance of automated dry clutch through the analysis of the influence of the main parameters on the frictional map. In: World Tribology Congress (2013)
13. Nowoisky, S., Knoblich, R., Gühmann, C.: Prüfstand für die Identifikation und den Funktionstest an automatisierten Schaltgetriebe. Virtuelle Instrumente in der Praxis; Begleitband zum 17. In: VIP-Kongress (XIV), pp. 150–156 (2012)
14. Funck, J.: Drahtloses Messsystem für Temperaturen in Kraftfahrzeugkupplungen (2012)

# Part VI
# Hybrid + E

# KlimaPro: Development of an Energy Efficient Operating Strategy for Carbon Dioxide Climate Systems Used in a Fully Electronic Bus

**Franz Groth and Michael Lindemann**

**Abstract** This article describes the stages of development towards an energy and fuel consumption optimized operating strategy for $CO_2$ air conditioning units in fully electronic buses. The focus of the development process is on the simulation. The focus of the development process is on the simulation. Based on air condition and power train data from the development platform, which was provided by the project partners, a physical vehicle model is created, which illustrates the system's most important power flows and describes the reciprocal effect between the subsystems regarding their impact on the energy consumption. Thus, for example, a change of the fan speed will have an effect on the respective air flow volume of the cooling circuit model's heat transmission system on the one hand and will cause a change of the energy consumption via the battery model connected to the generator in the longitudinal dynamics model on the other hand. Examinations of changes in subsystems and their effects on the entire system are possible as a consequence. The overall simulation model consists of the described physical vehicle model and a by means of a real vehicle digitally modeled control system, the conventional operation strategy. The vehicle model receives control signals such as driving cycles, ambient temperature, passenger load and further control commands (i.e. door opening, lowering the bus etc.). Out of these measurement signals the vehicle model is calculating the respective physical performance. All state variables, such as passenger load, inside and outside temperature, the battery state of charge, requirements of auxiliary units systems etc., will be delivered to the operating strategy. Out of these data, the variables for the air conditioning will be calculated and the measured results can be sent back to the vehicle model. A comparison of the simulation results between the impact of the conventional and the newly developed

Supported by ZIM - Central Innovation Programme SME.

F. Groth (✉) · M. Lindemann
Hochschule für Technik und Wirtschaft Berlin, Fachbereich 2 | Fahrzeugtechnik, Wilhelminenhofstraße 75A, 12459 Berlin, Germany
e-mail: f.groth@htw-berlin.de

M. Lindemann
e-mail: michael.lindemann@htw-berlin.de

 
C. Gühmann et al. (eds.), *Simulation and Testing for Vehicle Technology*,
DOI 10.1007/978-3-319-32345-9_13

operating strategy of the $CO_2$ climate system on the vehicle model supports the developing process. As a result, the operating strategy can be optimized with regard to the energy consumption. The development process of an overall simulation model and the obtained results will be presented. In addition, the potential of a demand-oriented power control concerning the energy consumption compared to conventional operating strategy will be demonstrated. Finally, it is shown which further steps are necessary in order to enhance a renunciation of fossil auxiliary heaters in the future and ensure at least the same range at the same time.

**Keywords** $CO_2$ air conditioner · Operating strategy · Energy optimization · Electric bus · Overall simulation model

# 1 Motivation

The project"KlimaPro" is a result of the discussion process within the context of "E-Bus-Pro" network [1]. Accordingly, the energy consumption combined with the battery management determines significantly the economical and potential uses electric busses. Auxiliary users and particularly the air conditioner are substantial energy consumers within this context. Examinations on possible saving potentials of air conditioning provided so far the use of efficient, optimized with heat pump function and $CO_2$ as a refrigerant. Clearly, air conditioners with $CO_2$ as a refrigerant are more efficient and environment-friendly than those that are based on the uses of R123a refrigerants. Achieved energy savings can be converted directly into a higher range and enhance a renunciation of auxiliary heaters based on fossil fuels. The application of improved regulation and controlling concepts offers another saving potential. Currently, only Wabco-Controlling can be classified as a market-relevant climate control which is used by almost all vehicle manufacturers. Although, it does not fulfill the Association for German Transportation Companies (VDV). Furthermore, the Wabco-Controlling can control only one electric auxiliary heater for the supply temperature heating [2].

The operating strategy is primarily aiming for the increase of range of busses by relieving specifically the traction battery. In order to achieve these objective two main approaches will be basically targeted:

- air conditioning adapted to the needs of operation
- energy-optimized operation within the composite with auxiliary units.

Overall are following development results aspired:

- In the climate system integrated battery cooling and—heating as well as use of the thermal waste heat from the electrical system
- Control electronic in comparison with the load capacity of the battery in real time

- Control electronic to priorities the energy resources for the drive and subordinate for the comfort (air conditioning)
- Fulfilment of the VDV-guidelines for air conditioning of vehicle interiors in the public transport
- Adaptive climate control depending on the concede amount of energy.

Despite the generic requirement of the control takes the development strictly along on the actually praxis circumstances of a concrete vehicle place. As a development platform serves a 12 m E-Bus from the company Skoda Electric a.s.. The build-in $CO_2$ air condition was provided from the Konvekta AG. It is possible to switch between heat- and cooling operation. For the heating is the cooling process reversed and a high proportion of the environmental heat from the ambient air is pumped in the room to be heated. An electrical auxiliary heater in the winter months is therefore waived in the future. As simulation tool is MATLAB Simulink in use.

## 2 Entire Vehicle Simulation Model

The dynamic complete vehicle simulation model is composed of a physical vehicle model and by means of a real vehicle modelled control of the conventional operation strategy. For the implementation of simulations is the vehicle model in accordance with boundary conditions exposed, which describe the actual driving status of the bus and the climate environmental conditions.

- driving status- boundary conditions (driving cycle, target vehicle speed, gradient)
- climate boundary conditions (ambient temperature, humidity, solar radiation) (Fig. 1).

The basic structure of the model parts will be described in the sections below. The explanation of thermodynamically model builds the center of attention.

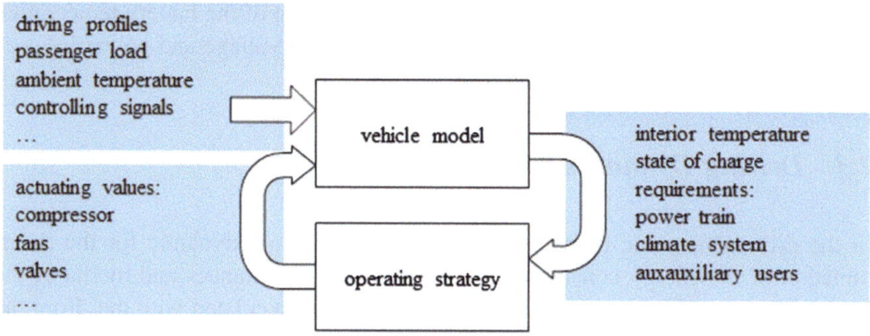


**Fig. 1** Entire vehicle simulation

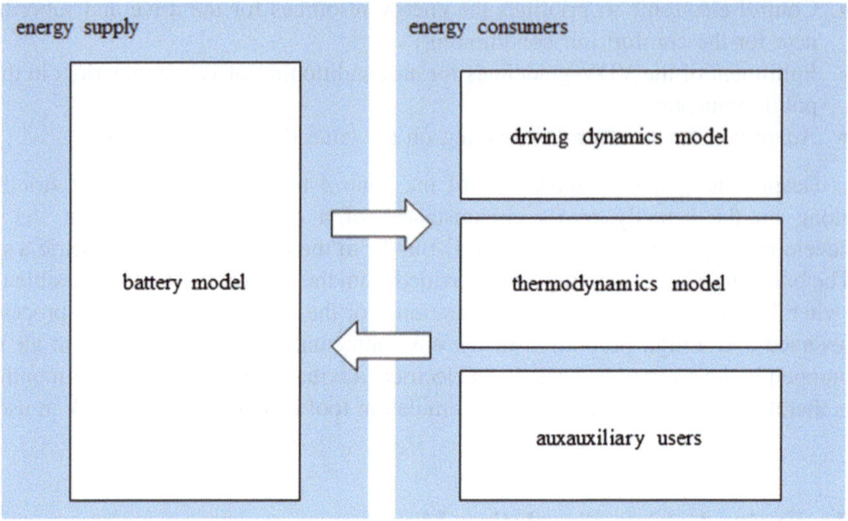


**Fig. 2** Vehicle model

## 2.1 Vehicle Model

The vehicle model simulates the total electrical powertrain and the thermodynamic, but furthermore comprised the physical route of the thermal control loop (thermal model of the vehicle interior and these components). It is structured by Fig. 2 in areas of energy supply und–consumer.

## 2.2 Battery Model

The supply side contains the battery model. This determined the actual available battery power and the state of charge. Therefore, are all performance requirements of the consumer side added up and calculated with the help of the battery temperature the necessary parameter as internal resistance, terminal voltage and power loss.

## 2.3 Driving Dynamics Model

In the driving dynamic model is the vehicle longitudinal dynamic for the determination of the energy consumption through driving resistances and inertia of the vehicle depicted. The integrated electric motor model calculated with the, from the driver requested torque, which through balance with the driving resistance, results in the vehicle speed and the driven distance.

## 2.4 Auxiliary Users

The block of the auxiliary users bundled not modelled performance requirements of the pneumatics (e.g. lowering of the bus on the bus stop, automotive headlights and other auxiliary units). The air condition itself is not a part of the auxiliary users, because for the optimization that needs a deeper view on it.

## 2.5 Thermodynamics Model

This model is divided into the interior and air conditioning model, which will be discussed in more details below.

**Vehicle Interior Model** Since the inner and outer thermal and moisture loads vary strong in time and quantity, also the air-conditioning load is constantly changing. However, to determine the current demand for air conditioning it's based on the 12-m-electric bus "Perun HE" by Skoda Electric a.s., a three-zone constructed state space model. In the vehicle longitudinal direction it is divided in three zones. The first zone describes the driver's area. The second and largest zone corresponds to the passengers and the third zone describes the thermal behavior of the battery. Accordingly to the division into three zones yields three different sizes of balance zones. The respective volume of the zones is assumed as mixed ideal, i.e. it will be expected a homogenous temperature distribution for each zone. Figure 3 shows a schematic representation of the vehicle with the parties on the balance sheets of heat and material flows. Heat flows are shown monochrome and material flows dashed.

The air conditioning load results per zone by balancing with the corresponding loads. As an example of the driver, the relationship is described as followed.

$$\dot{Q}_{bal-drv} = \dot{Q}_{clima-drv} + \dot{Q}_{amb-drv} + \dot{Q}_{sol-drv} + \dot{Q}_{sol-drv} \quad (1)$$

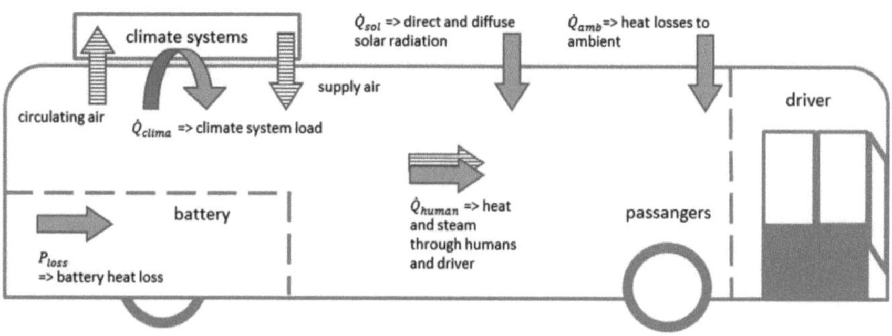


**Fig. 3** bus climate system load

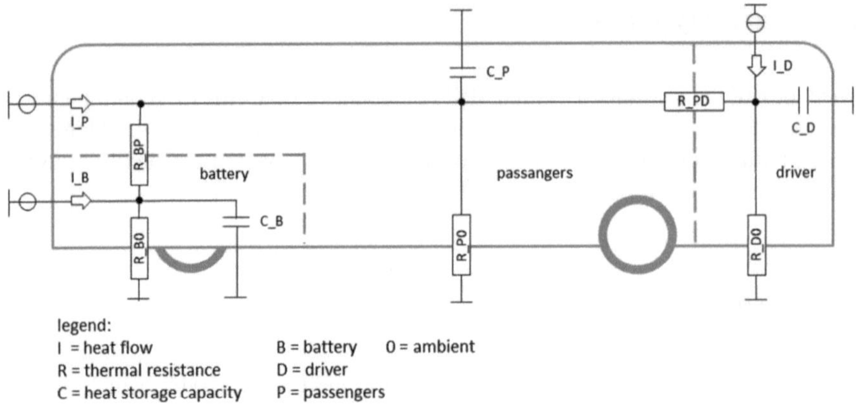


**Fig. 4** Thermal network

For the transmission of reported heat flows in the resulting room temperatures a thermal network out of capacities, resistors and structured sources was build. Therefore transverse sizes are to understand as temperatures and through sizes as heat flows. The resistors R (or conductance G) describe the Heat transfer capacity of the scenery or to neighboring areas. The Capacities (C), so the heat storage capacity of a zone is calculating from the specific heat capacity of the medium, the density and its volume. If only considers the thermal resistances, it would correspond to the stationary heat conduction, but because of the capacities and the related heat storage capacity per zone, a time-dependent temperature gradient and consequently a dynamic behavior of the thermal system. Figure 4 shows the thermal equivalent circuit. With it the indexes X0 describe the conductance to the surroundings and XY the conductance by zones to each other.

Fourier's law and from the electronics transmitted Kirchhoff rules can be applied in the equivalent circuit. The entire thermal system consists of three different sized capacity, to be understood as Energy storage as described. As state variables describe energy storage, arise from the capacity of three state variables and a system of the third order. Therefore three differential equations of 1st order are required. The resulting differential equations are as follows.

$$\text{Batterie: } \dot{U}_B = \frac{1}{C_B} \int [I_B - (G_{BO} + G_{BP})U_B + G_{BP}U_P]dt \tag{2}$$

$$\text{Fahrer: } \dot{U}_D = \frac{1}{C_D} \int [I_D - (G_{D0} + G_{PD})U_D + G_{PD}U_P]dt \tag{3}$$

$$\text{Passagiere: } \dot{U}_P = \frac{1}{C_P} \int [I_P - (G_{P0} + G_{PD} + G_{BP})U_P + G_{BP}U_B + G_{PD}U_D]dt \tag{4}$$

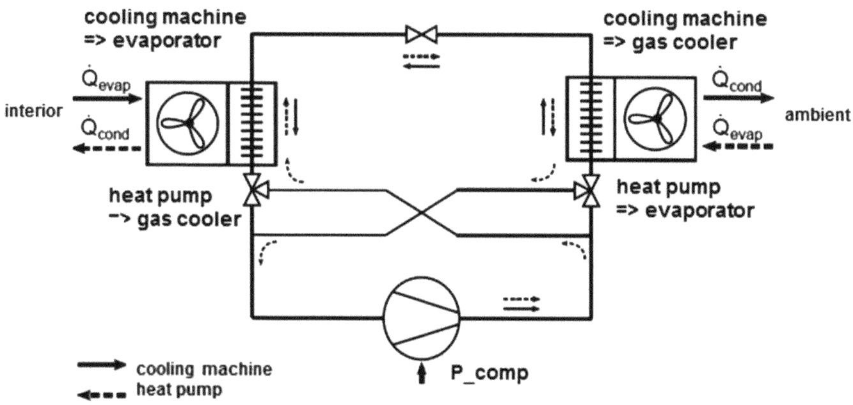


**Fig. 5** Invertible refrigeration cycle process [3]

The previously determined accounted heat flows go as inputs $I_x$ in the state space model, and provide at its output the ambient temperatures $U_x$.

**Climate System Model** The air condition model serves the figure of the refrigeration cycle process and determines the thermal usage of energy in dependence to the applied electrical power. A schematic representation of the process for the heating- and cooling mode is outlined in Fig. 5.

Is the liquid coolant on a low pressure level with an evaporating temperature below this of the cooling room, it withdraws the heat flow $\dot{Q}_{evap}$ and cooled the room as a consequence. The steam of the coolant is going to be compressed on a higher pressure- and temperature level in the compressor upon expenditure of the electrical power $P_{comp}$. Over the gas cooler the hot coolant steam gives the heat flow $\dot{Q}_{cond}$ to their environment off. In the expansion valve is the coolant again on evaporating pressure relaxed and repeated supplied to the evaporator. Through an accordingly use of valves, it is possible to heat with the cold circle process, therefore the process operates as a so-called heat pump. Thereby the emitted heat to the environment from the gas cooler is in use to heat the room. The thermal effective output is no longer $\dot{Q}_{evap}$, but $\dot{Q}_{cond}$. However, for the current account of the system always valid:

$$\dot{Q}_{cool} = \dot{Q}_{evap} + P_{comp} \tag{5}$$

At $CO_2$ air conditions becomes the critical point of the coolant exceeded by the compression. In this context it is called a transcritical refrigeration cycle process [4]. The subsequent heat dissipation to the ambient takes place in the supercritical state and does not lead to condensation. That is why it calls gas cooler and not condenser like it is usual in subcritical systems. Only by the expansion the refrigerant liquefies (Fig. 6).

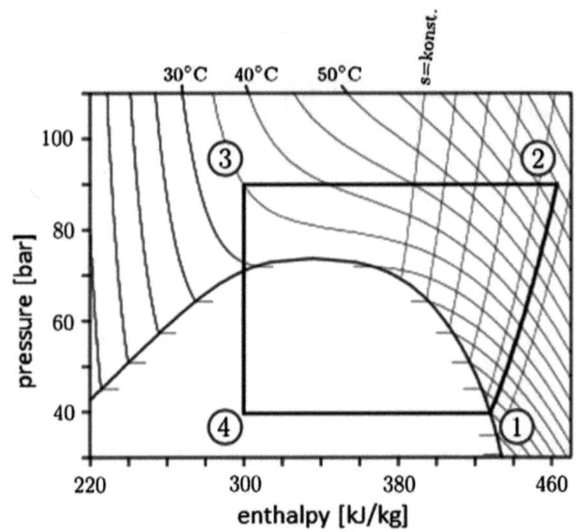


**Fig. 6** Cold cycle in $CO_2$ p, h-Diagram [5]

- 1-2: Isentropic compression in the compressor ($P_{comp}$)
- 2-4: heat emission with stable pressure in the gas cooler ($\dot{Q}_{cool}$)
- 3-4: Expansion in the expansion valve at a stable enthalpy
- 4-1: Heat absorption at stable pressure in the evaporator ($\dot{Q}_{evap}$).

*Calculation of the Efficiency Power* As a basis for calculating the process of the cold cycle, a compressor polinomical from Blitzer is utilized. Pressure difference between low pressure and high pressure as well as the frequency of compressor are necessary initial parameters and deliver the mass stream, as well as the introduced hydraulic benefit. With these values, the acceptance of a constant evaporation temperature of −5 °C chosen freely, the gas cooler outlet temperature of 25 °C and the substance data library are deposited in "TILMedia", all condition points can be determined in the p,h-diagram. The sought-after thermal efficiency ($\dot{Q}_{eff}$), arises according to the mode of operation after the following coherence:

$$\text{Effective output of the cooling machine: } \dot{Q}_{evap} = \dot{m}_{CO_2}(h_4 - h_1) \tag{6}$$

$$\text{Effective output of the heat pump: } \dot{Q}_{cool} = \dot{m}_{CO_2}(h_2 - h_3) \tag{7}$$

For the best energy-efficient use of the air-conditioning it is important to know in which operating point the arrangement with the slightest energetic expenditure, has the biggest benefit. Therefore is efficiency divided by effort. The result is generally referred as a coefficient of performance. In the case of the cooling machine it is spoken by EER (Energy Efficiency Ratio) and the head pump significate COP (Coefficient of Performance).

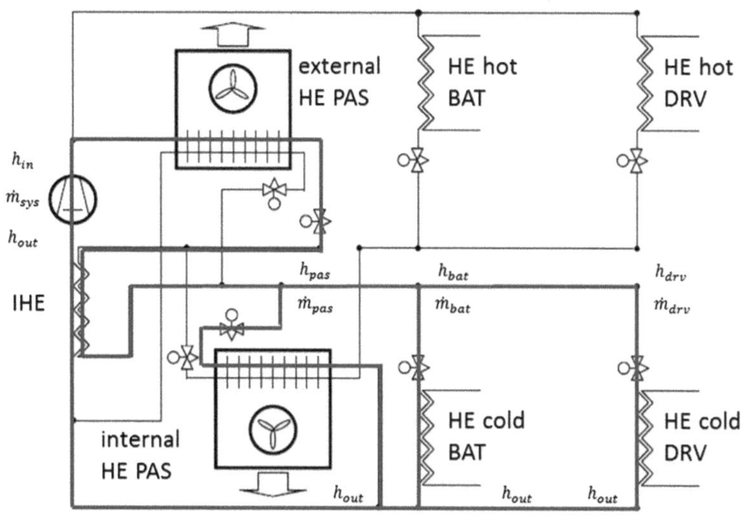


**Fig. 7** $CO_2$-climate system [6]

$$EER = \dot{Q}_{evap}/P_{comp} \tag{8}$$

$$COP = \dot{Q}_{cool}/P_{comp} \tag{9}$$

*Distribution of the Efficiency Power* The already generated efficiency power has to distribute to the heat exchangers (HE), which ones are presented in the system below. In Fig. 7 the example of a cooling machine show a possible combination of the treated three zones.

In the simulation of the air conditioning circle the terms shown in Gl.10 should be equal anytime. As already mentioned the first part arise the cooling capacity of the refrigerant. The middle part describes the heat transfer capacity of the heat exchanger and the third part refers to the heat flow of air or more generally spoken the heat flow of secondary fluid.

$$\dot{m}_1(h_{in} - h_{out}) = kA\Delta\vartheta_{log} = \dot{m}_2 c_2(\vartheta_{out} - \vartheta_{in}) \tag{10}$$

On bases the signal flow orientation of the approach with Simulink and the necessity of a generic model assumption are made that the simulation simplify the system circuit significantly. These simplifications are far legitimate, as that in the development of the operating strategy the electrical performance of the actors (Compressor, Ventilators) are of interest. Although the output power of the actors is depend on thermodynamic state of the system, but limits itself to a normal operation in the broadest sense, may with sufficient accuracy be assumed a linear behavior and mutual thermodynamic and nonlinear interactions of the various parts of the system are neglected.

To simplify the dependencies of the thermo-dynamical relations is as first to be assumed that the heat exchangers are able to transmit the heat flow from or to the refrigerant. The approach is to be welcomed regarding to a preferably generic, vehicle independent operating strategy. Therefore are parameters which are difficult to obtain like heat transfer and pressure loss coefficient are no longer required. Furthermore assumed it is established, that the enthalpies (h) after every heat exchanger are equal.

$$h_{in} = h_{bat-in} = h_{drv-in} = h_{pas-in} \tag{11}$$

$$h_{out} = h_{bat-out} = h_{drv-out} = h_{pas-out} \tag{12}$$

Consequently the mass flow is the only not determined variable in equation. Accordingly, depending to the magnetic valve position it leads to the effective power per heat exchanger ($\dot{Q}_{he-x}$) as an part of the entire effective output ($\dot{Q}_{eff}$).

For this it will introduced the new variable, named HFD (Heat Flow Distribution). It describes the percentage distribution of refrigerant mass flow after the compressor ($\dot{m}_{sys}$) into the other parts of circle.

With the law of conservation of energy the HFD has to be between zero and one and the sum of its parts has to be one, like the sum of its parts of mass flows ($\dot{m}_x$) has to be equal with the system mass flow.

$$\dot{m}_{sys} = \dot{m}_{bat} + \dot{m}_{drv} + \dot{m}_{pas} \tag{13}$$

$$HFD = 1 = HFD_{he-bat} + HFD_{he-drv} + HFD_{bat-pas} \tag{14}$$

$$\dot{Q}_{eff} = \dot{m}_{sys}(h_{in} - h_{out}) \tag{15}$$

$$\dot{Q}_{he-x} = \dot{Q}_{eff} * HFD_{he-x} \tag{16}$$

*Fans* Is now considered the air side arise the supply air temperature $\vartheta_{sup}$ through transforming the known equation Gl.10. The inlet temperature of fans $\vartheta_{air}$ correspond to the temperature of circulating or ambient air, depending which operating mode is actually chosen (a mixing temperature of both is also possible). $c_{air}$ correspond to specific heat capacity of air. It should also be noted that the lost heat of a fan has to be subtracted as electrical power from the effective power. For reasons of clarity is this fact in the following equation neglected.

$$\vartheta_{sup} = \frac{\dot{Q}_{he-x}}{\dot{m}_{fan} * c_{air}} + \vartheta_{air} \tag{17}$$

*Condensation* Depending on the relative humidity (φ) of the interior it is possible that during cooling of damp air, condensation forms it out. This will happen precisely when the temperature of supply air fall below the dew point temperature of the room.

To figure out the dew point ($\vartheta_{air-dew}$) is used the following equation [7].

$$\vartheta_{air-dew} = \varphi^{\frac{1}{8.02}}(\vartheta_{air} + 109.8) - 109.8 \tag{18}$$

The temperature difference between supply and dew point leads with the Eq. 19 to dehumidification performance part $\dot{Q}_{cond}$ and the difference to the heat exchanger cooling performance result to dry part $\dot{Q}_{dry}$ shown in Eq. 20.

$$\dot{Q}_{cond} = \dot{m}_{fan} * c_{air} * (\vartheta_{sup} - \vartheta_{dew}) \tag{19}$$

$$\dot{Q}_{dry} = \dot{Q}_{he} - \dot{Q}_{cond} \tag{20}$$

The dry part of the heat flow leaves this block und goes as the climate system performance $\dot{Q}_{clima}$ into the already described balance of climate system loads.

## 2.6 Control

The control block unites the control of the longitudinal dynamics model, the operating strategy of the climate and the performance management.

**Control of Longitudinal Dynamics Model** The driver model is a PID controller which controls the actual speed of the vehicle model with the regulating units accelerator pedal position and brake pedal position according to the target speed of the given driving cycle. The town part of the NEDC (New European Driving Cycle) serves as a base cycle. Other bus specific cycles can also be given as a standard.

**Performance Management** The available maximal performance of the battery is calculated in the battery model using the sum of the required performances and the thereby associated battery power. Therefore, provided that the maximum is not achieved, the requirements yield the availability. Only in case of exceedance, a hierarchical performance management is activated. The supply of the braking systems of, in this case, the requirements of the longitudinal dynamics model, stands in the first place of course. Within the thermic model, the conditioning of the battery is prioritised. Thereupon follows driver and subsequently the passengers.

**Operating Strategy of the Climate System** The characteristic line of comfort in the public transport for busses according to the VDV guidelines forms the basis for the ascertainment of the control circuit reference value. The set room temperature is therefore a function of ambient temperature (Fig. 8).

This target should be accomplished by the air conditioning system without the intervention of the driver. For his own area, the driver is able to switch between the automatic determination of the target value and the manually adjustable desired temperature. The thermic conditioning range of the battery is settled on 25 °C.

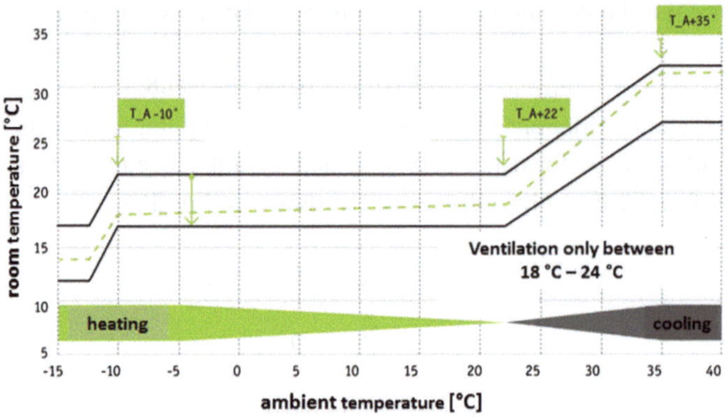


**Fig. 8** Behaglichkeitskennlinie nach VDV-Richtlinie [2]

The difference between the reference value and the output signal of the roadway, that means the difference between target and actual room temperature, results in the control deviation and serves as the decision factor for the ascertainment of the actuating variables in the controller. Initially, the controller is realized as a simple 3 points controller. For every area, it is decided whether heating, cooling or no thermic conditioning at all is needed. The thereby obtained information about the area requirements is, in a subsequent logic, transformed into the resulting requirement for the air conditioning system. That means, for example, in the simplest case, that a compressor is turned on when at least one area requirement is unequal zero. A once turned off compressor should, as the producer points out, not be turned on again for a certain time. The logic switch on of a component does not yet tell anything about the absolute value, the specific actuating variable. The transformation of the general actuating variable into the specific number of revolutions is momentarily statically resolved. This means that it can be differentiated between the maximal and minimal, respectively within the limits of an arbitrary rotational speed.

## 3 Results

### *3.1 Conventional Operating Strategy*

The conventional operating strategy was implemented for verification and testing the model. The city part of NEDC is used as the basic simulation-cycle. Across the simulation time following boundary conditions were set as constant.

- direct solar radiation = 590 W/m$^2$
- diffuse solar radiation = 128 W/m$^2$
- relative humidity of ambient = 50 %

**Table 1** results for various load cases

| Ambient temperature (°C) | Set temperatures/BAT = 25 °C | | |
|---|---|---|---|
| | Conventional os PAS & DRV = 18 °C | Conventional os PAS & DRV = 21 °C | New os => VDV |
| 30 | E_sy = 1,544 kWh/km<br>E_td = 0,555 kWh/km | E_sy = 1,542 kWh/km<br>E_td = 0,553 kWh/km | E_sy = 1,508 kWh/km<br>E_td = 0,518 kWh/km |
| 20 | E_sy = 1,500 kWh/km<br>E_td = 0,511 kWh/km | E_sy = 1,491 kWh/km<br>E_td = 0,502 kWh/km | E_sy = 1,525 kWh/km<br>E_td = 0,536 kWh/km |
| 10 | E_sy = 1,514 kWh/km<br>E_td = 0,525 kWh/km | E_sy = 1,514 kWh/km<br>E_td = 0,525 kWh/km | E_sy = 1,494 kWh/km<br>E_td = 0,504 kWh/km |
| 0 | E_sy = 1,524 kWh/km<br>E_td = 0,534 kWh/km | E_sy = 1,539 kWh/km<br>E_td = 0,550 kWh/km | E_sy = 1,525 kWh/km<br>E_td = 0,536 kWh/km |
| −10 | E_sy = 1,543 kWh/km<br>E_td = 0,554 kWh/km | E_sy = 1,554 kWh/km<br>E_td = 0,555 kWh/km | E_sy = 1,543 kWh/km<br>E_td = 0,554 kWh/km |

The set-temperatures are also constants and they do not depend on the eco characteristic curve. The actors operated at half rotational speed. Table 1 shows results for various load cases concerning to total energy requirement $E_{sy}$ and the requirement of the climate system $E_{td}$.

## 3.2 *Efficiency Optimization*

One opportunity to reduce the total energy consumption and simultaneously get the best-possible result of efficiency is the consideration of the optimal operating point.

Therefore were in use described coefficients of performance from Eqs. 8 and 9. The rotational speed of the actors will drive in the respectively best operating point. The boundary conditions correspond to Sect. 3.1.

Compared with the results of the conventional operating strategy is to recognized that the highest energy saving came to approximately 1.3 %. This seems to be a marginal outcome but currently the rational speeds are still constants, also when they are configured for the respectively best operating point. So it would be lead to better results if the rotational speeds will calculated automatically after the climate load. The studies indicated as well, that the efficiency of the compressor is much dependent on the high pressure and it is more important to control the entire system efficiency as only the component ones.

# 4 Perspective

Against the background of efficiency optimization it does not mean that a minimal rational speed is necessarily a minimal electrical effort or minimal electrical effort of one component is not immediately a minimal effort of the entire system. This is the reason that in future the rotational speed should be automatically controlled over various kinds of sophisticated mathematical algorithms. On the one hand, it is an aim to operate not only in the best efficiency of respective components but rather in the optimal operating point of the entire system. On the other hand it is necessary to produce in each moment just so much power like the climate system loads actually demand. Further it will be shown that the control of supply temperature also will have an impact on the required climate power. But also it is important to take a look on the situation if the ambient temperature conditions are closed to the set conditions. Because in the past it was very difficult to find an energy efficient path handling the switch-on inhibits. The topic automatic ventilation without cooling or heating and the consideration of a prospective analysis of driving cycles should offer the possibility not only to react anymore and be proactive. Expansion the reference variables to on air $CO_2$ content and the relative humanity also leave more detailed diagnosis. In addition to the rotational speed control refer the results from Sect. 3.2 the significance of optimal high pressure control. For heat pump mode it would be useful to discuss the possibility of a heat accumulator to use the battery lost heat. Ultimately, only a combination of those optimization activities will result in a sustainable solution.

**Acknowledgments** ZIM—Zentrales Innovationsprogramm Mittelstand. Bundesministerium für Wirtschaft und Energie aufgrund eines Beschlusses des Deutschen Bundestages.

# References

1. www.e-bus-pro.de. 02.5.2015/10:44
2. Klimatisierung von Elektrobussen. HTW Berlin, Nachhaltige Mobilität, Energiewende und Industrie 4.0. F. Groth, M. Lindemann
3. Entwicklung eines Simulationsmodells. zur Ermittlung des Energiebedarfs und der Lebenszykluskosten verschiedener Heiz- und Kühlsysteme in einem Elektro-Linienbus. Diplomarbeit D. Jefferies. Abgabe 2015
4. Thermodynamik: Grundlagen und technische Anwendung. H. Baehr, S. Kabelac. Springer
5. Luftgekühlter Flüssigkeitskühler mit dem Kältemittel $CO_2$. TU Braunschweig, Institut für Thermodynamik. Abschlussbericht aus 2004. N. Lemke, S. Försterling
6. http://www.konvekta.de/forschung-und-projekte/busklimaanlagen/ultralight.html. 19.12.2015/09:23
7. www.schweizer-fn.de/lueftung/feuchte/feuchte.php. 21.02.2016/13:54

# Model-Based Efficiency Improvement of Automotive Fuel Cell Systems

**Sebastian Bilz, Matthias Rothschuh, Katharina Schütte and Ralf Wascheck**

**Abstract** Fuel cells are seen as a promising potential technology for future automotive drive systems. The efficiency of fuel cells, their costs and dynamic behavior depend essentially on the surrounding system components. Different system topologies are possible for achieving an optimum interaction between system components and fuel cell stack and will be investigated as an ongoing major topic in the years to come. Development and operation of a fuel cell system for mass production is a challenge containing numerous trade-offs. One goal is an optimized operation of the fuel cell stack with the aiming on a maximum efficiency during the entire lifetime of the fuel cell system. Model-based approaches are used in early development stages to develop and verify the fuel cell system components. The investigations presented in this paper show detailed results for different fuel cell system topologies together with the interaction between system components and stack. The trade-off between optimum fuel cell stack supply and dynamic demands of the vehicle is also considered. The simulation framework developed is used for rapid and targeted initial calibration of new systems and for a detailed comparison of different topologies. It serves as a tool for ascertaining the optimum operating parameters for the fuel cell stack. Results provide a basis for demands, selection and positioning of different components in a fuel cell system.

S. Bilz · M. Rothschuh · K. Schütte (✉) · R. Wascheck
IAV GmbH, Rockwellstraße 16, 38518 Gifhorn, Germany
e-mail: katharina.schuette@iav.de

S. Bilz
e-mail: sebastian.bilz@iav.de

M. Rothschuh
e-mail: matthias.rothschuh@iav.de

R. Wascheck
e-mail: ralf.wascheck@iav.de



C. Gühmann et al. (eds.), *Simulation and Testing for Vehicle Technology*,
DOI 10.1007/978-3-319-32345-9_14

# 1 Introduction

Fuel cell hybrid vehicles are a promising drive concept for fulfilling current and future legislation requirements regarding greenhouse emissions. Nearly all major automotive manufactures are currently working on fuel cell driven vehicle concepts, investigating those vehicles in small-scale fleet programs or selling fuel cell vehicles in mass production.

Present development activities aim to reduce costs, increase durability and optimize fuel cell system efficiency, thus reducing hydrogen consumption. Currently available fuel cell vehicles operate in similar power ranges and hybridization concepts. The main differences are situated in the fuel cell system layout and its operating strategy. This paper presents possibilities for an optimized fuel cell system using model-based simulation tools.

## 1.1 *Fuel Cell System*

Today, PEM fuel cells (Proton Exchange Membrane) are used in automotive applications. Besides high power density, good dynamic properties and cold start capability, the high durability is another factor in favor of this technology. A fuel cell system designed for automotive applications has four sub-systems. The fuel cell system consists of the PEM fuel cell stack, hydrogen supply, air supply and cooling system. Figure 1 shows the schematic diagram of a fuel cell system suitable for automotive use.

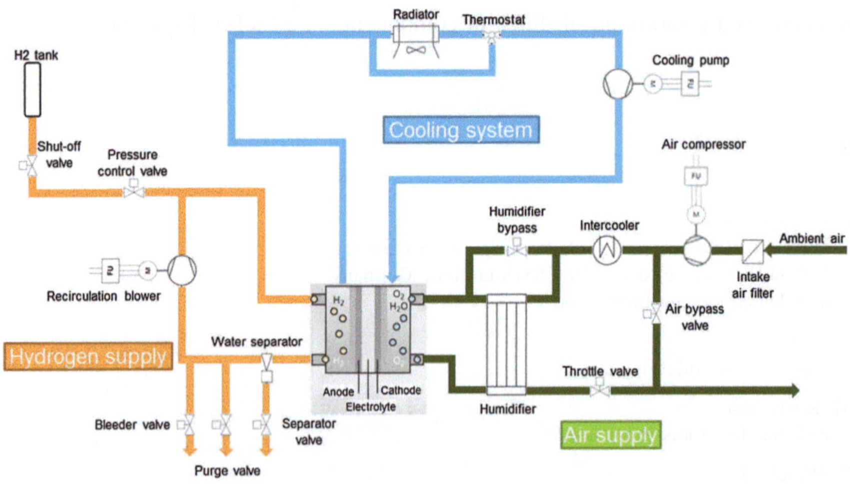


**Fig. 1** Fuel cell system

The fuel cell stack generates electric energy from hydrogen and oxygen taken from the ambient air.

The hydrogen needed for the electrochemical reaction is stored in the hydrogen tank at 700*bar*. The hydrogen passes through the shut-off valve and the pressure control valve to the stack inlet where it reacts with oxygen to water. The anode system operates at a leaner-than-stoichiometric level to enhance stack efficiency and durability. Hydrogen that does not react is recirculated by the recirculation blower and mixed with the mass flow of fresh hydrogen. Water generated as a by-product is collected in the water separator and discharged through the separator valve. The purge valve helps to increase hydrogen concentration and is opened cyclically due to inert gases which concentrate in the anode system caused by the partial pressure gradient between anode and cathode.

The process heat generated in the electrochemical reaction is dissipated through the cooling system. In contrast to a vehicle propelled by an internal combustion engine, more process heat has to be dissipated through the cooling system in a fuel cell vehicle.

The oxygen needed for the reaction is fed through the stack by an electrically driven compressor. The compressed air mass flow is conditioned in the intercooler and then humidified to enhance the conductivity of the membrane. A humidifier bypass controls humidity at the stack inlet. An additional throttle valve is fitted in the exhaust gas system to increase pressure. The operating point of the compressor can be shifted using the cathode bypass valve by discharging an additional mass flow that bypasses the stack.

Fuel cell stack durability and performance depend primarily on the selected operating conditions such as pressure, temperature, stoichiometry and relative humidity.

The following chapters take the air path as an example to illustrate how these parameters can be optimized in the interests of efficient system operation.

# 2 Model Setup

## *2.1 Fuel Cell Stack*

The model of the fuel cell stack shows the working principle of a PEM fuel cell (Fig. 2). It consists of the following sub-models:

- Fluid-dynamic model of the flow fields for anode and cathode
- Model of the thermal network
- Model of the polymer electrolyte
- Electrochemical model of the two electrodes

The flow field models are used to compute the thermodynamic state with the help of the volume chamber model, see also Sect. 2.2. These volume chambers are

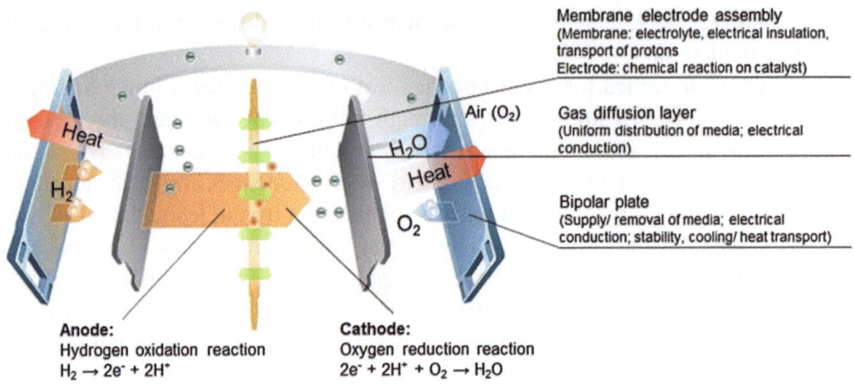


**Fig. 2** Fuel cell stack

the interfaces to the hydrogen or air supply. The thermal network connects them to the cooling circuit. Heat transport within the fuel cell stack and heat transfer into the coolant is computed in the thermal network.

The electrolyte of a PEM fuel cell consists of a semi-permeable membrane that electrically insulates the two electrodes and enables the proton exchange. A high water content of the membrane is essentially to uphold proton conductivity and maintain high stack efficiency. The water content indicates the number of water molecules per ionic end group and thus depends directly on the relative humidity in the membrane surroundings [1]. A relative humidity of $\phi = 1$ in the membrane surroundings results in a water content of 14.

The cathode air at the fuel cell stack inlet is humidified by a humidifier to keep the humidity level as close to saturation as possible. The anode side is supplied indirectly with water from the transport of water molecules through the membrane. Two essential effects are involved here: diffusion resulting from the concentration gradient on the anode and cathode side, and the electro-osmotic entrainment as a result of proton transport. The absorption and release of water by the membrane triggered by these effects is depicted dynamically in the sub-model of the polymer electrolyte.

The electrochemical model of the stack is based on the substitute diagram of a fuel cell (Fig. 3). Besides the voltage source, there is an ohmic resistance connected in series with an RC element.

The theoretical maximum voltage results from the Gibbs standard enthalpy of the total reaction $\Delta G_R^0$ of the fuel cell:

$$E^0 = \frac{\Delta G_R^0}{z \cdot F} = 1.229\,\mathrm{V} \tag{1}$$

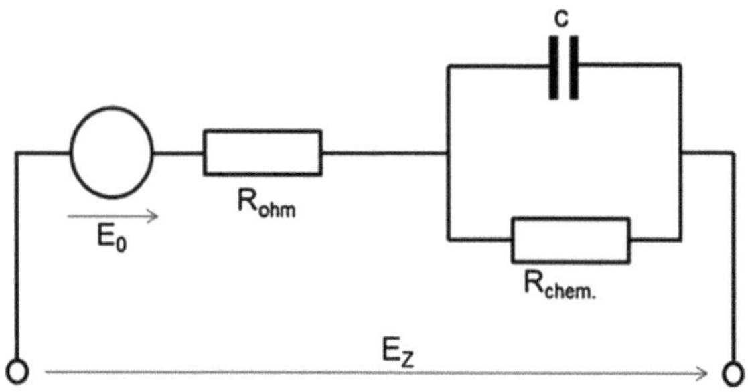


**Fig. 3** Substitute model

The valence $z$ is obtained from the total reaction of the fuel cell:

$$H_2 + \frac{1}{2}O_2 \rightarrow H_2O \tag{2}$$

$F$ is Faraday's constant.

Deviations from these standard conditions occur in reality. They are shown by the Nernst equation:

$$E_N = E^0 - \frac{R \cdot T}{z \cdot F} \ln\left(\prod_i a_i^{v_i}\right) \tag{3}$$

In the equation $a_i$ is the activity of the relevant substance and $v_i$ is the corresponding stoichiometric coefficient. Further losses caused by the abovementioned crossover effects and mixed potential are taken into consideration, resulting in the final open-circuit voltage.

If current is demanded from the fuel cell, additional losses occur as a result of the resistances mentioned above. The ohmic losses results from the electric conductivity of the used materials and the proton conductivity of the membrane.

There is also chemical resistance, caused primarily by the reaction kinetics. This is described by the Butler-Volmer equation:

$$i = i_0 \cdot \left[\exp\left(\frac{\alpha \cdot z \cdot F}{R \cdot T}\eta_{act}\right) - \exp\left(\frac{(1-\alpha) \cdot z \cdot F}{R \cdot T}\eta_{act}\right)\right] \tag{4}$$

Here $\alpha$ is the symmetry factor for the respective electrode reaction and $i_0$ is the exchange current density of the electrode. The equation is solved for over-voltage $\eta_{act}$ for each of the electrodes.

When high electric current is demanded, the consumption of reaction gases can no longer be covered by the supply system. Fuel depletion occurs at the electrodes, causing cell voltage to collapse. In the simulation model, this is depicted by the relationship:

$$\eta_{conc} = \frac{-R \cdot T}{z \cdot F} \cdot \ln\left(1 - \frac{i}{i_L}\right) \tag{5}$$

$I_L$ is the limit current which depends on the materials used.

Over-voltages result in the static cell potential:

$$E_Z^{stat} = E_{OCV} - \eta_{act} - \eta_{ohm} - \eta_{con} \tag{6}$$

In dynamic operation, the build-up and collapse of an electrochemical double layer at the contact area between membrane and electrode delays the adjustment of steady-state voltage, resulting in the following differential equation:

$$\dot{\eta}_{dyn}(t) = \frac{i}{c} - \frac{i}{c \cdot \eta_{chem}} \cdot \eta_{dyn}(t) \tag{7}$$

Fuel cell efficiency is obtained directly from the resulting cell voltage. It results from the achieved cell voltage divided by the thermal voltage.

$$\eta_Z = \frac{E_Z}{E_f^0} \tag{8}$$

Thermal voltage is defined as the complete conversion of chemical into electric energy. Figure 4 illustrates the current/voltage characteristic line with the occurring losses [2].

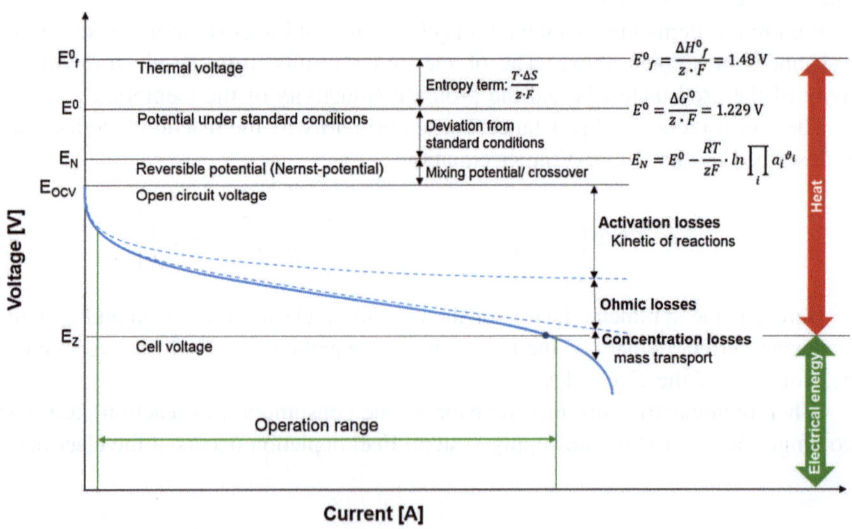


**Fig. 4** Characteristic line of a fuel cell [2]

The electrochemical sub-model is also used to compute the impact of the chemical reaction on the composition of the substance mixtures. Hydrogen is converted on the anode side, while on the cathode side oxygen reacts with protons and electrons to produce water (Fig. 2). This conversion is directly proportional to the current and is described by Faraday's law:

$$\dot{m}_i = \frac{I}{z \cdot F} M_i \tag{9}$$

$M_i$ is the molar mass of the substance, $z$ results from the chemical reaction of the electrode to $z = 2$ for hydrogen and water, and $z = 4$ for oxygen. In reality, larger quantities of process gases are supplied as there is no complete conversion and fuel depletion needs to be ruled out. The ratio of supplied and required fuel or oxygen is called stoichiometry:

$$\lambda_{an} = \frac{\dot{m}_{H_2}^{supplied}}{\dot{m}_{H_2}^{required}} \tag{10}$$

$$\lambda_{ca} = \frac{\dot{m}_{O_2}^{supplied}}{\dot{m}_{O_2}^{required}} \tag{11}$$

## 2.2 *Fuel Cell System Model*

The fuel cell system component models are described through the example of the cathode air supply. The same thermodynamic principles are applied to the anode gas supply.

The cathode air supply model is based on the thermodynamic description of a volume $V$ within a chamber of incoming and outgoing mass flows. Approaches of gas system modelling for current Diesel engines [3–5] are applied to a fuel cell system [6, 7].

Assuming a homogeneous distributed gas within a chamber, the energy balance for the chamber is described by the sum of incoming and outgoing enthalpy flows $\dot{H}$, considering incoming or outgoing heat flows $\dot{Q}$ by neglecting kinetic and potential energy:

$$\dot{U} = \sum_{i=1}^{n} \dot{H}_i - \sum_{i=1}^{m} \dot{H}_i + \dot{Q} \tag{12}$$

Assuming an isochoric process by application of the isochoric gas constant $c_v$, the change of the internal energy $\dot{U}$ within the chamber is defined as:

$$\dot{U} = c_v T \dot{m} + c_v m \dot{T} \tag{13}$$

The change of mass is derived by the sum of incoming and outgoing mass flows $\dot{m}$:

$$\dot{m} = \sum_{i=1}^{n} \dot{m}_{i,in} - \sum_{i=1}^{m} \dot{m}_{i,out} \tag{14}$$

Applying the ideal gas law by usage of the specific gas constant $R_s$ the pressure can be determined as:

$$p = \frac{m \cdot R_s \cdot T}{V} \tag{15}$$

The connection between volume chambers is modeled as a nozzle. Assuming an isentropic process and applying the equation of flow $\psi(\pi)$, the mass flow is represented by:

$$\dot{m} = A_{eff} \cdot \sqrt{\rho_1 p_1} \cdot \psi(\pi) \tag{16}$$

where $A_{eff}$ defines the effective surface area, $\rho_1$ describes the gas flow density and $p_1$ represents the pressure anterior of the nozzle.

The equation of flow is defined as:

$$\psi(\pi) = \begin{cases} \sqrt{\frac{2\kappa}{k-1}\left(\pi^{\frac{2}{\kappa}} - \pi^{\frac{\kappa+1}{\kappa}}\right)}, & \pi < \left(\frac{2}{k+1}\right)^{\frac{k}{k-1}} \\ \left(\frac{2}{\kappa+1}\right)^{\frac{k}{k-1}} \sqrt{\frac{2\kappa}{k+1}}, & \pi \geq \left(\frac{2}{k+1}\right)^{\frac{k}{k-1}} \end{cases} \tag{17}$$

depending on the pressure ratio $\pi$ and the ratio of specific heats $\kappa$.

The relation of compressor mass flow $\dot{m}_{comp}$, compressor pressure ratio $\pi_{comp}$ and compressor speed $n_{comp}$ is represented by a compressor flow map:

$$\dot{m}_{comp} = f(\pi_{comp}, n_{comp}) \tag{18}$$

The isentropic compressor efficiency $\eta_{comp}$ is also represented by an efficiency map taking pressure ratio and mass flow into account:

$$\eta_{comp} = f(\pi_{comp}, \dot{m}_{comp}) \tag{19}$$

The compressor dynamics are approximated as a first order system.

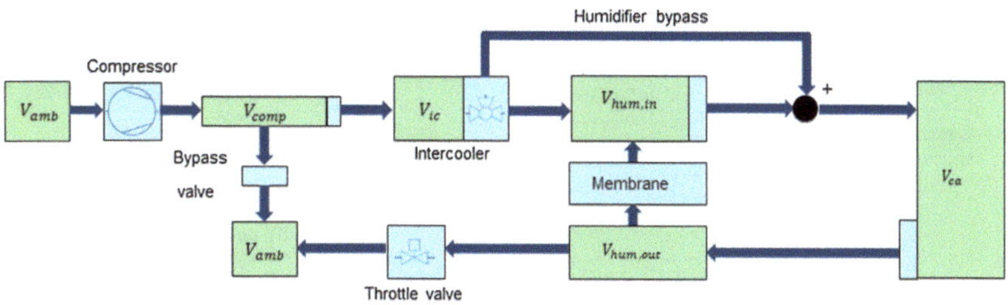


**Fig. 5** Cathode system model

Assuming an isentropic compression, the compressor power is defined as the sum of isentropic compressor power $P_{isen}$ and propulsive power $P_{mech}$:

$$P_{Verd} = P_{isentrop} + P_{mech} \tag{20}$$

The isentropic compressor power is obtained by:

$$P_{isentrop} = \frac{c_p \cdot T_1 \cdot \dot{m}}{\eta_{comp}} \cdot \left[ (\pi)^{\frac{k-1}{k}} - 1 \right] \tag{21}$$

Assuming constant compressor inertia $J$, the propulsive power is determined by:

$$P_{mech} = J \cdot \dot{\omega} \cdot \omega \tag{22}$$

where $\omega$ represents the angular velocity.

The cathode air supply model combines several volume chambers and nozzle models (Fig. 5). The required oxygen is taken from the ambient $V_{amb}$. The state equations for a gas composition of nitrogen $N_2$, oxygen $O_2$, water vapor $H_2O$ and hydrogen $H_2$ are applied to five volume chamber models: volume between compressor and intercooler $V_{comp}$, intercooler volume $V_{ic}$, humidifier inlet volume $V_{hum,in}$, fuel cell cathode volume $V_{ca}$ and humidifier outlet volume $V_{hum,out}$. Volume chamber models are connected by fixed or adjustable nozzle models which determine mass- and enthalpy flows (bypass valve, throttle valve, humidifier bypass). Pressure losses between volume chambers are calculated for each nozzle model.

## 2.3 Vehicle Environment and Hybrid Manager

The basic vehicle parameters refer to data of available fuel cell vehicles presented by various manufacturers. The fuel cell system was integrated in an existing hybrid vehicle simulation.

The simulated vehicle has the following parameters: (Fig. 6, Table 1).

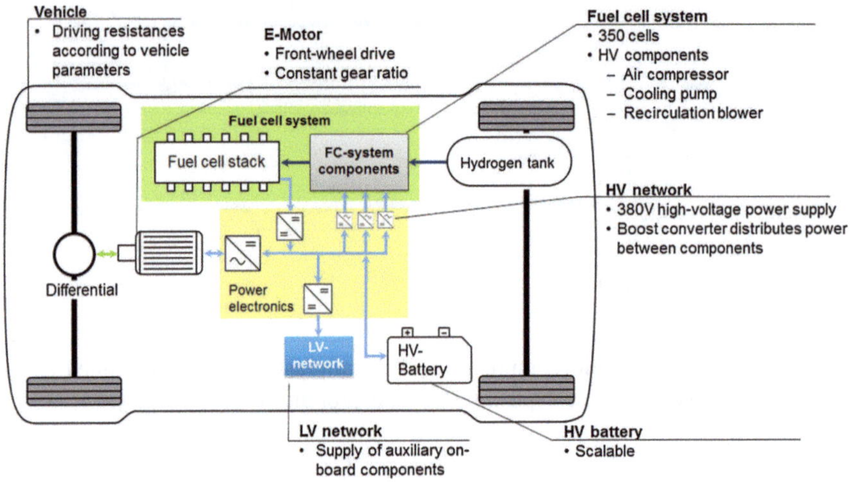


**Fig. 6** Fuel cell vehicle topology

**Table 1** Basic vehicle parameters

| Parameter | Value | Unit |
|---|---|---|
| Vehicle mass | 1950 | kg |
| Drag coefficient | 0.3 | – |
| Frontal area | 2.5 | $m^2$ |
| Dynamic wheel circumference | 1940 | mm |
| Roll resistant coefficient | 0.008 | – |
| Battery | 2.5 | kWh |
| Fuel cell system max power | 90 | kW |
| Electric motor power max | 100 | kW |

The operating strategy specifying the power split between battery and fuel cell system is based on IAV's ECMS control strategy (Equivalent Consumption Minimization Strategy) [8], adapted for a fuel cell hybrid vehicle.

The ECMS is a real-time energy flow coordinator for exploiting an additional energy source in a hybrid vehicle with regard to minimum fuel consumption. Boundary conditions include the current SOC (state of charge) of the battery and the system efficiency at the required operating point of the fuel cell. The optimization objective is the achievement of minimum hydrogen consumption for the drive cycle under the constraint of a balanced SOC.

## 3 Simulation Results for the Basic Vehicle Configuration

An energetic analysis of the fuel cell system in the WLTP cycle (Worldwide Harmonized Light-Duty Vehicles Test Procedure) confirms that the electric compressor causes the largest parasitic losses in the fuel cell system (Fig. 7). The investigated basic configuration employs an electric turbo-compressor.

The following results and optimizations are based on computations for a system being at operating temperature at the start of the cycle. Other consumption levels apply for cold starts and other operating parameters for the supplied media. No start/stop strategy is implemented during the simulation. The fuel cell supplies a minimum output of 5 kW.

The initially selected operating parameters for cathode gas supply have already been adapted to the map of the turbo-compressor and the requirements of the fuel cell stack.

The operating parameters of the fuel cell cathode system are shown in Fig. 8.

Figures 9 and 10 show the operating behavior of the fuel cell system chosen by the operating strategy. At the start of the cycle and during idle operation, the fuel cell operates at low-load ranges and the battery is charged due to the fuel cell system‘s maximum efficiency at low loads. Electric drive energy is generated primarily by the fuel cell system. When faced with high load demands and dynamic power peaks occurring primarily at the end of the cycle, electric propulsion is supported by power withdrawn from the HV battery, thus discharging the battery. Altogether there is a balanced SOC over the cycle.

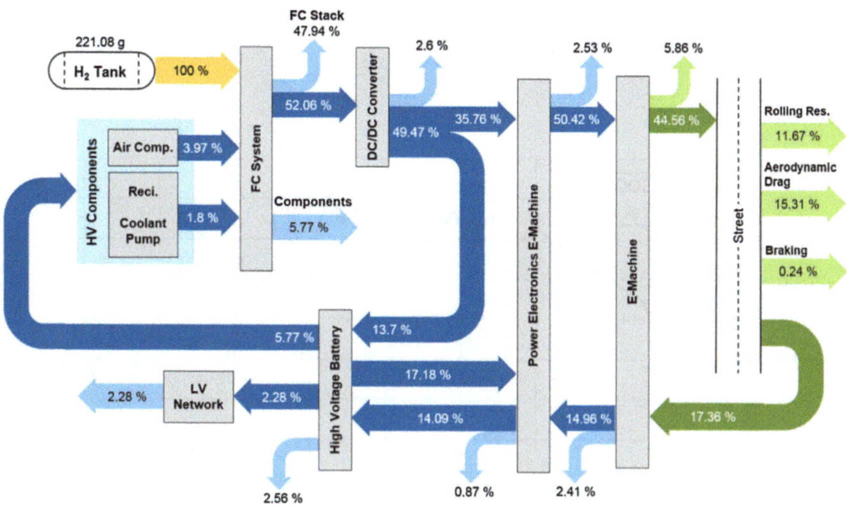


**Fig. 7** Energy flows in WLTP

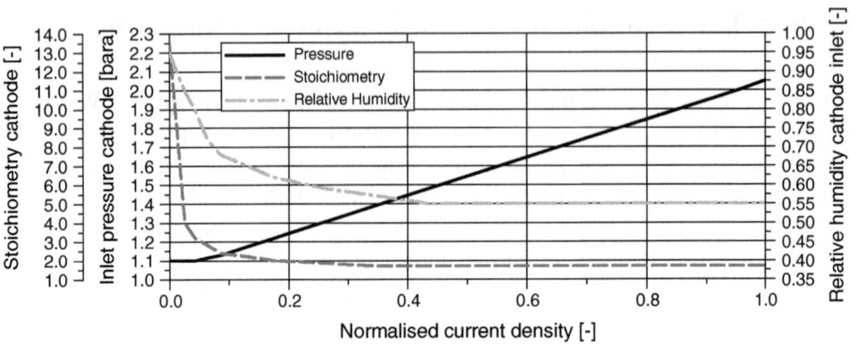


**Fig. 8** Operating parameters of cathode in basic configuration

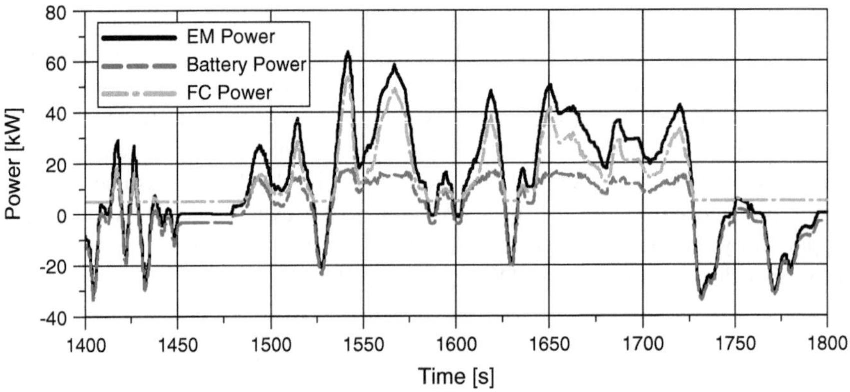


**Fig. 9** Power split WLTP

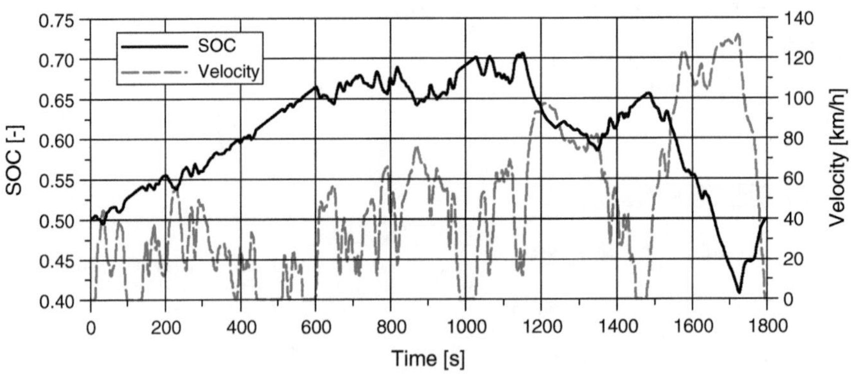


**Fig. 10** SOC and velocity WLTP

# 4 Optimizing the Operating Parameters Using the Example of the Air Path

## 4.1 Components

Besides the electric turbo-compressor different compressor types are conceivable to be used for automotive fuel cell applications [9]. The differences emerge particularly in the type of compression and are reflected in the respective compressor map. In addition to compressor output, stack humidity is also influenced by operating pressure so that other effects on system performance have to be considered.

While the turbo-compressor is limited in its map towards the surge limit with relatively low efficiency at the surge line, mechanical compressors such as a Roots-type superchargers have a wider operating range, particularly in the range of small mass flows.

In the framework of this paper, the operating parameters of the cathode path will therefore be optimized for two other possible compressors compared to the basic configuration with simple turbo-compressor. Conventional, commercially available compressors are used in each case.

## 4.2 Optimization Problem Definition

The modular modelling concept allows the replacement and analysis of different compressor types. Simulation results and analysis show that the initially formulated cathode operating conditions result in an operating line that does not match the system limits of the roots compressor map (Fig. 12). Stable fuel cell operation under initially formulated operating conditions is possible up to a current density of $0.75i_{max}$. For a wide operating range new operating conditions have to be defined for different compressor types.

Operating parameters regarding to the cathode air supply are cathode humidity, cathode pressure and cathode stoichiometry. For an effective parameter definition an optimization problem is defined and solved by applying a genetic algorithm to the optimization problem. Genetic algorithms are a useful search heuristics for solving complex, non-linear optimization problems [10]. Genetic algorithms, evolutionary algorithms and other numerical optimization methods are implemented in IAV's Engineering Toolbox [11]. Detailed information about implemented algorithms is provided in [12, 13].

The aim of the optimization problem is the operation of the fuel cell system with its maximum efficiency $\eta_{Sys}$ considering a constant current density $i$: Maximization problem:

$$Max\,\eta_{Sys}(p,\phi,\lambda,)_{i=konstant};\,i_{min}\le i\le i_{max} \tag{23}$$

With regard to following constraints:

- Permanent control offset stoichiometry control: 1 %
- Permanent control offset pressure control: 1 %
- Permanent control offset humidity control: 1 %
- Compressor operating point meets system limits
- Amount of water in membrane 12–14

The set of feasible solutions is restricted by system limits for different components:

$$1.05 < p < 3[bara] \tag{24}$$

$$1.05 < \lambda < 10 \tag{25}$$

$$5 < \phi < 95 \tag{26}$$

The control input for the anode pressure is set to a fixed pressure rise of $200mbar$ between cathode and anode, whereas in the basic simulation model the control input for the cathode pressure is depending on the anode pressure control input.

Figure 11 shows the dependence of the fuel cell system efficiency from cathode stoichiometry and cathode pressure for a constant current density $0.5i_{max}$ and constant humidity. The maximum system efficiency $\eta_{Sys} = 0.458$ is achieved for a cathode stoichiometry of $\lambda = 1.64$ and a cathode pressure of $p = 1.4bara$. An increase in cathode stoichiometry and pressure results in an improvement of fuel cell efficiency (Sect. 2.1), certainly the compressor power increases, resulting in optimized operating parameters for a maximum system efficiency. The system efficiency map boundaries are limited by fulfillment of the optimization problem constraints.

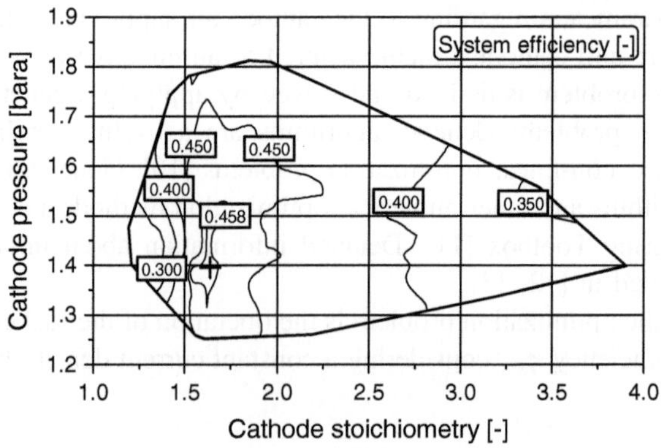


**Fig. 11** System efficiency map

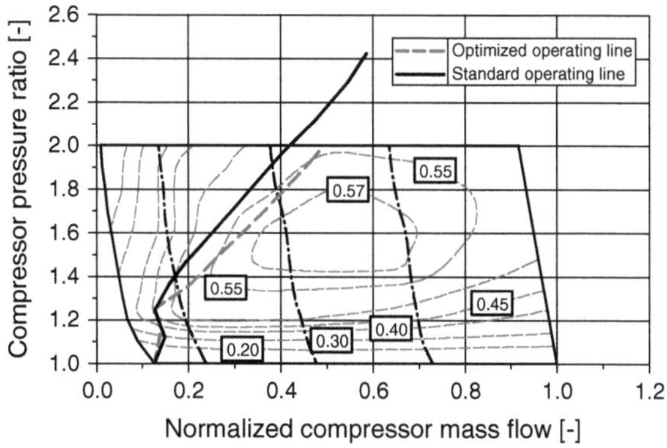


**Fig. 12** Compressor map roots compressor

## *4.3 Simulation Results*

### 4.3.1 Roots Compressor

Figure 12 shows the resulting operating line of the cathode parameter optimization for the roots compressor compared to initially defined operating parameters. In contrast to the initially standard operating line the optimized operating line meets the compressor limitations, which allows a fuel cell operation through the complete operating range.

Compared to the basic configuration employing an electrical turbo charger the system efficiency at medium loads can be increased by usage of a roots compressor. This is caused by a bypass mass flow minimization and increased cathode pressure, enabled by a wider operating range. For high loads the system efficiency is decreased due to poor compressor efficiencies compared to an electrical turbocharger.

In conclusion, an electrical turbocharger with an enhanced operating range benefits the system's efficiency. For low and medium loads the possibility of a decreased bypass mass flow is advantageous, for high loads the enhanced turbocharger efficiency positively effects the system's efficiency.

### 4.3.2 Variable Trim Compressor

The Variable Trim Compressor (VTC) developed at IAV for efficiency improvement of internal combustion engines, offers the possibility for an enhanced compressor operating range [14]. Caused by continuous pinch in compressor flow, the compressor trim varies apparently, resulting in an enhanced operating range towards the surge limit. The shift of the surge line relies on compressor speed and design and achieves between 20 and 30 %. Besides, the compressor efficiency is

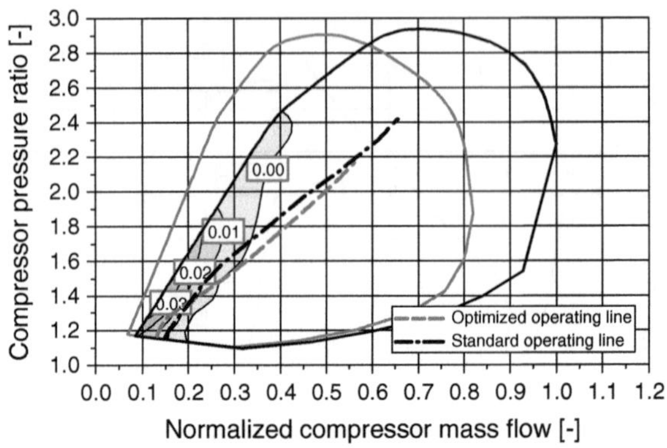


**Fig. 13** Compressor map VTC compressor

improved for low mass flows. Figure 13 shows the shift of the surge limit and the efficiency improvement achieved by a VTC compressor. In addition, the operating line for initially defined cathode operating parameters and the optimized operating line after parameter optimization is displayed. The aim of the optimization is maximum system efficiency at varying cathode operating parameters: pressure, stoichiometry and relative humidity.

### 4.3.3 Optimization Results

Figure 14 shows the simulation results of the fuel cell efficiency for the basic configuration, the optimized roots compressor and the optimized VTC compressor. At low loads the roots compressor achieves the highest fuel cell efficiency due to an increased cathode pressure. At high loads the fuel cell efficiency decreases determined by reduced cathode stoichiometry and pressure.

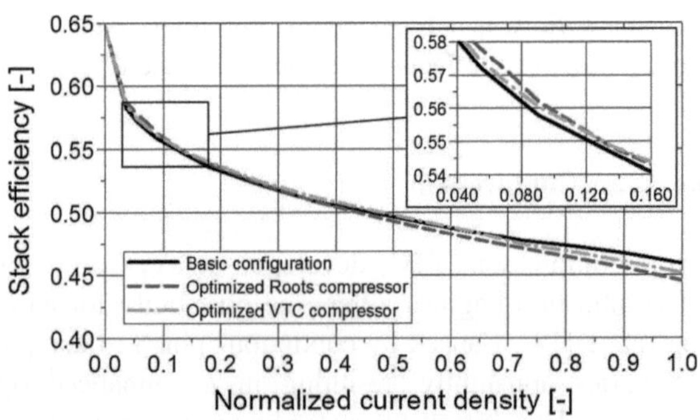


**Fig. 14** Fuel cell efficiency

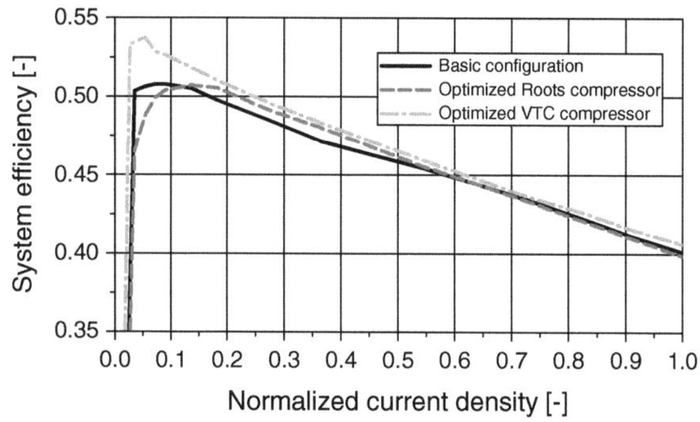


**Fig. 15** System efficiency

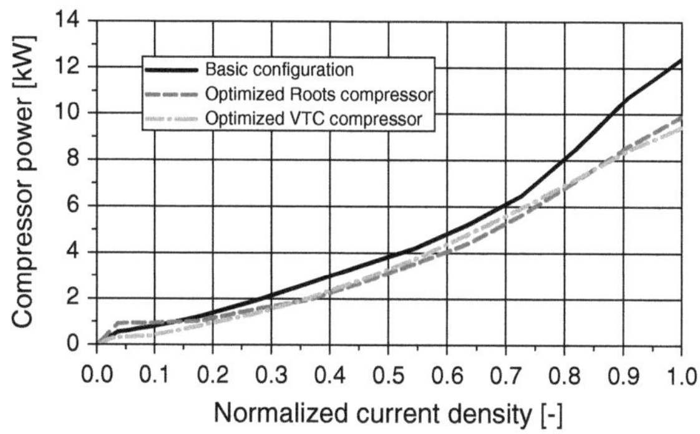


**Fig. 16** Compressor power

The VTC compressor's efficiency exceeds the efficiency of the basic configuration slightly due to an improved cathode pressure which is enabled by the surge line shift. At high loads cathode pressure and stoichiometry are reduced effecting in a decreased fuel cell efficiency which is compensated by reduced compressor power resulting in maximum system efficiency.

Simulation results of the system efficiency for three analyzed options are displayed in Fig. 15. Especially at low loads the system efficiency is improved by the VTC compressor. The main reason is situated in the reduced bypass mass flow which is achieved by the shift of the surge line. At medium loads VTC compressor and roots compressor exceed the basic configuration efficiency. At high loads the system efficiency of the roots compressor decreases due to poor compressor efficiency compared to turbocharged options. The VTC compressor obtains the maximum efficiency at all operating points and provides the opportunity for reduced vehicle hydrogen consumption.

Figure 16 shows simulation results of the compressor power for three analyzed options. The basic configuration compressor power exceeds the roots and VTC compressor power at almost all operating points. Especially at low loads the VTC

**Table 2** Simulation results WLTP

| | Basic configuration | Roots compressor | VTC compressor |
|---|---|---|---|
| $H_2$ consumption (g) | 221.1 | 227.2 | 214.4 |
| Compressor energy/$H_2$ equivalent | 1254 MJ<br>8.78 g | 1.739 MJ<br>12.2 g | 0.744 MJ<br>5.21 g |
| FC energy (MJ) | 16.45 | 16.88 | 15.98 |
| Average FC-system efficiency (%) | 49.38 | 46.85 | 51.47 |
| Average FC efficiency (%) | 57.05 | 57.27 | 57.07 |
| Average compressor power (kW) | 0.7 | 0.97 | 0.43 |

compressor shows a significantly reduced power due to reduced bypass mass flow and increased compressor efficiency. At high loads roots and VTC compressor achieve a decreased compressor power caused by operation parameter optimization.

Table 2 shows vehicle simulation results of the WLTP cycle for three analyzed options. The hydrogen consumption for the WLTP cycle can be reduced by 3 % caused by increased system efficiency and optimized operating parameters. The average fuel cell system efficiency is increased by approximately 2 %. Parameter optimization results in the highest average fuel cell efficiency for the roots compressor, while the average system efficiency is the lowest of all analyzed options.

In conclusion the VTC compressor is the most promising option for cathode air supply of PEM fuel cell systems, regarding to system efficiency and variability in operating parameter definition.

## 5 Summary and Outlook

The paper validates the modularity of the developed fuel cell vehicle model.

The example illustrated in this paper, individual components of the air path were tested by replacing the component models and showing their effect on the overall system.

Using the simulation environment, diverse optimization tasks can be performed for the whole powertrain. New concepts can be assessed in terms of their potential in the simulation framework before proceeding laboratory tests.

Continuous examinations should investigate additional component configurations, aiming for example at the possibility of including an expander in the energy studies. Furthermore, an operating parameter adjustment is necessary for different operating modes of the fuel cell vehicle. For example the start up at different temperatures has to be analyzed due to the effects of warming up the fuel cell stack.

The current central issues for development work in fuel cell vehicles are durability over service life, costs and package. These criteria should be incorporated in the component models when pursuing further investigations and extending the simulation environment. The implementation and analysis of multi-criteria optimization considering the above-mentioned parameters can be enforced.

## References


1. Springer, E., Zawodzinski, T.A., Gottesfeld, S.: Polymer elektrolyte fuel cell model. J. Electrochem. Soc. **138**(8), 2334–2342 (1991)
2. Mench, M.: Fuel Cell Engines. Wiley, Hoboken (2008)
3. Schollmeyer, M.: Modellbasierte Ladedruckregelung für PKW-Dieselmotoren. Fakultät für Elektrotechnik und Informatik der Gottfried Wilhelm Leibniz Universität Hannover, Hannover (2010)
4. Friedrich, I.: Motorprozess-Simulation in Echtzeit – Grundlagen und Anwendungsmöglichkeiten. Fakultät V – Vekehrs- und Maschinensysteme der Technischen Universität Berlin, Berlin – (2007)
5. Allmendinger, K.: Modellbildung und modellbasierte Estimitation thermodynamischer Prozessgrößen am Beispiel eines Dieselmotors. Fachbereich Elektrotechnik und Informatik der Universität-Gesamthochschule Siegen, Siegen (2002)
6. Hans, R., Panik, F., Reuss, H.-C.: Modular modeling of a PEM fuel cell system for automotive applications. In: 14. Internationales Stuttgarter Symposium, Springer Fachmedien, Wiesbaden (2014)
7. Reuter, J., Beister, U.-J., Liu, N., Reuter, D., Eybergen, B., Radhamohan, M., Hutchenreuther, A.: Control of a fuel cell air supply module (ASM). SAE Int. (2004)
8. Siebenpfeiffer, W. (ed.): Energieeffiziente Antriebstechnologien Hybridisierung – Downsizing – Software und IT. Springer Vieweg, Wiesbaden (2013)
9. Yu, W., Sichuan, X., Ni, H.: Air compressors for fuel cell vehicles: an systematic review. SAE Int. (2015)
10. Bürger, S.: Multikriterielle Optimierung von Hybridantriebskonzepten hinsichtlich Verbrauch und Dynamik. Shaker Verlag, Aachen (2012)
11. IAV GmbH.: IAV Engineering Toolbox: Der Entwicklungshelfer: Engineeringportal, Optimierung Auswertung. https://www.iav.com/sites/default/files/handouts/2015/de/engineeringtoolbox.pdf. Retrieved 18 Feb 2016
12. Putzert, F.: Moderne Evolutionäre Verfahren/Genetische Algorithmen für den Einsatz in technischen Aufgabenstellungen und Bewertungsstrategien zum Vergleich der Güte und Effektivität von multikriteriellen Optimierungsverfahren. Fakultät für Mathematik der Technischen Universität Chemnitz, Chemnitz (2012)
13. Stöcker, M.: Untersuchung von Optimierungsverfahren für rechenzeitaufwändige technische Anwendungen in der Motorenentwicklung. Fakultät für Mathematik der Technischen Universität Chemnitz, Chemnitz (2007)
14. IAV GmbH.: Variabilitäten am Turbolader – Lösungen für steigende Systemanforderungen. https://www.iav.com/sites/default/files/handouts/2014/de/variabilitaeten-am-turbolader.pdf. Retrieved 25 Feb 2016

# Hybrid Vehicle Simulation Addressing Real Driving Emissions Challenges

**Theodora Zacharopoulou, Grigorios Koltsakis and Klaus von Rüden**

**Abstract** The upcoming Real Driving Emissions (RDE) legislation requirements impose new challenges for future powertrains. In parallel, hybridization increases the importance for the engine and exhaust system thermal management. In this new technological and legislative environment, the application of model-based methodologies could support the tough optimization problem of exhaust system design and control. The purpose of the present study is to present a practical simulation approach for the complete vehicle-powertrain-aftertreatment chain, focusing on Hybrid Electric Vehicle applications. The methodology is supported by a complete vehicle simulation platform (*velodyn*), with emphasis on the prediction of engine-out and tailpipe emissions via appropriate engine and aftertreatment modeling. The exhaust aftertreatment devices simulation is based on well-established and validated mathematical models of the physico-chemical phenomena using the software *axisuite*®. The engine-out emissions prediction is based on a semi-empirical approach with sufficient degrees of freedom to describe the effect of the engine thermal state on baseline engine emissions. The models are coupled in order to allow a closed-loop simulation of the complete powertrain and apply various aftertreatment control strategies depending on the driving scenaria. Next, the impact of hybridization on emissions is discussed and studied on the basis of complete vehicle simulation, addressing the exhaust system thermal management

T. Zacharopoulou (✉) · G. Koltsakis
Laboratory of Applied Thermodynamics, Aristotle University of Thessaloniki,
University Campus, 54124 Thessaloniki, Greece
e-mail: theodorz@auth.gr

G. Koltsakis
e-mail: grigoris@auth.gr

K. von Rüden
IAV GmbH, Carnotstraße 1, 10587 Berlin, Germany
e-mail: klaus.von.rueden@iav.de

 
C. Gühmann et al. (eds.), *Simulation and Testing for Vehicle Technology*,
DOI 10.1007/978-3-319-32345-9_15

with the support of advanced 3D thermal loss calculations. It is demonstrated that the presented methodologies could substantially support the design and optimization of integrated engine and aftertreatment controls to comply with Real Driving Emissions requirements.

**Keywords** Hybrids · RDE · Emissions · Aftertreatment · Simulation

## 1 Introduction

The upcoming Euro 6c Emission Regulation will implement Real Driving Emissions (RDE) as an additional type-approval requirement, in order to cover a much wider range of engine operating modes [1–5]. The RDE legislation for passenger cars is introducing the road as a new environment for emissions testing and certification. This will impose significant challenges on the design, development and calibration of powertrains, vehicles and aftertreatment.

It is obvious that the validation of all the test cases needed for the RDE legislation would be almost impossible to realize on real vehicles for an OEM because of the high cost of the test trips and the restricted availability of prototype vehicles. In addition, the number of climatic engine test beds and climatic chassis dynos, which would be needed in order to develop the altitude calibration with real engine hardware, would be hardly affordable [6].

Therefore, given the complexity and the high cost for conducting such time-consuming measurements with PEMS, the need for developing validated simulation models with aftertreatment control systems in order to study the behavior of the legislated emissions in real-driving conditions becomes imperative.

Recently, there have been model-based research performed towards this target, based on recent experimental studies with PEMS [4, 7–10], including HiL/SiL [5], EiL [11], EiL for hybrid applications [12] and simulation tools for diesel applications [13, 14] including its EGR control [15].

In this direction, this study presents the development of a complete hybrid vehicle model, consisting of the internal combustion diesel engine, the SCR-based aftertreatment system enhanced with appropriate control systems and the vehicle submodels.

To demonstrate the implementation concept of the proposed approach, the developed simulation platform is applied to the cases of conventional and hybrid electric vehicles in real world driving conditions in terms of pollutant emissions, in order to comply with the Euro 6 limits and the upcoming RDE compliance [13, 16, 17].

# 2 Methodology and Modeling Approach

## 2.1 Internal Combustion Engine Model

The proposed methodology for the raw exhaust regulated emissions (CO, HC, $NO_x$) and the exhaust flow properties (temperature, mass flow) prediction of the engine utilizes steady-state maps, generated from steady-state dynamometer measurements, as functions of the engine speed and torque. Appropriate corrections are implemented to account for transient operating modes: a submodel accounting for the cold start extra emissions (CSEE) during engine warm-up. A block diagram of the developed semi-empirical ICE model is shown in Fig. 1, whereas the respective corrections are discussed in the following subsection.

### 2.1.1 Cold Start Extra Emissions Correction

During the engine cold start, CO and HC engine-out emissions are higher due to incomplete combustion. To account for these effects, the model applies emission correction factors on the map-based steady-state values. The emission correction factors are calculated as functions of the coolant temperature. Since the coolant temperature is not a direct input parameter of the model, it is estimated by applying a simplified energy balance using the cumulated consumed fuel as input variable, as explained in the following equations.

$$f_1\left(\dot{m}_{fuel}\right) \rightarrow f_2(Q_{heat\_losses}) \rightarrow f_3(T_{coolant}) \rightarrow f_4(C_{CSEE_{CO}}, C_{CSEE_{HC}}) \tag{1}$$

$$Q_{heat\_losses} = \frac{1}{3} * 43.1 * \dot{m}_{fuel} \tag{2}$$

$$T_{coolant} = a_1 * Q^3_{heat_{losses}} + a_2 * Q^2_{heat\_losses} + a_3 * Q_{heat\_losses} + T_{amb} \tag{3}$$

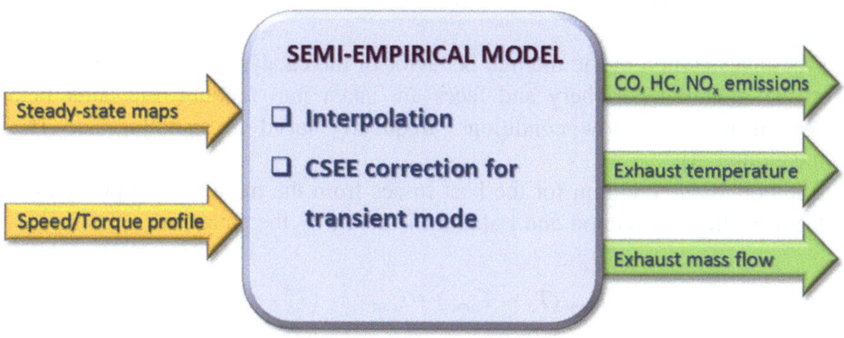


**Fig. 1** Internal combustion engine semi-empirical model

$$C_{CSEE_{CO}} = b_1 * T_{coolant}^2 + b_2 * T_{coolant} + b_3 \quad (4)$$

$$C_{CSEE_{HC}} = c_1 * T_{coolant}^2 + c_2 * T_{coolant} + c_3 \quad (5)$$

The CSEE correction becomes more critical for hybrid operation mode, when the ICE is switched-off and undergoes a cooling process. During the ICE-off cooling periods the coolant temperature is estimated by implementing an equation that calculates the heat losses to the environment. The above correction functions are semi-empirical and therefore have to be tuned to each engine by using actual experimental data.

When the coolant temperature is lower than 85 °C the CSEE factor for CO and HC is greater than 1. While the ICE starts the coolant temperature increases and the CSEE factor decreases. During the engine shut-off period a small decrease of the coolant temperature is observed. As soon as the coolant temperature reaches 85 °C the CSEE factor becomes equal to 1 and no correction is implemented.

## 2.2 *Aftertreatment Model*

### 2.2.1 Main Balance Equations

The modeling tool used for the simulation of the flow-through catalytic converter is axisuite® [18–20]. The model has been extensively used in the past to simulate diesel and gasoline exhaust systems, including HC adsorption (DOC) [21, 22] and selective catalytic reduction (SCR) [23] catalysts. The main differential and algebraic equations for the heat and mass transfer in the converter are given in Fig. 2.

The model is further equipped with intra-layer washcoat discretization to account for the coupling between reaction and diffusion phenomena inside the washcoat layer, as presented in Fig. 3.

### 2.2.2 Heat Transfer Boundary Conditions

For the investigation of the thermal behavior of the catalyst bricks, the heat losses from the monolith periphery and faces are taken into account, the latter being critical at near-zero flow conditions, frequently faced in start-stop and HEV applications [24].

The boundary condition for the heat losses from the monolith periphery to the ambient by free convection and radiation is given by the equation.

$$\lambda_{S,r} \frac{\partial T_S}{\partial r} = C_{amb} \cdot h_{amb}(T_S - T_{amb}) + \varepsilon_{rad} \cdot \sigma \cdot (T_S^4 - T_{amb}^4) \quad r = 0.5 D_f \quad (6)$$

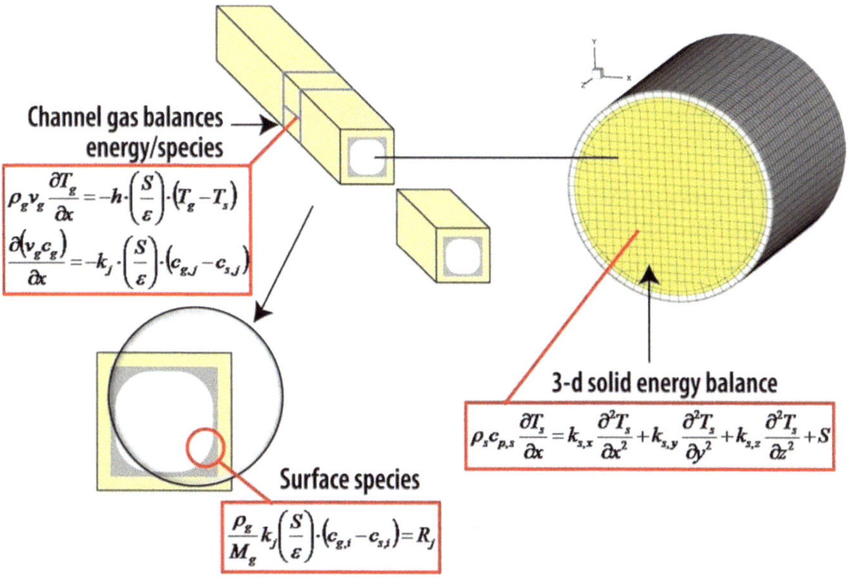


**Fig. 2** Flow-through catalytic converter schematic and basic model equations

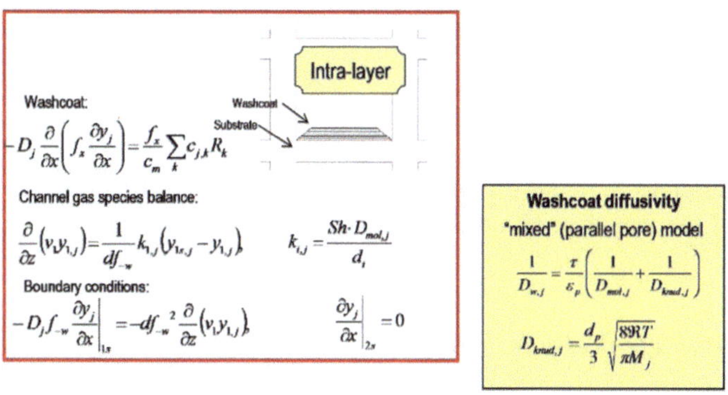


**Fig. 3** Intra-layer modeling equations

Heat losses by free convection and radiation from the front and rear faces of the device are taken into account as boundary conditions as shown below.

$$\lambda_{S,z} \frac{\partial T_S}{\partial z} \underset{z=0}{} = C_{amb,f} \cdot h_{amb,F} \cdot \left(T_S - T_{\infty,in}\right) + C_{rad,f} \cdot (1 - \varepsilon)\varepsilon_{rad} \cdot \sigma \cdot (T_S^4 - T_{\infty,in}^4) \tag{7}$$

$$\lambda_{S,z} \frac{\partial T_S}{\partial z}\bigg|_{z=L} = C_{amb,r} \cdot h_{amb,F} \cdot \left(T_S - T_{\infty,out}\right) + C_{rad,r} \cdot (1-\varepsilon)\varepsilon_{rad} \cdot \sigma \cdot \left(T_S^4 - T_{\infty,out}^4\right) \tag{8}$$

The terms for natural convection and radiation, which are included in the above expressions, are explained in detail below:

- Natural (free) convection

In case of near-zero flow conditions, natural convection from the monolith faces becomes important. The convection coefficient from the monolith faces $h_{amb,\ F}$ is computed using the equations for natural convection for a vertical plate. For the 3D simulation, the local Nusselt number is computed based on the following equations:

$$Nu_x = \left(\frac{Gr_x}{d}\right)^{\frac{1}{4}} g(\mathrm{Pr}) \tag{9}$$

$$g(\mathrm{Pr}) = \frac{0.75\sqrt{\mathrm{Pr}}}{\left(0.609 + 1.221\sqrt{\mathrm{Pr}} + 1.238\,\mathrm{Pr}\right)^{\frac{1}{4}}} \tag{10}$$

$$Gr_x = \frac{g\,\beta(T_s - T_\infty)x^3}{v^2} \tag{11}$$

- Radiation

The emissivity $\varepsilon_{rad}$ refers to the substrate material. The term $(1-\varepsilon)$ refers to the fraction of the face area of a monolith occupied by solid material.

## 2.3 *Complete Vehicle Model*

The previously described engine and DOC-SCRF model were embodied into a complete vehicle model, which was developed using *velodyn* [25], a flexible simulation platform, suited for real-time, control oriented simulation of complete motor-vehicle powertrains within the Simulink/MATLAB® environment.

*velodyn* works on the basis of so called "carrier blocks" that incorporate the signals from any Simulink block into a bus structure. The driving scenario, the environmental data, the parameters for the common driving resistances to be considered in the calculation and the characteristics of the vehicle (transmission gear ratios, differential ratio etc.) are specified in the appropriate blocks of the *velodyn* model. The engine speed and the requested torque are calculated from *velodyn* according to the requested power, vehicle speed and engine output power. The conventional diesel vehicle was used for this work is shown in Fig. 4.

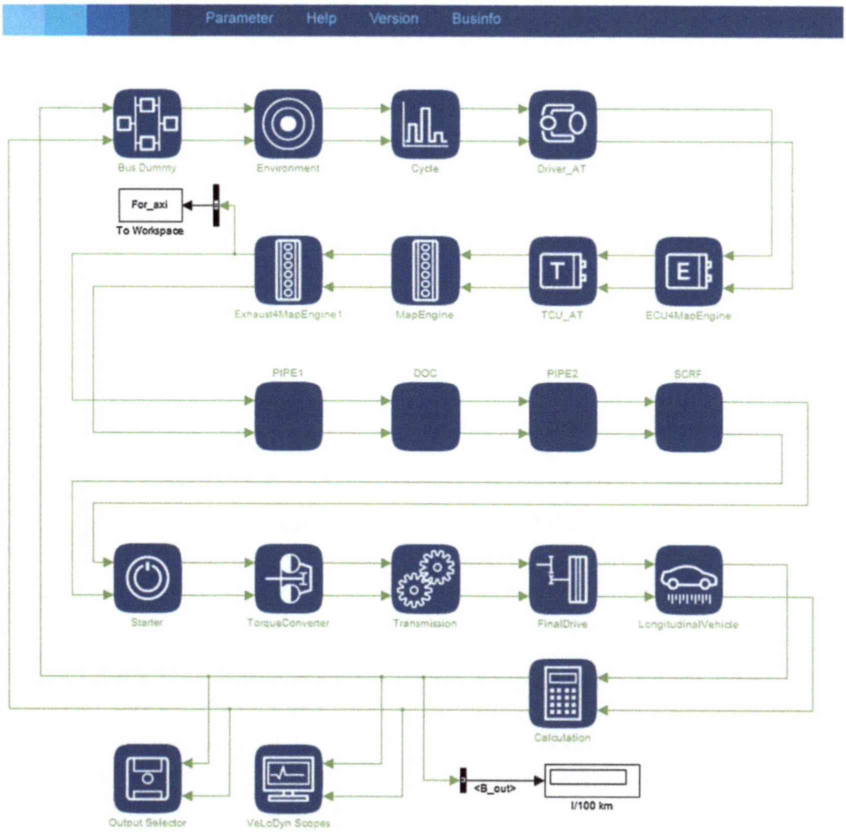


**Fig. 4** Complete conventional vehicle model coupled with DOC and SCRF aftertreatment devices

### 2.3.1 ICE Model Characteristics

The semi-empirical ICE model integrated to the vehicle for the purposes of this study was a Euro 5 diesel engine with a displacement of 1.4 ℓ and 190 Nm of maximum torque.

### 2.3.2 Pipe Models Characteristics

In order to take into account the heat losses of the exhaust gas to ambient between the engine-out and the inlet of the aftertreatment devices exhaust pipe models upstream the DOC and upstream the SCRF devices were integrated to the overall vehicle model using *axiheat* [20] module of *axisuite*® simulation tool. *axiheat* computes the heat losses to the ambient by all possible heat transfer modes (conduction, convection and radiation). The mass flow rate used in the heat transfer calculation is the time-dependent mass flow rate defined in the exhaust gas scenario, or computed at the outlet of the upstream exhaust component.

The pipes implemented upstream the DOC and the SCRF devices were 0.8 m and 0.2 m long respectively, having 60 mm of diameter and 1.5 mm of wall thickness.

### 2.3.3 DOC Model Characteristics

The DOC device used for the model was a cordierite monolith of 1.5 ℓ, cell density equal to 400cpsi and Pt-zeolite washcoat loading equal to 150 g/ℓ.

### 2.3.4 SCRF Model Characteristics

The SCRF device used for the model was a cordierite monolith of 2.5 ℓ, cell density equal to 180 cpsi and Cu-zeolite washcoat loading equal to 150 g/ℓ. Initial storage was 0.15 mol/mole site for $S1NH_3$ and 0.4 mol/mole site for $S2NH_3$, taking into account also a non-uniform distribution along the device.

### 2.3.5 AdBlue® Injection Control Model Characteristics

The semi-empirical ICE model and the *axisuite®* devices are coupled with the *velodyn* model within the Simulink platform using the appropriate libraries. By coupling *velodyn* and *axisuite®* and by performing a closed-loop function according to the desired $NH_3$ storage for the SCRF device as shown in Fig. 5, a dynamic model is finally configured, capable to simulate the complete powertrain of the vehicle.

The control strategy applied for the purposes of the current study was imposing $NH_3$ injection for SCRF inlet temperatures equal or higher than 180 °C in order for the stored $NH_3$ to remain within the 0.5–0.6 g/ℓ limit values.

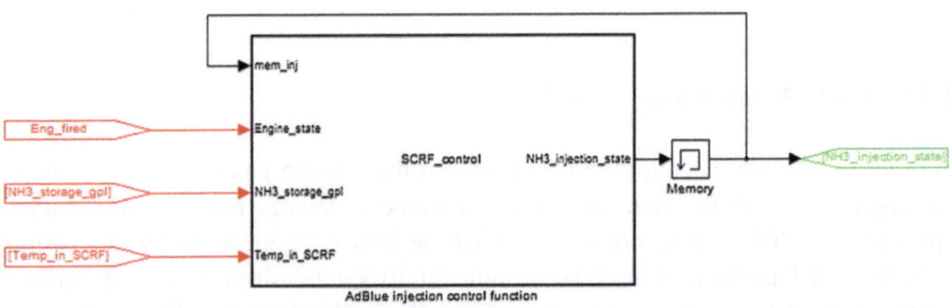


**Fig. 5** AdBlue® injection control function in Simulink

## 2.4 Hybridization

After the development of the conventional vehicle model also a parallel hybrid model was built. The ICE of the HEV model operates in 3 modes, the charge-by-engine mode, the torque-assist mode and the boost mode. The charge-by-engine mode simulates the behavior of the battery charge by using the ICE in a load point increase mode. During the torque-assist mode, the ICE operates on the characteristic line of the optimal fuel consumption in order to achieve better efficiency of the HEV. At last, boost mode is activated only for particularly high driving torque and the engine operates in full load curve.

The developed diesel parallel hybrid model with an automated transmission coupled with the previously described SCR-based deNO$_x$ aftertreatment system is illustrated in Fig. 6.

It is obvious that carrier blocks additional to the conventional vehicle are embodied into the complete model, such as Transmission Control Unit (TCU), Connectivity Control Unit (CCU), Hydraulic Control Unit (HCU), Battery Management Unit (BMU) and Electric Motor.

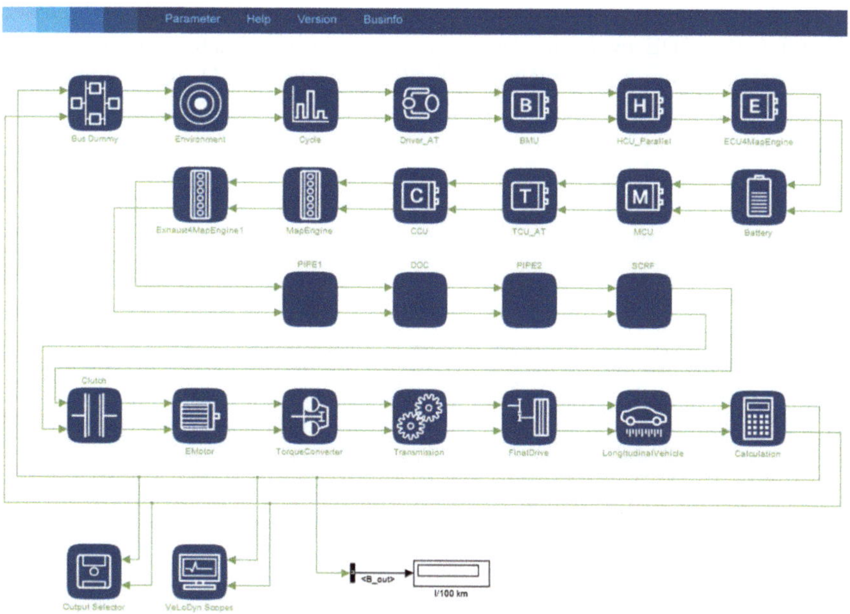


**Fig. 6** Complete hybrid vehicle model coupled with DOC and SCRF aftertreatment devices

### 2.4.1 HEV Model Characteristics

The developed model was a parallel hybrid vehicle and its powertrain consists of an ICE (1.4 ℓ of displacement and 190 Nm of maximum torque), an electric motor (14 kW of maximum power) and a Li-Ion battery (1.5 kWh of capacity). Additional mass compared to the previously described start-stop vehicle of 98 kg is considered (48 kg for the battery and 50 kg for the e-motor).

# 3 Results and Discussion

## *3.1 RDE for Baseline Vehicle*

For the purposes of this study two different driving cycles were used: NEDC and a mixed real-driving cycle RDE (average vehicle velocity: NEDC: 33.4 km/h, RDE: 55.5 km/h). The profile of the real-world cycle, shown in Fig. 7, was obtained from measurements with PEMS conducted by TÜV NORD under the European Commission Project "Development of a method for assessing real world emissions of hybrid diesel light duty vehicles" [26].

In this section, the raw results of the investigation are presented and discussed; regarding the RDE cycle, it was processed and divided into "moving average windows" according to the regulation for RDE, using the $CO_2$ mass emitted over

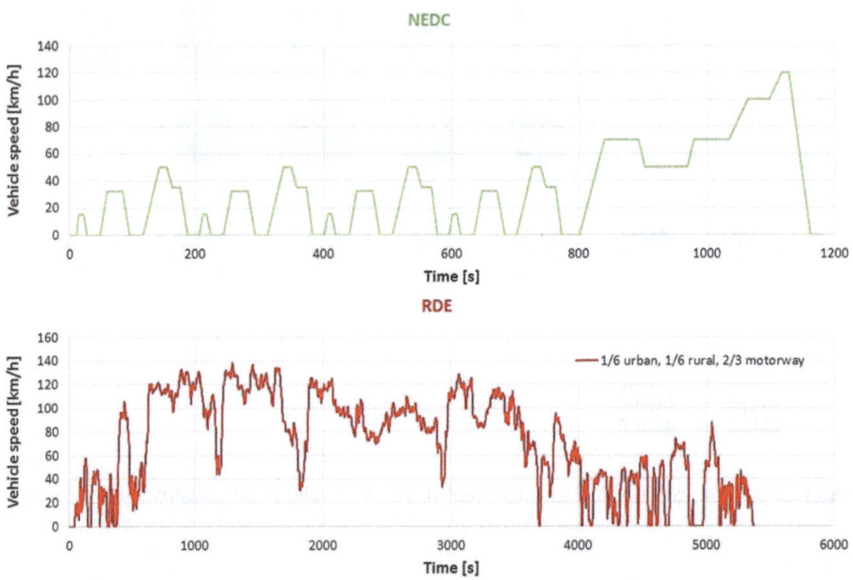


**Fig. 7** Velocity profiles of NEDC and RDE cycle

NEDC as a reference value. In this way, it is possible to relate real-world data to cycle values by selecting a fixed reference magnitude pertaining to the test cycle and subsequently windowing RDE data according to this magnitude. Since the windows share a common characteristic with the reference cycle, it would be possible to compare windowed emissions to the known cycle results or emission limits and have a meaningful estimation of the influence of real-world driving conditions.

In Fig. 8, the SCRF inlet and exit temperatures and the exhaust mass flow rate during NEDC are illustrated for the case of the conventional vehicle model.

As first, the engine-out and tailpipe instantaneous $NO_x$ emissions predicted by the model during NEDC with cold start are shown in Fig. 9 in case of the conventional vehicle.

The engine-out and tailpipe cumulative $NO_x$ emissions for the same NEDC scenario are illustrated in Fig. 10, compared to the Euro 6 T/A value.

Similarly, the SCRF inlet and exit temperatures and the exhaust mass flow rate during RDE are plotted on Fig. 11 for the case of the conventional vehicle model.

Subsequently, the engine-out and tailpipe instantaneous $NO_x$ emissions predicted by the model during the respective window of the cold start of the real-driving cycle, using the equivalent $CO_2$ mass of the NEDC as common reference magnitude with cold start are shown in Fig. 12 in case of the conventional vehicle.

The engine-out and tailpipe cumulative $NO_x$ emissions for the RDE cold-start window are illustrated in Fig. 13, compared to the Euro 6 T/A value.

Tailpipe $NO_x$ emissions values remain safely within the Euro 6 threshold, both in case of NEDC and RDE cycle.

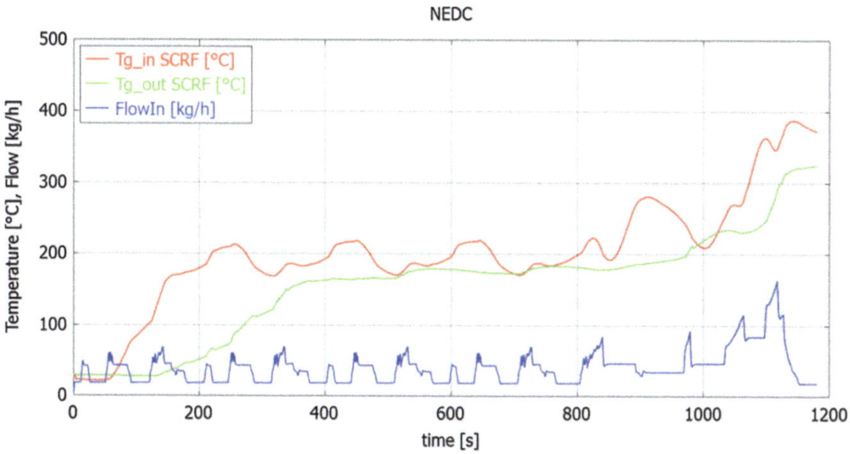


**Fig. 8** SCRF inlet and exit temperatures and exhaust mass flow for NEDC (cold-start)—conventional vehicle

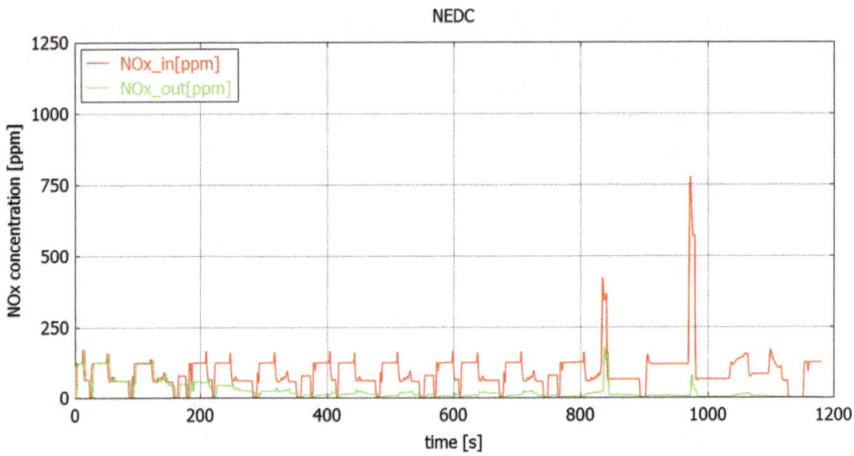


**Fig. 9** Engine-out and tailpipe instantaneous $NO_x$ emissions for NEDC—conventional vehicle

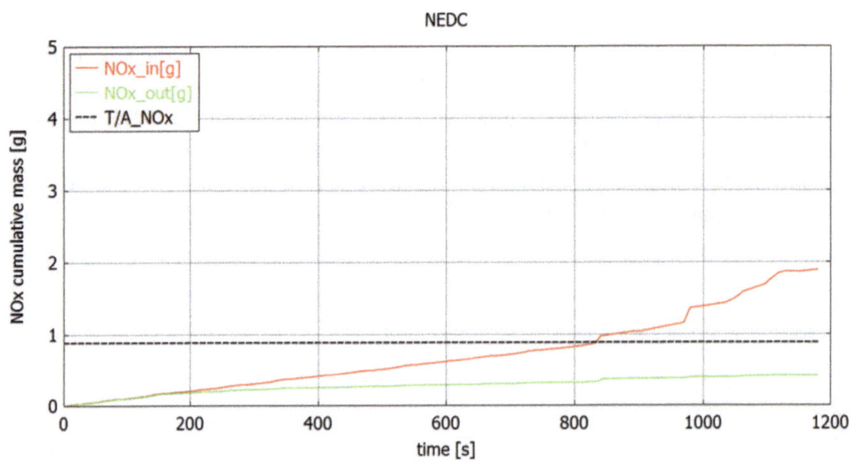


**Fig. 10** Engine-out and tailpipe cumulative $NO_x$ emissions for NEDC—conventional vehicle

## 3.2 *Hybridization Impact on RDE*

The HEV model with combined power of the ICE and an e-motor that was previously described was implemented in order to predict the tailpipe emissions of NEDC with cold start and the respective window of the cold start of the real-driving cycle.

In Fig. 14, the SCRF inlet and exit temperatures and the exhaust mass flow rate during NEDC are illustrated for the case of the hybrid electric vehicle model.

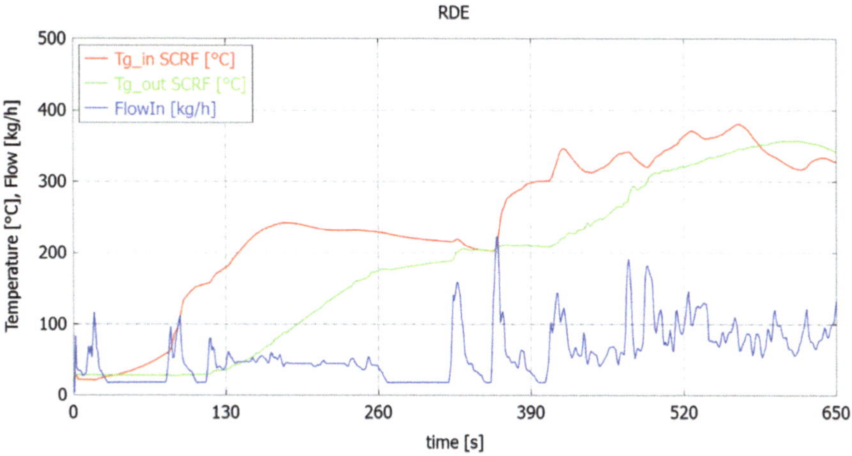


**Fig. 11** SCRF inlet and exit temperatures and exhaust mass flow for NEDC equivalent $CO_2$ mass real-driving cycle RDE window (cold-start)—conventional vehicle

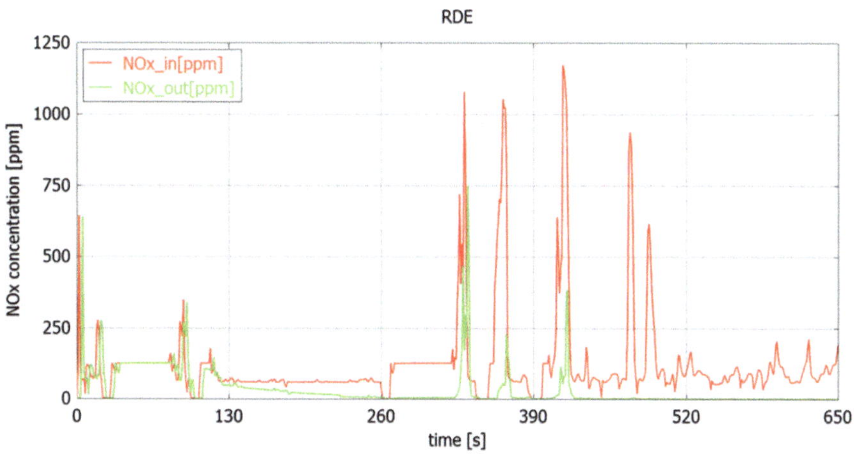


**Fig. 12** Engine-out and tailpipe instantaneous $NO_x$ emissions for NEDC equivalent $CO_2$ mass real-driving cycle RDE window (cold-start)—conventional vehicle

It is obvious that the SCRF device undergoes a cooling process during the ICE-off (e-motor on) periods for the hybrid vehicle, when there are zero-flow conditions, which is predicted by the SCRF thermal model that was previously described.

In case of the HEV the engine-out and tailpipe instantaneous $NO_x$ emissions predicted by the model during NEDC with cold start are presented in Fig. 15.

The cumulative $NO_x$ emissions for the same NEDC scenario are illustrated in Fig. 16, compared to the Euro 6 T/A value.

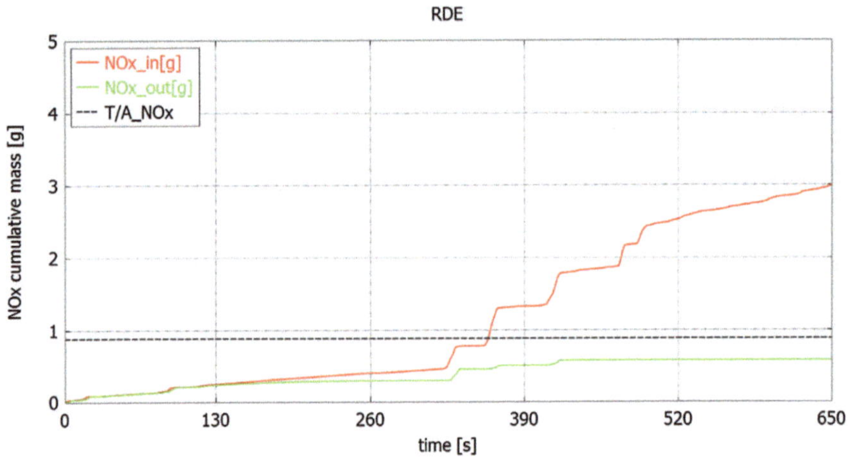


**Fig. 13** Engine-out and tailpipe cumulative $NO_x$ emissions for NEDC equivalent $CO_2$ mass real-driving cycle RDE window (cold-start)—conventional vehicle

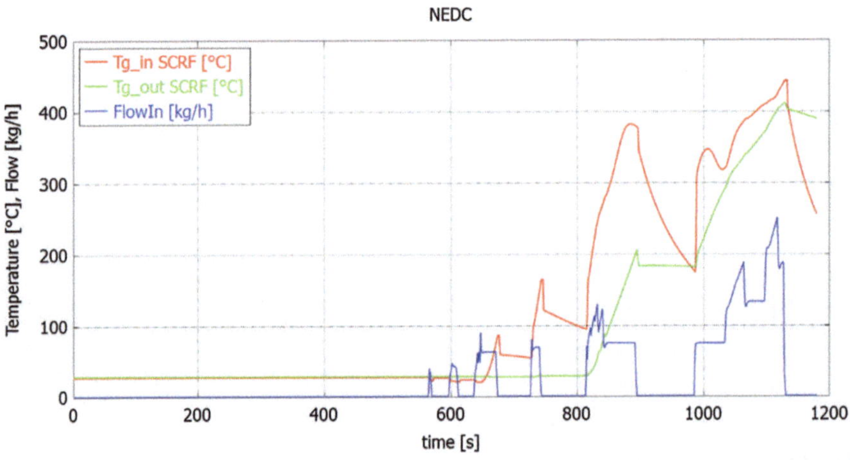


**Fig. 14** SCRF inlet and exit temperatures and exhaust mass flow for NEDC (cold-start)—hybrid vehicle

The cold-start real-driving cycle window of RDE profile, having NEDC equivalent $CO_2$ mass, is presented for the hybrid vehicle case as an example in Fig. 17. Apart from the speed-torque profile, SCRF inlet and exit temperature and exhaust mass flow are plotted on the same graph.

Similarly to the NEDC scenario, the cooling of the SCRF device during the ICE-off (e-motor on) periods for the hybrid vehicle, when there are zero-flow conditions, is predicted by the previously described A/T thermal model.

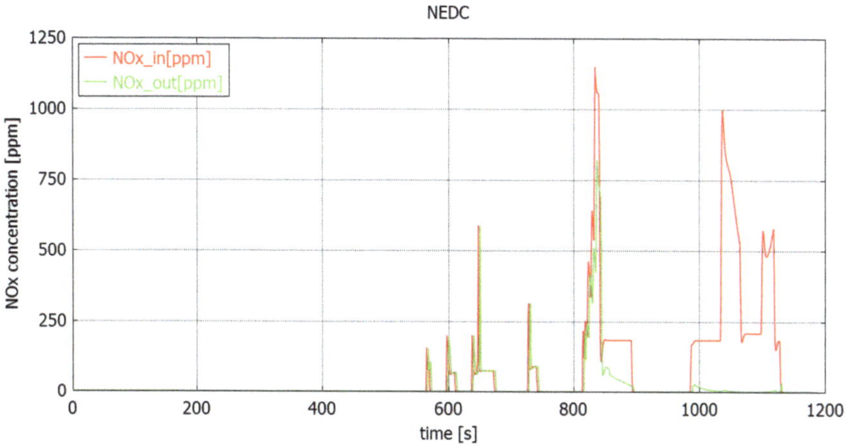


**Fig. 15** Engine-out and tailpipe instantaneous $NO_x$ emissions for NEDC—hybrid vehicle

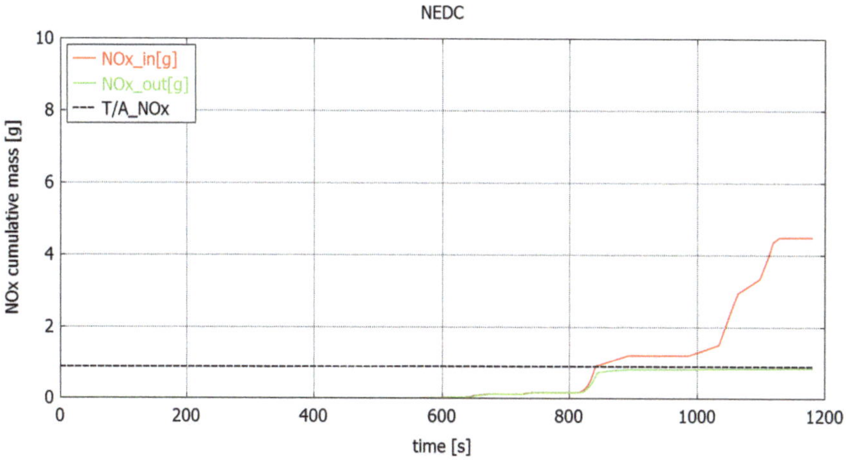


**Fig. 16** Engine-out and tailpipe cumulative $NO_x$ emissions for NEDC—hybrid vehicle

The engine-out and tailpipe instantaneous $NO_x$ emissions predicted by the model during the respective window of the cold start of the real-driving cycle, using the equivalent $CO_2$ mass of the NEDC as common reference magnitude with cold start are shown in Fig. 18 in case of the hybrid vehicle.

The cumulative $NO_x$ emissions for the same NEDC scenario are illustrated in Fig. 19, compared to the Euro 6 T/A value.

Although tailpipe $NO_x$ emissions for the hybrid vehicle during NEDC appear just within the type-approved value, the case of RDE seems to become even more critical for $NO_x$ emissions, after applying the same $NH_3$ injection control strategy.

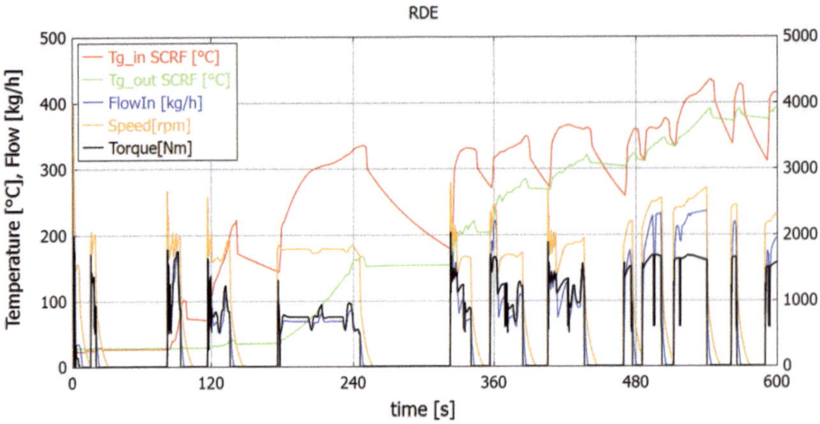


**Fig. 17** RDE cycle profile and SCRF inlet and exit temperatures for NEDC equivalent $CO_2$ mass (cold-start)—hybrid vehicle

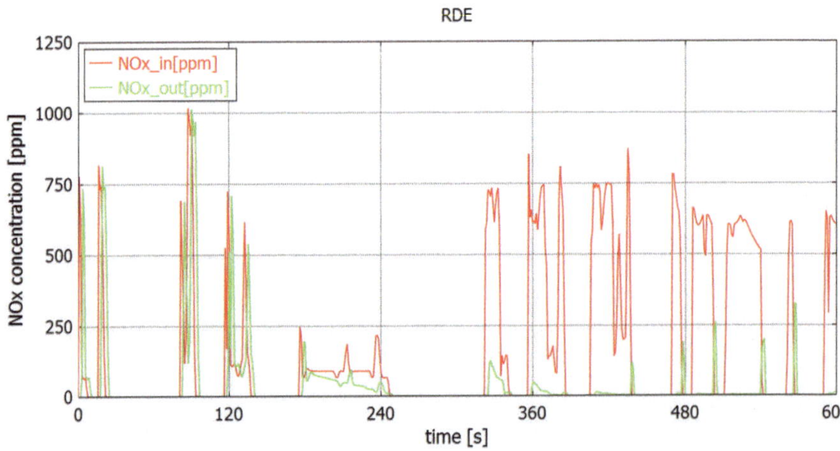


**Fig. 18** Engine-out and tailpipe instantaneous $NO_x$ emissions for NEDC equivalent $CO_2$ mass RDE cycle window (cold-start)—hybrid vehicle

Compared to the conventional vehicle model results, which appear safely under the Euro 6 legislative limit both for the NEDC scenario and the respective RDE window, hybridization seems to imply the risk of exceeding the T/A value when implementing the same $NH_3$ injection control strategy, especially in case of the RDE.

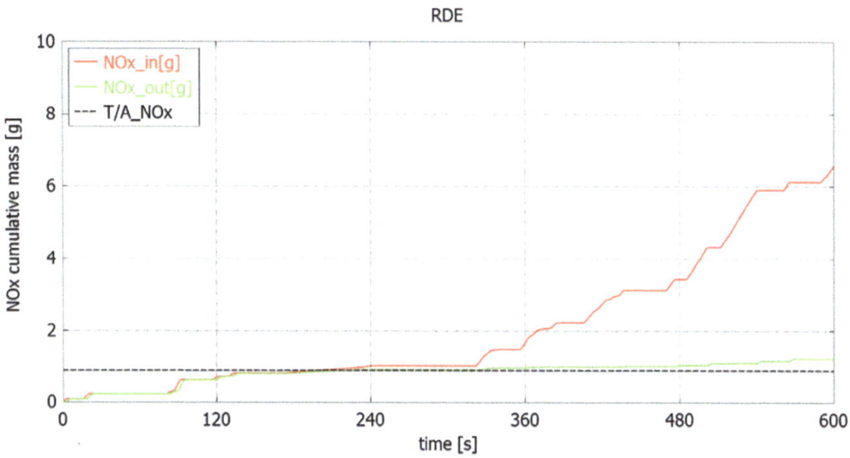


**Fig. 19** Engine-out and tailpipe cumulative $NO_x$ emissions for NEDC equivalent $CO_2$ mass RDE cycle window (cold-start)—hybrid vehicle

# 4 Conclusions and Future Work

A methodology was proposed and demonstrated for simulating the regulated tailpipe emissions for a conventional diesel vehicle and an equivalent HEV under transient real-driving conditions. The methodology aims to predict the trends in engine-out and tailpipe emissions, exhaust temperature and exhaust mass flow, taking into account the cold or warm ICE-starting conditions, under realistic driving modes.

According to the results, tailpipe emissions of the conventional vehicle model remain within the limits of Euro 6 standards even under real-driving conditions. On the other hand, hybridization seems to introduce the risk of exceeding T/A values for $NO_x$ emissions, when implementing the same $NH_3$ control strategy for the SCRF device. In particular, real-driving conditions may induce a higher risk for tailpipe $NO_x$ of HEV application, since engine-out $NO_x$ emissions appear more sensitive on the high temperatures of RDE and, consequently, the ICE-starting and driving conditions, regarding the SCRF light-off, being critical for cold-starts.

It is demonstrated that the presented methodologies could substantially support the design and optimization of integrated engine and aftertreatment controls, like $NH_3$ injection control, to comply with Real Driving Emissions requirements and hybridization.

In this direction, further study on the thermal management could include the implementation of various warm-up strategies and the investigation of a possible external heating application.

Likewise, different strategies for HEV could be easily tested, proposing optimized solutions towards the future even more stringent regulations for hybrid real-world driving conditions.

## References


1. EU.: Commission Regulation (EC) No 715/2007 of the European Parliament and of the Council. Official J. Euro. Union, 1–16 (2007)
2. EU.: Commission Regulation (EC) No 692/2008. Official J. Eur. Union, 1–136 (2008)
3. EU.: Commission Regulation (EU) No 459/2012. Official J. Eur. Union, 16–24 (2012)
4. Martin Weiss, P.B., Hummel, R., Steininger, N.: A complementary emissions test for light-duty vehicles: assessing the technical feasibility of candidate procedures. In: European Commission, Joint Research Centre, Institute for Energy and Transport (IET) (2013)
5. Fortuna, T., Koegeler, H.-M., Kordon, M., Vitale, G.: DoE and beyond—evolution of the model-based development approach. ATZ Worldw. **117**, 30–35 (2015)
6. Platner, S., Kordon, M., Fakiolas, E., Atzler, H., Model-based production calibration—the efficient way of variant development. MTZ Worldw. **74**, 18–25 (2013)
7. Wang, Z., Wu, Y., Zhou, Y., et al.: Real-world emissions of gasoline passenger cars in Macao and their correlation with driving conditions. Int. J. Environ. Sci. Technol. **11**, 1135–46 (2013)
8. Bonnel, P.: Cleaner road vehicles: How are European regulations addressing real driving emissions? In: Green Week Conference, Brussels (2013)
9. Bosteels, D.: Real driving emissions and test cycle data from 4 modern European vehicles. In: IQPC 2nd International Conference Real Driving Emissions, Düsseldorf (2014)
10. Norbert Ligterink, G.K., van Mensch, P., Hausberger, S.: (TUG), Martin Rexeis (TUG). Investigations and real world emission performance of Euro 6 light-duty vehicles. TNO 2013 R11891 (2013)
11. Maschmeyer, H., Kluin, M., Beidl, C.: Real driving emissions—a paradigm change for development. MTZ Worldw. **76**, 16–21 (2015)
12. Disch, C., Koch, T., Spicher, U., Donn, C.: Engine-in-the-Loop as a development tool for emissions optimisation in the hybrid context. MTZ Worldw. **75,** 40–46 (2014)
13. Striok, S., Wancura, H., Platner, S., Schöggl, M.: Real driving emissions a paradigm shift in vehicle application. MTZ Worldw. 75, 4–9 (2014)
14. Sellers, R., Ward, A.: Diesel engine and aftertreatment strategies to meet real driving emissions legislation. In: Bargende, M., Reuss, H.-C., Wiedemann, J. (eds.) 14 Internationales Stuttgarter Symposium. Springer Fachmedien Wiesbaden, pp. 1443–58 (2014)
15. Körfer, T., Schnorbus, T., Miccio, M., Schaub, J.: Emission-based EGR strategies in diesel engines for RDE requirements. MTZ Worldw. **75**, 10–17 (2014)
16. EU. Commission Regulation (EU) No 630/2012. Official J. Eur. Union, 14–26 (2012)
17. Klaus, Land A.: Potential for reducing emissions from road traffic and improving air quality. ENVI Committee - Public Hearing on Air Quality Policy. 2014
18. Exothermia.: Axicat user manual - multi-dimensional flow-through catalyst model (2015)
19. Exothermia.: Axitrap user manual - catalyzed diesel particulate filter simulation (2015)
20. Exothermia.: Axiheat user manual - exhaust pipe heat transfer (2015)
21. Pontikakis, G.N., Koltsakis, G.C., Stamatelos, A.M., Noirot, R., Agliany, Y., Colas, H., Versaevel, P.h., Bourgeois, C.: Experimental and modeling study on zeolite catalysts for diesel engines topics in catalysis. 329–335 (2001)
22. Stamatelos, A.M. Koltsakis, G.C, Kandylas, I.P., Pontikakis, G.N.: Computer aided engineering in diesel exhaust aftertreatment systems design. In: Proceedings of the Institution of Mechanical Engineers, Journal of Automobile Engineering, vol 213, pp. 545–560 (1999)
23. Tsinoglou, D.N., Koltsakis, G.C.: Modelling of the selective catalytic NOx reduction in diesel exhaust including Ammonia storage. In: Proceedings of the IMechE, Part D: J. Automobile Engineering, vol 221, pp. 117–133 (2007)

24. Koltsakis, G.C.. Samaras, Z.„ Karvountzis-Kontakiotis, A., Zacharopoulou, T., Haralampous, O.: Implications of engine start-stop on after-treatment operation. SAE Int. J. Engines 4 (1),1571–1585 (2011)
25. IAV.: VeLoDyn user manual. Vehicle longitudinal dynamics simulation (2012)
26. Jan Hammer, H.S., Vicente, F., Francisco Posada, S., Zissis, S., Theodora, Z.: Development of a method for assessing real-world emissions of hybrid diesel light duty vehicles. Report—By order of European Commission, DG Environment (2015)

[illegible]

# Part VII
# Model-Based Development

# Model-Based Control Design for Comfort Enhancement During Drive Off Maneuvers

**Simon Eicke, Steffen Zemke, Ahmed Trabelsi, Matthias Dagen and Tobias Ortmaier**

**Abstract** Model-based control provides a high potential to reduce costs of testing and application in vehicle software development as well as an improvement of control quality. In this paper a physical model for drive off behavior of vehicles with manual transmission is presented. The model includes wheel and vehicle longitudinal dynamics along with clutch and powertrain behavior. The declared parameters are identified using a combination of evolutionary and deterministic optimization algorithms. Based on the model two algorithms for engine torque control are designed to enhance driving comfort during drive off maneuvers. On the one hand the driver gets assistance with regard to comfort and constant acceleration and on the other hand unpleasant oscillations are prevented. Both strategies show good results in vehicle tests under real-life conditions.

**Keywords** Model-based control · Power hop · Powertrain · Clutch · Drive off

S. Eicke (✉) · M. Dagen · T. Ortmaier
Institute of Mechatronic Systems, Leibniz Universität Hannover, Hannover, Germany
e-mail: simon.eicke@imes.uni-hannover.de

M. Dagen
e-mail: matthias.dagen@imes.uni-hannover.de

T. Ortmaier
e-mail: tobias.ortmaier@imes.uni-hannover.de

S. Zemke · A. Trabelsi
IAV Automotive Engineering, Gifhorn, Germany
e-mail: steffen.zemke@iav.de

A. Trabelsi
e-mail: ahmed.trabelsi@iav.de



C. Gühmann et al. (eds.), *Simulation and Testing for Vehicle Technology*,
DOI 10.1007/978-3-319-32345-9_16

# 1 Introduction

Ride comfort during drive off maneuvers on vehicles with manual transmission is primarily determined through the driver's experience or skill, since the force acting on the powertrain is regulated by his chosen dosage of the clutch pedal. In particular, in case of frequent change of vehicle, the driver must first get used to the clutch behavior because each vehicle is individually in this regard. On the one hand the interaction between clutch and accelerator pedals at drive off has to be in a certain way so that the vehicle is accelerating as desired, while the engine does not stall. This complex task of the driver can be actively supported by control algorithms on the engine control unit. On the other hand especially on slippery road surfaces very high slip occurs during drive off in case of high torque demands by the driver. Despite existing traction control systems, a phenomenon referred to as "Power Hop" partly arises. This vertical oscillatory motion of the driven axle results in unpleasant jerking of the entire vehicle and thus reduces driving comfort of the occupants [1–4]. Figure 1 illustrates measurements of occurring oscillations as stated in [5]. High values in longitudinal acceleration prove the perceptibly negative comfort influence.

Based on vehicle or powertrain models new control strategies can be tested and parameterized in simulations. In this way cost intensive tests under real-life conditions can be reduced. In this paper, a physical model of the vehicle's behavior during drive off maneuvers is presented. The model parameters are identified using an optimization algorithm based on measured data. With the help of this model a controller is designed which supports the driver during drive off and clutch

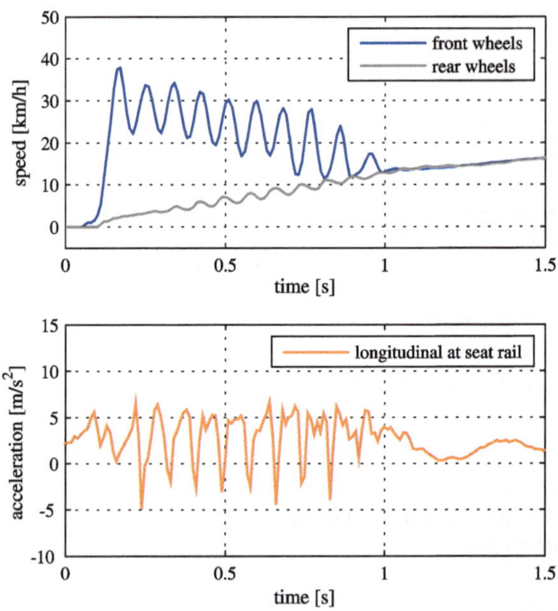


**Fig. 1** Measured wheels speeds and longitudinal acceleration during "Power Hop"

engaging with a suitable torque intervention. Thereby, a comfortable maneuver with a constant acceleration is achieved. The simulation results are validated in vehicle tests under real-life conditions. In addition, a model-based controller is designed for a reduction of the power hop phenomenon. The algorithm is characterized by a small number of adjustable parameters, thus keeping the application effort low. Simulations and measurement results of real driving tests confirm the increase in comfort for this control algorithm as well.

The content is structured as follows: at first the physical model is introduced (Sect. 2). In the Sect. 3 the identification algorithm and results are presented. Afterwards, the development of the control strategies is described in Sect. 4. Finally, the measurement results of tests in vehicles under real life conditions are given in Sect. 5. In the last section a brief summary is given and a conclusion is drawn.

## 2 Modeling

In this section a physical model for vehicle dynamics during drive off maneuvers is introduced. It will be used for the model-based control design and provides also a simulation environment to reduce test effort under real-life conditions.

### 2.1 *Wheel Dynamics*

Since the driving force which accelerates a vehicle is transferred by contact area between wheel and road surface, the physical mechanism plays an important role. Figure 2 shows a free body diagram of a wheel with all acting forces, torques and characteristic dimensions.

The wheel with radius $r_W$ and inertia $I_W$ is driven by torque $T_d$ while frictional losses due to the bearings and the rolling resistance are summed up by the torque

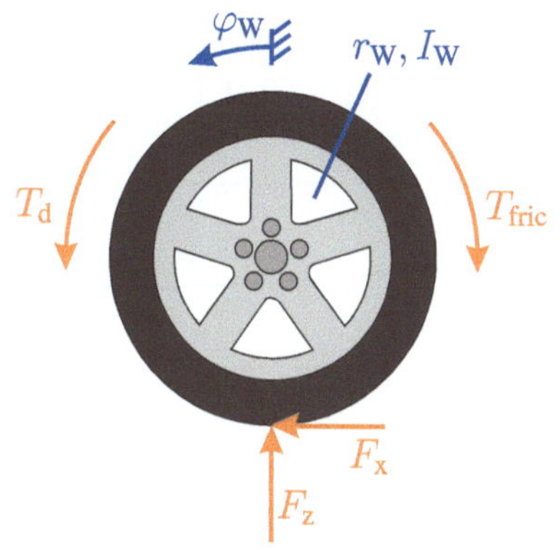


**Fig. 2** Free body diagram of a wheel

$T_{\text{fric}}$. The drive torque $T_{\text{d}}$ might also be negative in case of a braking maneuver. In the contact point between road and surface the acting forces in horizontal and vertical direction are represented by $F_{\text{x}}$ and $F_{\text{z}}$, respectively. The rotational degree of freedom of the wheel is symbolized by $\varphi_{\text{W}}$. The principle of angular momentum leads to

$$I_{\text{W}}\ddot{\varphi}_{\text{W}} = T_{\text{d}} - T_{\text{fric}} - r_{\text{W}}F_{\text{x}}. \tag{1}$$

The longitudinal force $F_{\text{x}}$ depends on the wheel load $F_{\text{z}}$ and can be calculated by considering the actual friction coefficient $\mu$, which is a function of the longitudinal slip $\lambda$ by

$$F_{\text{x}} = \mu(\lambda)F_{\text{z}}. \tag{2}$$

The functional relation between the slip and the friction coefficient can be approximated by Pacejka's Magic Tire Formula [6] as stated in the following equation

$$\mu = D\sin[C\,\text{atan}(B(1-E)\lambda + E\,\text{atan}(B\lambda))]. \tag{3}$$

The constants $B, C, D, E$ can be adjusted and allow a representation of different tire-road conditions. In Fig. 3 typical slip curves for high and low $\mu$ surfaces are shown.

The required longitudinal slip value $\lambda$ can be calculated by the quotient

$$\lambda = \frac{|\dot{\varphi}_{\text{W}}r_{\text{W}} - \dot{x}|}{\max(\dot{\varphi}_{\text{W}}r_{\text{W}}, \dot{x})} \tag{4}$$

where $\dot{x}$ describes the vehicle's velocity in longitudinal direction. The derivation of the latter as well as of the wheel load $F_{\text{z}}$ is presented in the following section.

## 2.2 *Longitudinal Vehicle Dynamics*

In addition to the longitudinal force some further dependencies have to be considered in order to describe the vehicle dynamics in longitudinal direction. Figure 4 shows the acting forces on a vehicle on a slope.

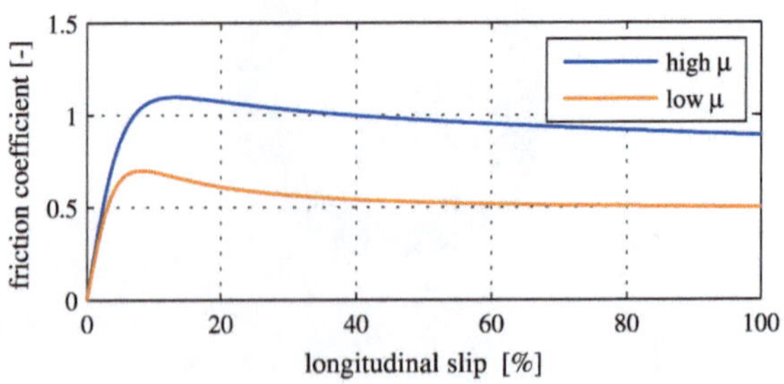


**Fig. 3** Slip curves for different road conditions

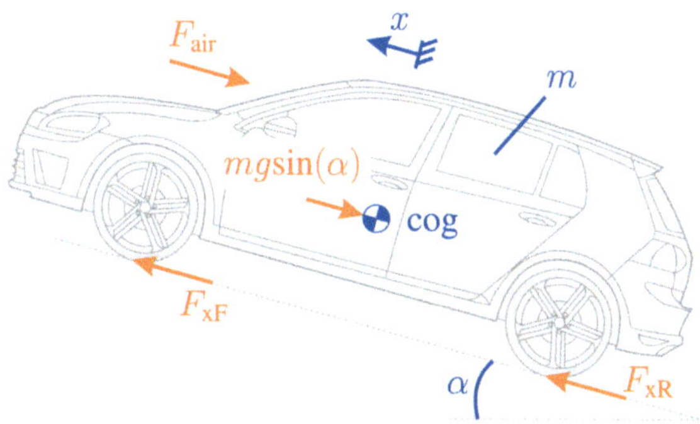


**Fig. 4** Free body diagram of vehicle with forces in longitudinal direction

Application of the principle of linear momentum in longitudinal direction yields

$$m\ddot{x} = F_{xF} + F_{xR} - F_{air} - mg\sin(\alpha). \tag{5}$$

In this context the longitudinal force is separated into two parts acting on the front axle $F_{xF}$ and at the rear axle $F_{xR}$ as a general formulation. Because the vehicle with mass $m$ might stand on a slope with angle $\alpha$, a force caused by the gravity g acting in the center of gravity (cog) has to be included. The degree of freedom in longitudinal direction is further denoted by $x$.

The wheel load $F_z$ consists in the first instance of a stationary part which is primarily determined by geometrical quantities. In Fig. 5 the influencing variables are depicted.

Using again the principle of angular momentum at the front and rear tire contact point, it follows

$$F_{zF,stat} = mg(\frac{l_R}{l_{wb}}\cos(\alpha) - \frac{h}{l_{wb}}\sin(\alpha)) \tag{6}$$

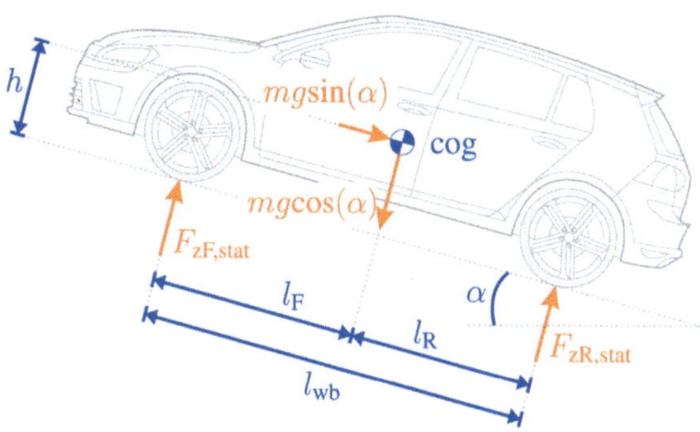


**Fig. 5** Geometric dimensions for stationary wheel load

and

$$F_{zR,stat} = mg(\frac{l_F}{l_{wb}}\cos(\alpha) + \frac{h}{l_{wb}}\sin(\alpha)) \tag{7}$$

for the stationary wheel load at the front and rear axle respectively. The wheelbase $l_{wb}$ and the distances from the center of gravity to the road surface $h$ and to front $l_F$ and rear axle $l_R$ are decisive parameters here. It can be seen that under the assumption of a constant road slope, the stationary wheel load is nearly constant for a single vehicle and might have only slight changes because of differences in the laden state.

In a second instance the wheel load has a dynamic part caused by possible acceleration or braking of the vehicle. Figure 6 shows a vehicle with acting D'Alembert's inertia forces due to dynamics in longitudinal direction. It can be easily deduced for the dynamic wheel load of the rear axle $F_{zR,dyn}$ by using the principle of angular momentum at the contact point of the front axle

$$F_{zR,dyn} = m\ddot{x}\frac{h}{l_{wb}}. \tag{8}$$

The front axle is relieved by the same amount as the rear axle is burdened. Therefore, the dynamic wheel load at front axle $F_{zF,dyn}$ is equal to Eq. (8) but with negative sign.

## 2.3 Powertrain Dynamics

Since the overall model should use the engine torque as its input, the dynamics in the powertrain from the engine to the wheel are taken into account. In Fig. 7 a model of the powertrain with three inertias representing engine $I_E$, gearbox $I_G$ and wheel $I_W$ is illustrated.

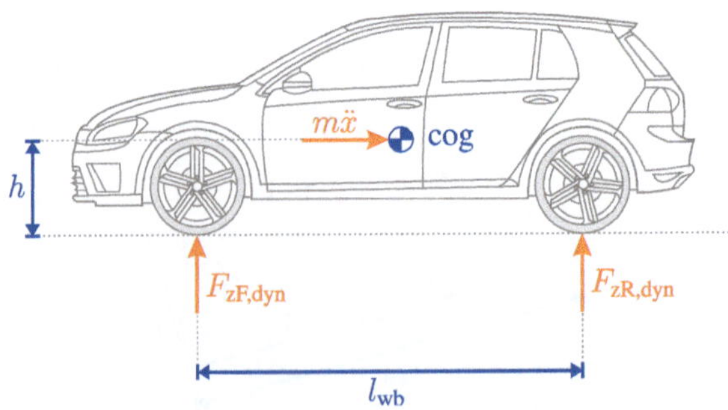


**Fig. 6** Dynamic wheel load influenced by D'Alembert's inertia forces

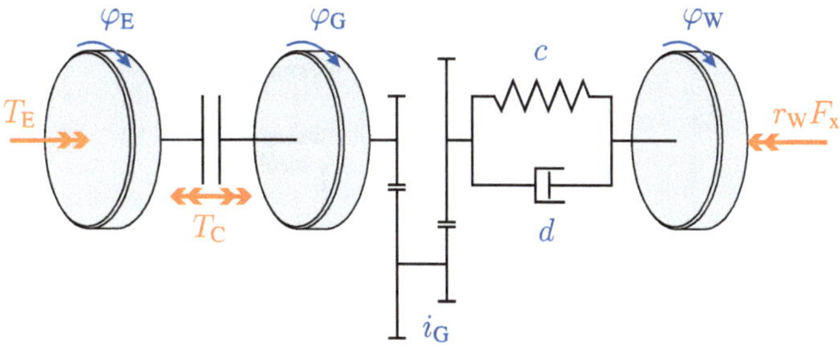


**Fig. 7** Model of the powertrain

Application of the principle of angular momentum once again for each inertia yields

$$I_E \ddot{\varphi}_E = T_E - T_C, \tag{9}$$

$$I_G \ddot{\varphi}_G = T_C - \frac{1}{i_G}\left[c\left(\frac{\varphi_G}{i_G} - \varphi_W\right) + d\left(\frac{\dot{\varphi}_G}{i_G} - \dot{\varphi}_W\right)\right], \tag{10}$$

$$I_W \ddot{\varphi}_W = c\left(\frac{\varphi_G}{i_G} - \varphi_W\right) + d\left(\frac{\dot{\varphi}_G}{i_G} - \dot{\varphi}_W\right) - r_W F_x. \tag{11}$$

The inertias have each a degree of freedom $\varphi_E, \varphi_G, \varphi_W$ respectively which results in three differential equations to describe the system. The engine torque $T_E$ is an input of the model while the clutch torque $T_C$ and the wheel torque $r_W F_W$ have to be calculated as explained in Sects. 2.1, 2.2 and 2.4. The actual gearbox ratio $i_G$ as well as stiff-ness $c$ and damping factor $d$ represent the influence of transmission and driveline flexibility respectively.

## 2.4 Clutch Dynamics

As introduced in the previous subsection the clutch torque $T_C$ plays an important role for power transmission in the drivetrain. Since in conventional vehicles the torque is standardly not available by measurements, an appropriate model which generates this information depending on other given data is presented in the following. Figure 8 shows the function and interaction of dry clutch components as often used in passenger cars.

It can be seen that the force is transmitted from the crankshaft via friction pads and the clutch disk to the gearbox primary shaft. The actual clutch torque depends on the contact pressure which is generated by the washer spring and the flat spring of the clutch disk. It can be calculated by

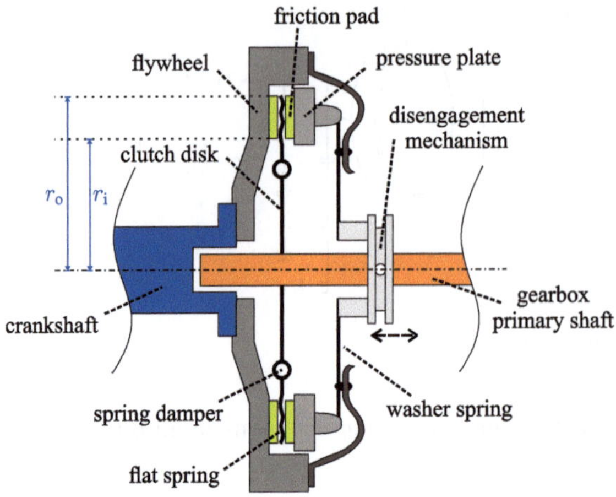


**Fig. 8** Schematic sketch of a dry clutch

$$T_C = zF_C r_C \mu_C, \tag{12}$$

where $F_C$ denotes the contact force, $\mu_C$ the friction coefficient and $z$ the number of friction surfaces which is generally equal to two for dry clutches. The average clutch radius $r_C$ can be calculated with regard to the inner $r_i$ and outer radius $r_o$ by

$$r_C = \frac{2}{3}\frac{r_o^3 - r_i^3}{r_o^2 - r_i^2}. \tag{13}$$

As stated in [7] the contact force is a function of the disengagement distance of the clutch and the friction coefficient might change with differences in relative speed between the friction partners.

# 3 Identification

## 3.1 Identification Algorithm

In order to identify the model parameters measurements are taken in different driving situations to allow a comparison between simulation and real data. The squared error between those values is summed up over time. To find appropriate model parameters an optimization algorithm is used to minimize the total error by adjusting the parameters. In this contribution a combination of the Particle Swarm Optimization (PSO) as introduced in [8–11] and a gradient-based optimization as described in [12, 13] is used. Both methods can handle the nonlinearities of the models while PSO helps avoiding local minima using an evolutionary approach. The subsequent application of the gradient-based technique provides deterministic solutions for the minimum in accordance to the defined criteria.

## 3.2 Identification Results

In the following major results for the model are presented. To identify the clutch parameters different drive off maneuvers with a vehicle with manual transmission are taken. The data is generated by standard vehicle sensors for engine speed, wheel speeds and clutch pedal position. It is assumed that the clutch pedal affects linearly the disengagement mechanism of the clutch. In Fig. 9 a comparison between measured and simulated engine's and wheel's angular velocity is depicted. In the related maneuver the clutch was slowly closed at about 1 s and was kept constant for about 0.5 s afterwards. Subsequently the clutch was slightly opened again before it is finally closed and the angular velocities draw near. Thereby, the model gets good stimulation for improved identification. In addition, accurate results for those exceptional maneuvers can confirm high model quality and adequate level of detail.

Engine torque and clutch pedal position are used as inputs for the introduced model. It can be seen that the simulated curves correlate very good with its equivalents. Especially the renewed increase in engine speed after about 2.2 s is matched well. In Fig. 10 the identified curves for the contact force and the friction coefficient are depicted. The clutch pedal position in the left plot equals 0 % for no activation of the clutch pedal in accordance with closed clutch and maximum

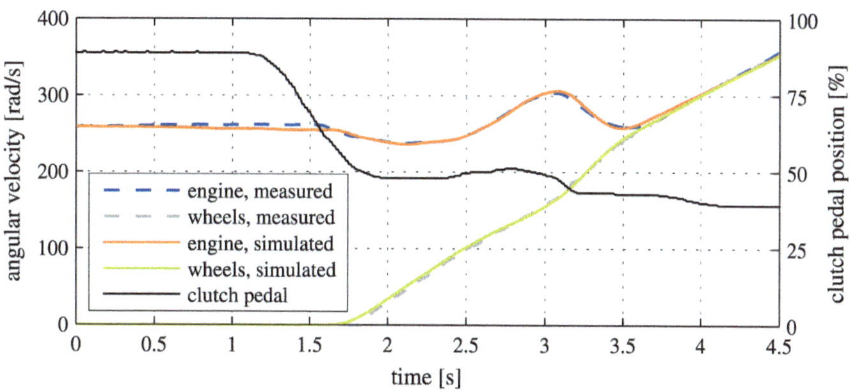


**Fig. 9** Identification results for engine and wheel speed during drive off maneuver

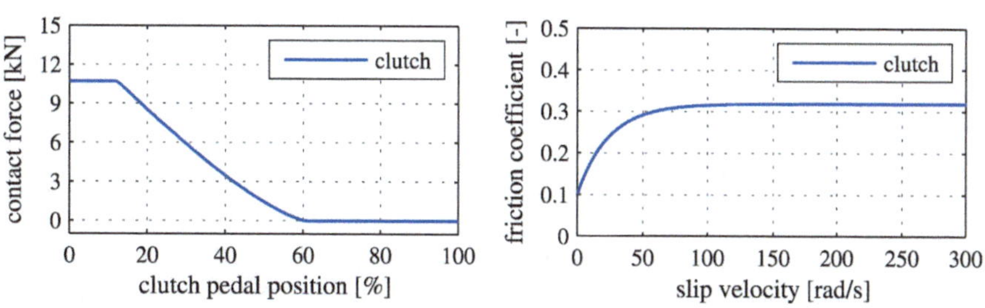


**Fig. 10** Identified clutch parameters: contact force (*left*) and friction coefficient (*right*)

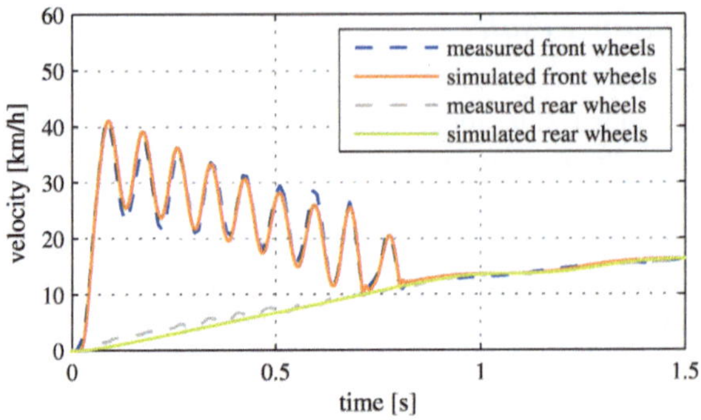


**Fig. 11** Identification results for drive off with occurring power hop

contact pressure. In contrast a value of 100 % means a completely pressed down pedal and thus an opened clutch without torque transmission. It can be recognized that with pedal position less than approximately 60 % the clutch begins to transmit power. After that point the contact force increases until it reaches a maximum and remains constant. The plot on the right illustrates the dependency of the friction coefficient on the slip velocity. It comes clear that the friction coefficient increase with growing slip until it saturates after about 100 rad/s on a constant level. Grabbing of clutch can be avoided by such a damped clutch characteristic. This result is in accordance with [14].

In Fig. 11 the results for a snap start maneuver with occurring power hop oscillations are shown. In contrast to the previous maneuver in Fig. 9, in this case the clutch was closed abruptly while engine speed was held on a level of about 2500 rpm before.

Typical vibrations of the driven axle arise for about 0.8 s as intended. The inputs for the simulation are again only engine torque and clutch pedal position. The good match between the measured and simulated data of the front and rear wheels attests again a high quality of the introduced model.

## 4 Control Design

The presented model can be used as a part of model-based control algorithms or can be utilized for simulation-based control design and parameter setting of common PID-controllers. The following control examples will briefly address possible purposes.

### *4.1 Drive Off Control*

With regard to the mentioned difficulties for the driver during drive off maneuvers and clutch engagement, control algorithms can be helpful. Since in vehicles with manual transmission the clutch is mechanically actuated by the driver a direct control intervention in this mechanism is complicated and unfavorable. On the contrary a software algorithm for the engine control unit (ECU) provides an easy and cheap feasibility for improvements.

The presented model offers information about inner torques in the clutch and thus can help to counteract the decreasing engine speed and in particular a stalling engine. Furthermore, the compensation torque can ensure the same constant acceleration after closing the clutch as during clutch engagement phase. The consideration of the current clutch torque for individual drive off maneuver makes this behavior suitable for both comfort and sport requirements.

Thereby, the driver can concentrate on the handling of the clutch pedal without need for compensations with the accelerator pedal. For this purpose the control algorithm adds a compensation torque to the driver's demand $T_{\mathrm{driver}}$

$$T_{\mathrm{res}} = T_{\mathrm{driver}} + T_{\mathrm{comp}} \tag{14}$$

in order to achieve a comfortable driving behavior by the resulting torque $T_{\mathrm{res}}$. The compensation torque $T_{\mathrm{comp}}$ is calculated as a sum of the modeled clutch torque $T_{\mathrm{C}}$ and an additional part of a PID-controller $T_{\mathrm{PID}}$ for engine speed to consider parameter uncertainties, e.g., temperature influences,

$$T_{\mathrm{comp}} = T_{\mathrm{C}} + T_{\mathrm{PID}}. \tag{15}$$

## 4.2 *Power Hop Control for Snap Starts*

As stated in [5] engine torque control also provides the opportunity to reduce the power hop phenomenon. Therefore, an additional control algorithm is designed which focuses on the described oscillations. Since in this case external influences as the uncertain friction coefficient between tire and road surface plays a major role, a robust control algorithm is chosen. Following [15, 16] a sliding mode controller based on the presented model is designed. Instead of the slip value as the controlled variable as stated in [16], the speed of the driven axle $v_{\mathrm{W}}$ is chosen to be included in the sliding variable $s$

$$s = v_{\mathrm{W}} - v_{\mathrm{W,ref}} \tag{16}$$

where the index "ref" denotes the reference input for the sliding mode control. This is advantageous even at low speeds since the slip is unsteady in this range and the scope of the function should be the drive off maneuver from stand still. In order to achieve good command response the sliding variable and its derivative has to equal zero. Using the model from Sect. 2 it can be stated

$$\dot{s} = \dot{v}_W = \left( \frac{(T_d - T_{fric} - r_W \mu F_z) r_W}{I_W} \right) \stackrel{!}{=} 0. \tag{17}$$

It comes clear that the sliding control depends on model-based estimations for friction torque $T_{fric}$ and longitudinal force $\mu F_z$. Regarding [16] the final controlled torque $T_{ctl}$ with maximum uncertainty for longitudinal force estimation can be written as

$$T_{ctl} = T_{driver} \left( \frac{1 - (1 + \alpha) \mathrm{sat}(s, \Delta)}{2} \right) \tag{18}$$

which is dependent on the sliding window $\Delta$ and an adjustable parameter $\alpha$ in order to influence the intensity of control intervention. To avoid chattering of the manipulated variable instead of the common signum-function a saturation function defined as

$$\mathrm{sat}(s, \Delta) = \begin{cases} \mathrm{sgn(s)} & \mathrm{if} |s| \geq \Delta \\ \frac{s}{\Delta} & \mathrm{if} |s| < \Delta \end{cases} \tag{19}$$

is used. In extension to this approach the presented model could be used to provide more accurate estimations of the unknown variables in Eq. (17).

# 5 Experimental Results

The presented model-based control algorithms are implemented for real-time testing using rapid prototyping software and hardware in common passenger cars. Exemplary results are presented below.

## 5.1 Drive Off Control

In order to evaluate the presented drive off function a maneuver with slow clutch engagement with both activated and deactivated control was carried out. The engine speed was kept constant by a limiter function during stand still. Afterwards the accelerator pedal was set to a constant value of 30 % to allow a comparison under at most same conditions. Figure 12 shows the measured results for a drive off maneuver with deactivated control algorithm. It can be recognized that the vehicle acceleration drops at about 2.2 s after the clutch is closed. This point might be unpleasant for the vehicle occupants, especially because the driver intends constant acceleration by a constant accelerator pedal position.

In contrast, in Fig. 13 the measurements with activated control are depicted. It becomes evident that the vehicle preserves constant acceleration after the clutch has

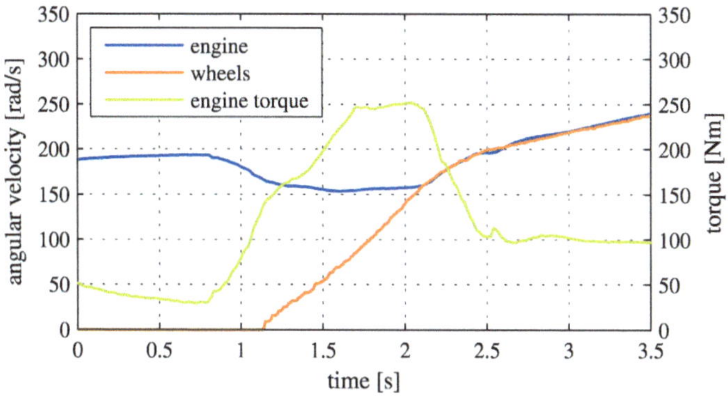


**Fig. 12** Measurements during drive off with deactivated control

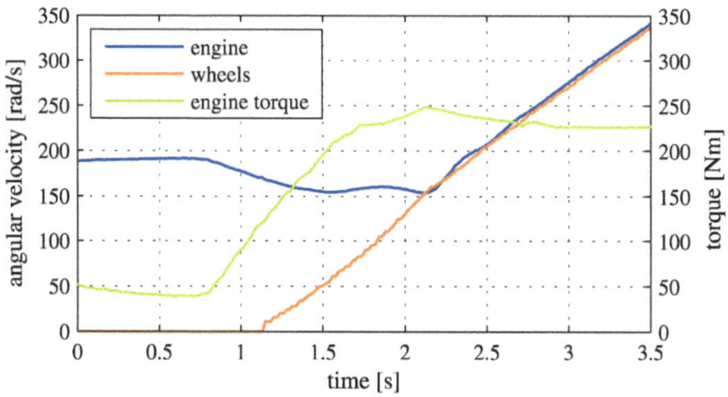


**Fig. 13** Measurements during drive off with deactivated control

been closed. The controller thus provides intuitive behavior for the driver and helps him to drive off from stand still with self-determined constant acceleration.

## 5.2 *Power Hop Control*

Snap start maneuvers were executed on a low $\mu$ surface to assess the developed power hop control. A snap start maneuver means an abrupt release of the clutch in stand still whereby the engine speed was kept on a higher level of about 3000 rpm before, see [17]. The accelerator pedal position was set to 70 % in the following. In Fig. 14 the measurements of the driven axle show persistent oscillations when the additional control is deactivated. This results in unpleasant vehicle vibrations and reduces passenger comfort. In addition the vehicle acceleration is low because of high slip and related low friction coefficient.

On the other hand Fig. 15 depicts the results with activated power hop control as stated in Eq. (18). It can be seen that the oscillations are reduced within 1.3 s by a

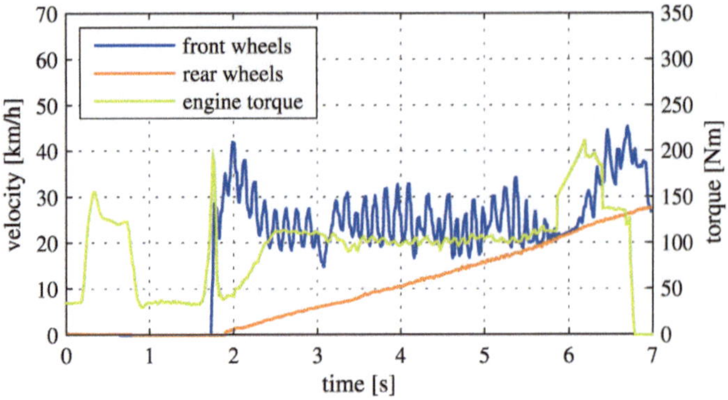


**Fig. 14** Measurements during snap start with deactivated control

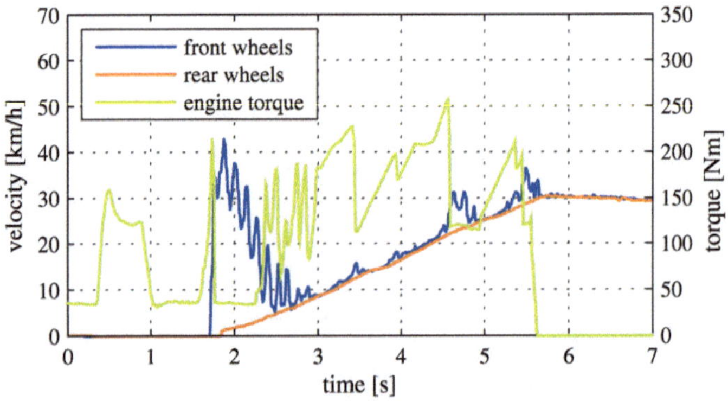


**Fig. 15** Measurements during snap start with activated control

sharp reduction of engine torque. Attempts of newly arising oscillations are reduced immediately. In consequence the comfort is increased clearly. Furthermore the acceleration of the vehicle can be enhanced on this road surface since the slip is quite low.

## 6 Conclusion

The presented physical model of vehicle's behavior during drive off shows great representation of reality and thus allows usage in vehicle dynamic simulations. This might help to reduce costs during development and testing phase of new functions or products. Furthermore, the introduced control strategies for drive off maneuvers and power hop control offer good results with regard to comfort enhancement and assistance of the driver. On the one hand the drive off control supports the driver via engine torque control to prevent stalling engine and to facilitate constant acceleration. On the other hand the power hop control reduces engine torque so that unpleasant oscillations can be reduced.

The described model might also be helpful for further intelligent control strategies in combination with a model based observer as it provides inner states of the system.

## References


1. Engel, D., Meywerk, M., Wojke, B.: Self-excited full-vehicle oscillations in dynamic packaging examination. ATZ Worldw. **111**(5), 52–57 (2009)
2. Brandhoff, H., Pfau, W., Tögel, F.-J.: Extrembelastungen im Pkw-Antriebsstrang durch Missbrauch. ATZ Automobiltechnische Z. **101**(12), 1006–1015 (1999)
3. Fan, J., Hierlwimmer, P., Berthold, B., et al.: Analyse des Systemeinflusses auf die Spitzenbelastung im Pkw-Handschaltgetriebe. ATZ Automobiltechnische Z. **101**(10), 812–819 (1999)
4. Thorausch, M., Kirchner, E., Scheuermann, M., et al.: Assessment method for passenger car transmissions under abusive loads. ATZ Worldw. **111**(1), 40–46 (2009)
5. Eicke, S., Zemke, S., Trabelsi, A., Dagen, M., Ortmaier, T.: Experimental investigation of power hop in passenger cars. SAE, Technical Paper 2015-01-2185 (2015)
6. Pacejka, H.: Tire and Vehicle Dynamics, 3rd edn. Elsevier Ltd., Oxford (2012)
7. Wohnhaas, A., Hötzer, D., Sailer, U.: Modulares Simulationsmodell eines Kfz-Antriebsstrangs unter Berücksichtigung von Nichtlinearitäten und Kupplungsvorgängen. VDI Berichte Nr. 1220 (1995)
8. Kennedy, J., Eberhart, R.: Particle swarm optimization. In: Proceedings of IEEE International Conference on Neural Networks, pp. 1942–1948 (1995)
9. Eberhart, R., Kennedy, J.: A new optimizer using particle swarm theory. In: Proceedings of the Sixth International Symposium on Micro Machine and Human Science, pp. 39–43 (1995)
10. Shi, Y., Eberhart, R.: A modified particle swarm optimizer. In: Proceedings of the IEEE International Conference on Evolutionary Computation, pp. 69–73 (1998)
11. Behrendt, H., Ortmaier, T., Trabelsi, A., Zemke, S.: Analyse und modellbasierte Regelung von KFZ-Antriebsstrangschwingungen. Simulation und Test für die Automobilelektronik III, pp. 145–161. Expert Verlag, Renningen (2010)
12. Scales, L.E.: Introduction to Non-Linear Optimization. Macmillan Publishers LTD, London (1985)
13. Nelles, O.: Nonlinear System Identification. Springer, Berlin (2001)
14. Albers, A., Herbst, D.: Rupfen – Ursachen und Abhilfen. 6. LuK Kolloquium, LuK GmbH & Co. KG (1998)
15. Shtessel, Y., Edwards, C., Fridman, L., Levant, A.: Sliding Mode Control and Observation. Springer, Berlin (2014)
16. de Castro, R., Araujo, R.E., Diamantino., F.: Wheel slip control of EVs based on sliding mode technique with conditional integrators. IEEE Trans. Indus. Electron. **60**(8) (2013)
17. Kember, S., Powell, N., Poggi, M., Ellis, D., Sung, J.: Modelling of snap start behaviour in an automotive driveline. MSC software support library (1999)

# Successful Integration of a Model Based Calibration Methodology for Non-standard Corrections and Protection Functions

**Nikolaus Keuth, Guillaume Broustail, Kieran Mcaleer, Marijn Hollander and Stefan Scheidel**

**Abstract** Due to increasing requirements on the performance and emissions at non-standard conditions the amount of correction functions are increasing in order to utilize the maximum performance of the engine within its limitations. To be able to calibrate all those functions more capacity of an altitude test bench would be needed. On one hand those test benches are limited and on the other hand their operation is very expensive. The target of the AVL approach is to minimize the usage of a real test-benches, to maximize performance, keep emissions and increase dataset quality due to frontloading based on a semi-physical engine model. Those semi-physical model are able to simulate the thermodynamic behavior and the emissions under non-standard ambient conditions. Due the physical model parts also a model based system fault simulation can be performed in order to simulate engine failures and validate the component protection calibration. These results are established by the usage of these semi physical models on a hardware in the loop test bed in combination with Cameo automation and optimization (AVL Virtual Test Bed). The main advantage of using a Virtual Test Bed is, that the calibration environment for the application engineer is similar to the test-bench. In this way the calibration engineer does not need to have specific modeling knowledge. By using the Virtual Test Bed as environment the calibration can be done faster and with better engine performance as result. This can mainly be achieved due to the fact that

N. Keuth (✉) · G. Broustail · K. Mcaleer · M. Hollander · S. Scheidel
AVL List GmbH, 8020 Graz, Austria
e-mail: nikolaus.keuth@avl.com

G. Broustail
e-mail: guillaume.broustail@avl.com

K. Mcaleer
e-mail: kieran.mcaleer@avl.com

M. Hollander
e-mail: stefan.scheidl@avl.com

S. Scheidel
e-mail: marijn.hollander@avl.com



C. Gühmann et al. (eds.), *Simulation and Testing for Vehicle Technology*,
DOI 10.1007/978-3-319-32345-9_17

during the calibration process for instance the real hardware limitation do not need to be taken in account. In combination with the automation with AVL Cameo and optimization with AVL Cameo this leads to a faster and better result. The successful integration of this methodology will be described.

# 1 Introduction

Calibration engineers are currently facing numerous new challenges arising from increasingly stringent legal requirements. Together with the introduction of EURO VI, EPA10, Stage IV and Tier4 a Not-To- Exceed area is introduced, where a certain level of tail pipe emissions needs to be assured. These tail pipe emissions need to be measured stationary in the WNTE (World harmonized Not-To-Exceed) area (see Fig. 1) combined with specific ambient conditions.

In addition to the WNTE the In-Service-Conformity (ISC) is introduced with EURO VI. The conformity of in-service vehicles or engines of an engine family shall be demonstrated by testing vehicles on the road operated over their normal driving patterns, conditions and payloads. The ambient conditions where the conformity factor needs to be demonstrated are equal to the one of WNTE, see Fig. 1. The engineer needs to elaborate a calibration that on the one hand assures the fulfillment of the emission requirements at the various ambient conditions and on the other hand still delivers the desired engine performance and fuel consumption. In this process, he is facing two major challenges:

WNTE Area Grid for diesel engines with a rated speed < 3000rpm:

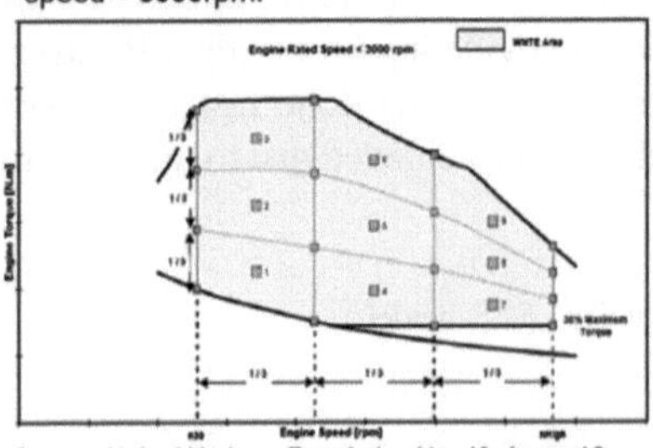

Source: United Nations, Regulation No. 49, Annex10

Emission Limits: Source: Commission Regulation (EU) No 582/2011

- CO 2000 mg/kWh
- THC 220 mg/kWh
- NOx 600 mg/kWh
- PM 16 mg/kWh

- Gaseous pollutants shall not exceed the limits measured over a grid cell
- Particulate pollutants shall not exceed the limit measured over the whole 3 selected grids

- Ambient conditions: Atmospheric pressure ≥ 82.5 kPa (~1680m)
  Temperature ≥ 266 K (-7 °C)
  Temperature ≤ determined by the following equation:
  $T = -0.4514 * (101.3 - p_b) + 311$ — T is the ambient air temperature in K; $p_b$ is the atmospheric pressure in kPa
- Engine coolant temp. ≥ 343 K (70 °C).

**Fig. 1** WNTE limitations and testing conditions

1. Finding the optimal compromise between emission, performance and fuel consumption targets under all ambient conditions in calibrating the complex non-standard engine functions.
2. Achieving this with a very limited access to non-standard engine tests beds. Such test beds that are able to reproduce extreme temperatures and altitudes are usually rare and come with high costs.

## *1.1 Connecting the Virtual and the Real Worlds*

In order to tackle these challenges, massive frontloading activities must be undertaken to succeed in the future and meet upcoming legal requirements. Therefore, the Austrian company for the development of powertrain, instrumentation and test systems AVL strives to break new ground in the field of model-based development. The integration of elements such as simulation models, data storage systems, execution environments, capabilities, processes and automation systems, to form dynamic, consistent and interdependent tool networks and functions will become more and more essential for success. The AVL strategy is to connect existing as well as new elements, whether they are virtual (simulated) or real, over the whole development process (Fig. 2).

To meet the requirements of EURO VI, EPA10, Stage IV and Tier4 and also to increase efficiency and robustness and reduce costs, a shift to upfront testing and calibration is necessary. This can be realized through the connection of the virtual and the real worlds. Simulation models are connected to each other to make up functional prototypes and as the development process proceeds, they are combined with real hardware components. Thus, test beds are enhanced with simulation and offer a plethora of new possibilities [1] (Fig. 3).

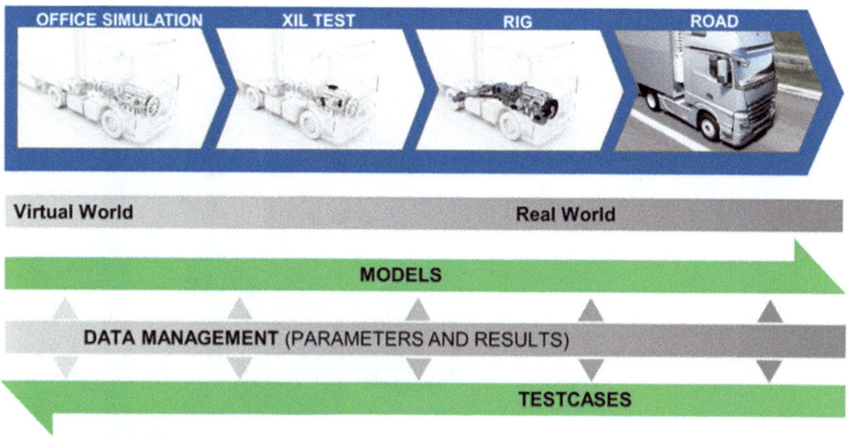


**Fig. 2** Connecting the virtual and the real worlds

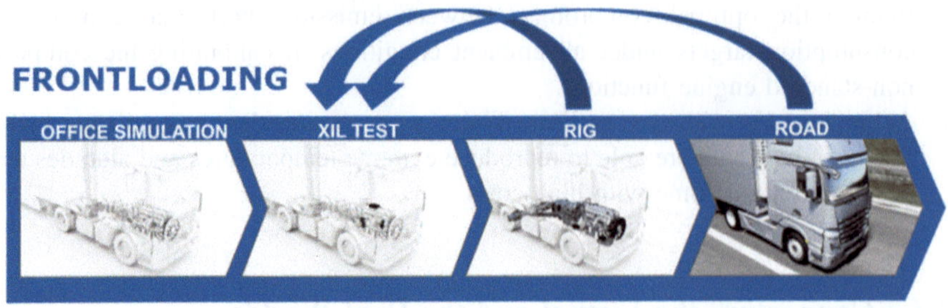


**Fig. 3** Frontloading tasks through model-based development

A typical application of this AVL methodology is to pre-calibrate ECU-functions on an AVL virtual test bed. One of the use cases performed on a hardware-in-the-loop system is to pre-calibrate the non-standard engine functions integrating a design of experiments optimization approach using AVL CAMEO.

## *1.2 AVL Virtual Test Bed*

The AVL virtual test bed, illustrated in Fig. 4 is an advanced Hardware-in-the-Loop test bed enhanced with advanced semi-physical powertrain models. It comes with test automation, central host data storage and xCU calibration software, and features the same interface as on a conventional engine test bed.

The engineer is therefore given a familiar environment, which enables a seamless shift from conventional to virtual test bed. Furthermore, he can import both

**Fig. 4** AVL virtual test bed

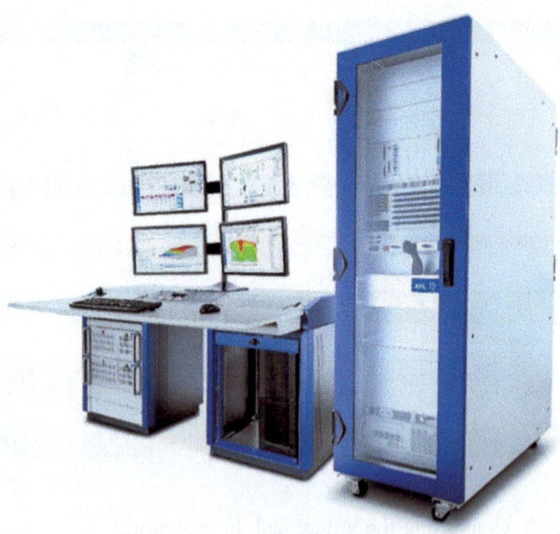

conventional and virtual test bed results from the data storage host, and analyze these results in a common post-processing layout. In this way, the calibration engineer does not have to have modeling or HIL specific skills in order to work with a virtual environment.

## 1.3 Comparison Virtual Versus Real Measurement Data

The key factor to be able to calibrate successfully on a virtual test bed is the quality of the engine models used. For the actual use case presented here, AVL MoBEO and CRUISE M models are used.

This approach is based on semi-physical models, combining the advantages of physical and empirical modelling approaches. The physical layer increases the range of application thanks to better extrapolation capability and makes it possible to easily modify some hardware specifications. The empirical layer enhances the accuracy within the covered parameter-space and increases the calculation speed.

The parameterization of the models has been done based on stationary real test bed measurements in standard ambient conditions. When changing the ambient conditions during calibration, the physical components ensure that the engine

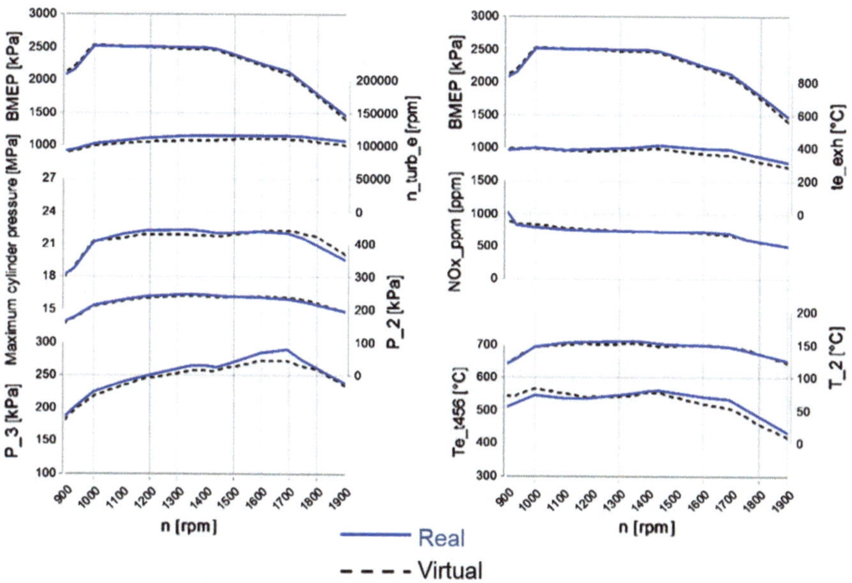


**Fig. 5** Validation measurements, real (*dotted*) versus virtual (full) test bed at −10 °C ambient temperature at sea-level

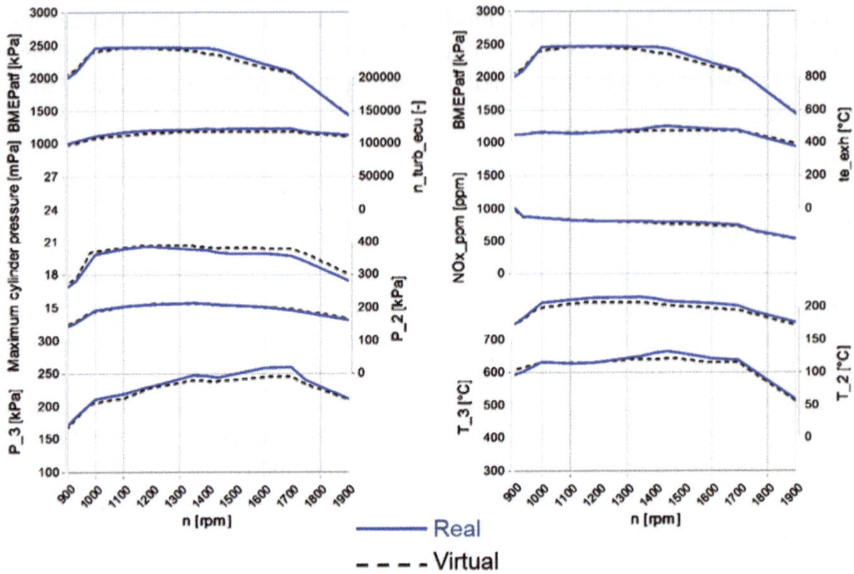


**Fig. 6** Validation measurements, real (*dotted*) versus virtual (full) test bed at 25 °C ambient temperature and 1640 m

model is extrapolated with the desired accuracy. Figures 5 and 6 show validation measurements from a climate test bed versus the measurements performed on the virtual test bed.

## 2 Test and Optimization Strategy

In order to optimize the ambient correction functions, an efficient test strategy has to fulfill the following requirements:

- Capability to find the optimal setting for the parameters
- Minimizing the testing time needed to achieve optimal results
- Providing assistance for the generation of engine maps

As the variation parameters have strong interactions between each other, manual testing or automated "one factor at a time" variations would not necessarily lead to optimal settings. Full factorial testing can lead to settings close to the optimum, but requires a high number of measurement points. Therefore, a "design of experiments" (DoE)-approach, combined with empirical modeling, model based optimization and map generation, is used. For every step of the workflow, the software-package AVL CAMEO is utilized.

## 3 CAMEO DoE Screening

The variations are distributed by a D-optimal test-plan, as the responses of interest for the optimization (NOx-emissions, torque, fuel consumption, turbine speed, various pressures and temperatures) are likely to be modeled accurately using polynomial equations. To perform the DoE-measurements, a CAMEO 2-layer screening test is used. This test-type approaches each variation point in steps or a ramp from a safe starting point, either until the target point is reached or until one of the parameterized limits is violated. In case of a limit violation, using the "re-approach" limit-reaction-strategy, the borderline of the drivable area is closely approached.

The parameterized limits can be divided into two different groups:

Hard limits—like peak firing pressure, exhaust temperature, turbine speed etc.—are engine limits that can cause damage on the engine and therefore are not to be exceeded on a real engine test bed.

Soft limits—like NOx-emissions, EGR-valve-position, pressure drop over the engine, etc.—can be used additionally to restrict the variation space to a reasonable range.

## 4 Advantage of Virtual Testing

Using a real engine, CAMEO screening methodology has to be set up taking into account all engine hard limits. Depending on the engine behavior, the order of variation parameter adjustment, the limit reaction strategy and the selected starting point for the screening method, a different test space arises. As shown in Fig. 7, the

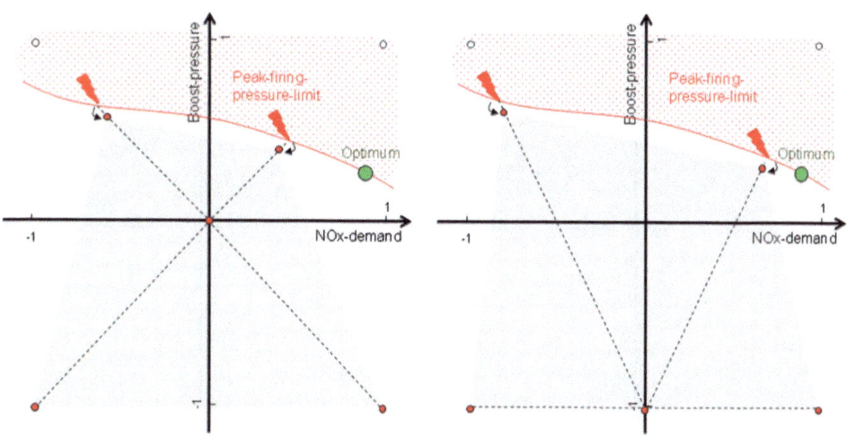


**Fig. 7** Schematic representation of the resulting test space depending on the starting point of the screening methodology, taking the real engine limits into account as hard limits

test space can highly differ from the initially planned variation space, especially for investigations under full-load-conditions.

Consequently, a drivable optimum (i.e. input-settings within the initially planned variation range and all limit-channels within the allowed range) may be located outside the test space.

To enlarge the test space on a real engine test bed, several procedures are discussed [2]. These procedures require a far more complex test-setup and the necessity to include pre-knowledge of the system behavior and still cannot ensure that the optimum will be within the test space.

Performing the DoE on a virtual test bed, the hard-limits of the real engine can be used like soft-limits: As no damage has to be prevented, the limit values can be exceeded and only be used to limit the test space to a reasonable range. Figure 8 shows the schematic representation of the benefit of virtual testing:

During the test, the limits are exceeded and hence the optimum can be found within the test space.

As for the calibration the real engine limits still have to be considered, all limit channels are modeled and used as hard constraints during the CAMEO optimization. Figure 9 shows the resulting test-space in the input-space using the real engine limits and the extended limits on the actual virtual test bed.

A comparison of the models and the optimization results using the two strategies is given in Fig. 10.

In the given example, the extension of the limits led to an optimization result with approximately 1 % lower fuel consumption. This improvement in fuel consumption found in the optimization based on the empirical models was validated on the virtual test bed. The settings for this optimum are clearly outside the test-space of the test-run considering the engine limits as hard-limits; i.e. it is not possible to find these optimal settings on a real test bed using a standard screening methodology.

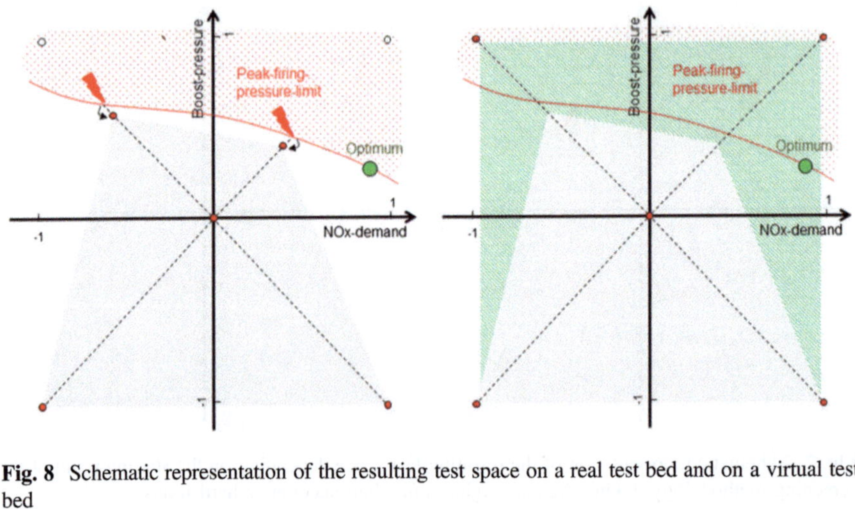


**Fig. 8** Schematic representation of the resulting test space on a real test bed and on a virtual test bed

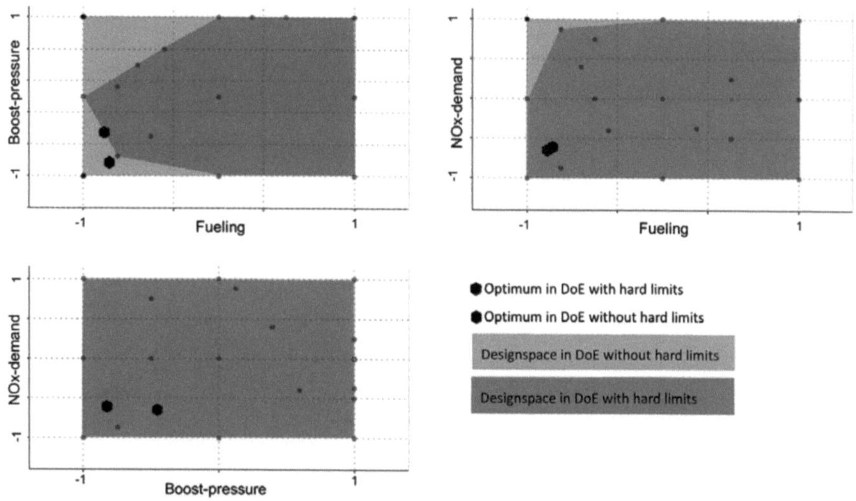


**Fig. 9** Measurement point distribution in the input space and the optimized settings respectively

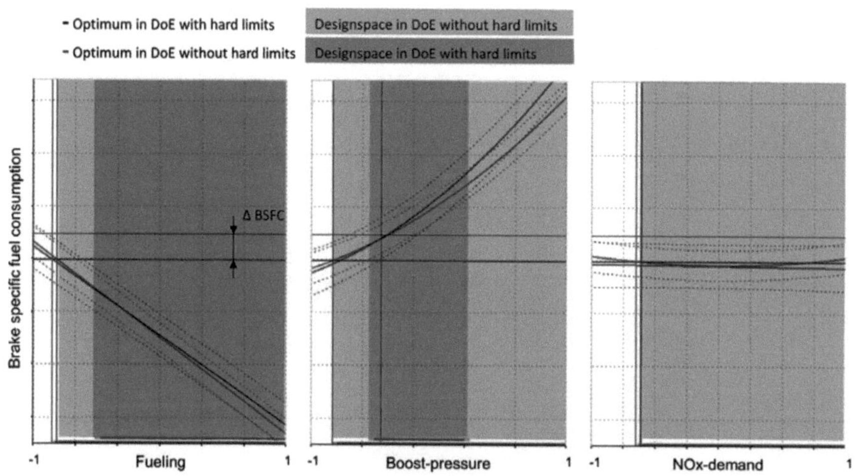


**Fig. 10** Intersection plot of fuel consumption model

# 5 Model Based Optimization and Map Generation

The measurement-data generated on the virtual test bed (using the DoE-procedures as described above) was used for model based optimization of the parameter maps in CAMEO. As a first step—after importing the data—empirical models for all channels needed for the optimization were created. This includes the two optimization targets, fuel consumption (which has to be minimized) and torque (which

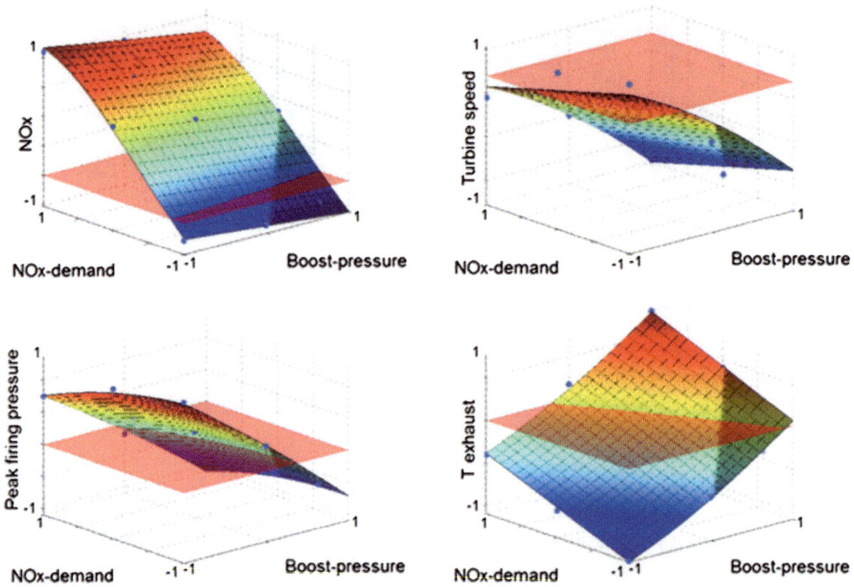


**Fig. 11** 3-D plots of empirical models for engine limits

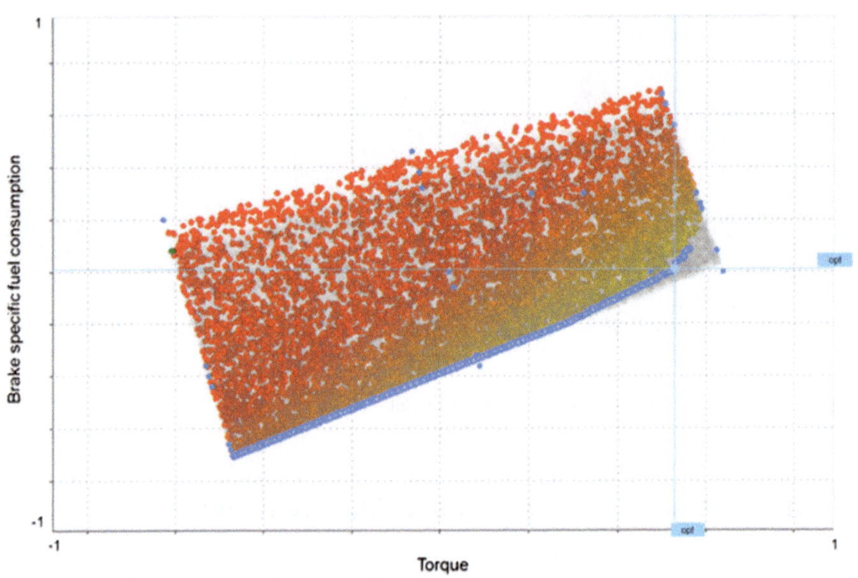


**Fig. 12** Pareto-front as result of two-criterial optimization

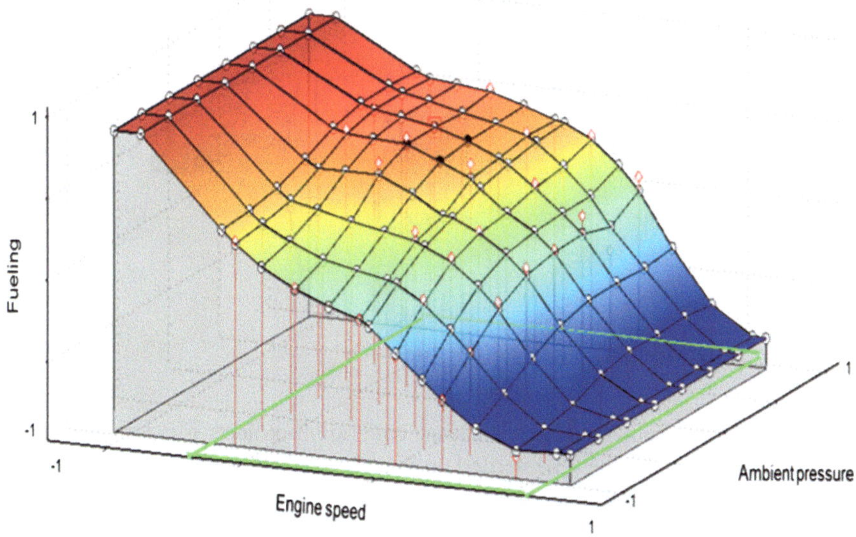


**Fig. 13** Generation of ECU-maps based on the optimization results

has to be maximized), and all limitations (like emissions, pressures, temperatures etc.). Figure 11 shows the 3D-plots of four engine limit models with respect to two of the input parameters.

Based on the two target functions and the values for the limit-channels, a two-dimensional optimization was performed. The result of this optimization is not a single solution but a selection of different best combinations for the two targets, i.e. a so-called pareto-front as shown in Fig. 12. By applying this procedure, the calibration engineer can choose a reasonable compromise between power output and fuels consumption.

This optimization workflow was performed for different engine speeds and under different ambient conditions. Finally, the in such a way generated optimized setting had to be transformed into calibratable maps. For this integration of the results in smooth ECU-maps, CAMEO map-generator was used.

Figure 13 shows the generation of the fuel-limitation-map.

## 6 Conclusions

Enabling frontloading, with model-based development new means of increasing efficiency and quality are generated. Development tasks that are today carried out only in latter process stages using real hardware components can be performed much earlier through the consistent integration of virtual and real worlds (Fig. 14).

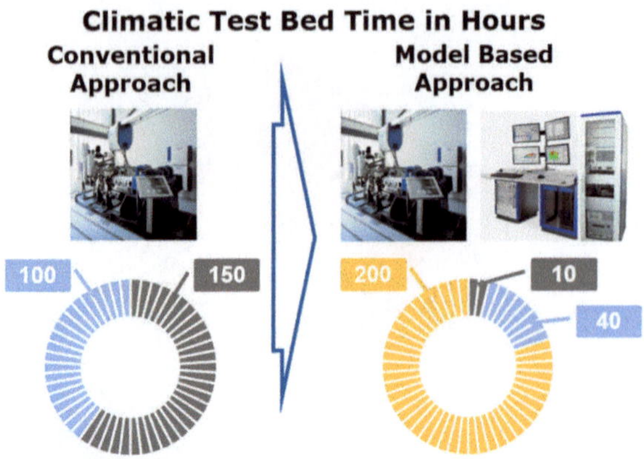


**Fig. 14** Saving resources with model-based development

Based on numerous commercial engine calibration projects it is known that for this particular use case the time needed on the real test bed can be reduced by 80 %, compared to the conventional approach. These savings can be used for enhanced validation or further frontloading leading to an increased dataset quality.

Combining the virtual test bed with a DOE approach leads to further increase of the engine performance within the limitations of hardware constraints and emission legislation. Only with a truly interdisciplinary approach the enormous challenges arising from upcoming legal requirements such as EURO VI, EPA10, Stage IV and Tier4 or the increasing complexity and speed can be mastered today and in the future.

## References


1. Fritz, S., Hötzendorfer, H., Koller, M.: Design of experiments in large diesel engine optimization. MTZ Indus. **04**, 40–45 (2014)
2. Voigt, K., Puntigam, W.: Simulation or Test, Integrated or Open? (MTZ—Fachtagung) VPC.plus Hanau (2014)

# Optimal Steady-State Base-Calibration of Model Based ECU-Functions

**Yijiang Xie**

**Abstract** In the ECU, there are a number of submodels which are used to calculate immeasurable signals, for example, EGR mass flow, exhaust gas temperature, etc. Usually these submodels consist of physical equations and empirical parts, that are modeled with parameters, curves and maps, which can only be calibrated experimentally. Currently the data set for the steady-state calibration of these submodels is generated by grid measurements on the engine testbench. There is a great potential for reduction of the cost and time needed for the calibration of these submodels with the utilization of the DoE approach and physical connections between them. This paper presents an algorithm which should calibrate all of the existing submodels as a network in the "Air System Model" at Bosch-ECU, namely automated and with minimal calibration cost at engine testbench. The algorithm should find the most informative combination of inputs, such as engine speed, fuel quantity, actuator position of throttle valve, EGR valve and variable turbine geometry (VTG), for the calibration of the network of submodels in air system. The process is implemented in the framework of "Sequential Experimental Design". After the initial experiment, where the submodels are fed with the equally and loosely spaced inputs within a predefined range, initial statistical models can be built with a Gaussian process model. At each iteration of the process, before the measurement is conducted, a function of the information content with respect to the combination of the inputs is derived. This can be done in 3 steps.

1. At first the relevant system variables for the calibration, such as air mass flow and gas temperature after mixture with EGR, will be predicted depending on the combination of the inputs, current statistical models for the relevant submodels and the physical structure of the system
2. Based on the predicted system variables an extended Kalman filter can be employed to estimate the variance of the measurement points for calibration of the submodels

Y. Xie (✉)
Robert Bosch GmbH, 70469 Stuttgart, Germany
e-mail: yijiang.xie@de.bosch.com



C. Gühmann et al. (eds.), *Simulation and Testing for Vehicle Technology*,
DOI 10.1007/978-3-319-32345-9_18

3. The information content of the predicted measurement points for the calibration of the submodels is calculated and summed up. It is defined by the reduction of the uncertainty of the unmeasured region in each Gaussian process model by adding the predicted measurement point to the current data set.

With the above derived function, the most informative combination of inputs can be found and used in the experiment at this iteration. The statistical model for each is updated with the generated data set and can be used in next iteration. The process continues till a desired quality of calibration is reached, which is described in an "Automatic Stopping Criterion". At the end the statistical models for the submodels are converted into maps or curves and stored in the ECU. As a use case, the algorithm was applied to choose the most informative measurement points in a grid measurement for the calibration of three submodels, cylinder charge, pressure upstream of the throttle valve and EGR mass flow. As a result, with less than 30 % of the grid measurement points, a slightly better calibration quality for the three submodels was achieved.

**Keywords** Steady-state base-calibration of ECU-functions · Sequential experimental design · Gaussian process regression · Mutual information · Extended Kalman filter

# 1 Introduction

The diesel engine is characterized by a high efficiency of 40 % in vehicles and even 50 % for vessels. To meet the increasingly stringent demands on both emissions and driveability, nowadays most diesel engines are equipped with turbocharging and system for exhaust gas recirculation (EGR), which have considerably increased the complexity and expense of the calibration of the air system.

## *1.1 Air System of the Diesel Engine*

As it is shown in Fig. 1, the air system is composed of the fresh air path and the exhaust path, which are connected through a pipe for the EGR. On one side, the fresh air first flows through the air filter to the compressor where it is compressed and then cooled by the intercooler. Then the fresh air is mixed with the cooled exhaust gas and flows into the cylinder. On the other side of the engine, the exhaust gas flows through the exhaust manifold to the turbine, where the exhaust gas is decompressed and the enthalpy is extracted. After the turbine the exhaust gas is cleaned by aggregates for exhaust aftertreatment, for example, particulate filter, and selective catalytic reduction (SCR).

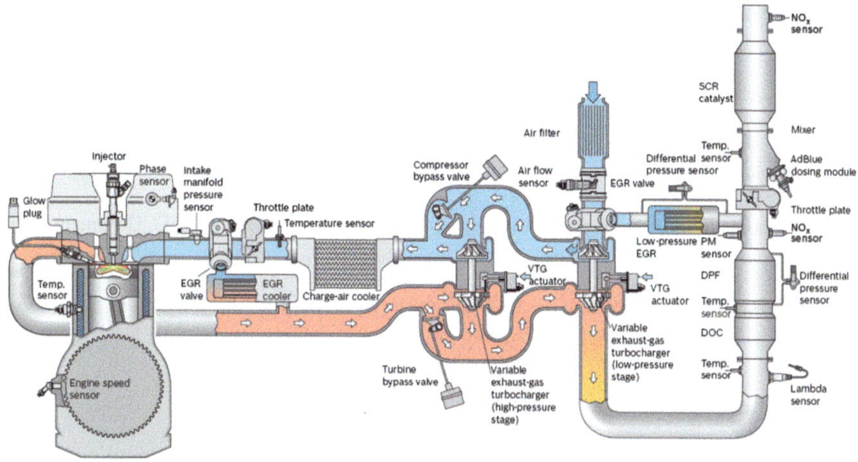


**Fig. 1** Components of the air system of diesel engine

## *1.2 Steady-State Base-Calibration of ECU-Functions*

In order to control or regulate the air system of the diesel engine properly, many functions in the ECU are designed and implemented. There, the entire air system of the diesel engine is divided into submodels and modeled by means of the filling and emptying method, which is mainly based on the ideal gas equation of state, throttle equation, mass balance and enthalpy balance. In the submodels, not only are the normal system variables (temperature, pressure, etc.) calculated but the immeasurable variables such as volumetric efficiency and EGR mass flow are provided as well. Typically, this kind of submodels are grey-box models and are composed of a physical part and an empirical part which is represented by a parameter, curve or map. Such empirical models are also used to describe relations that are difficult to depict with physical equations and thus can only be experimentally calibrated. A perfect example is the submodel for the modeling of EGR mass flow. Usually in ECU-functions for air system the EGR mass flow is modeled with the throttle equation below:

$$\dot{m}_{\mathrm{EGR}} = \sqrt{\frac{2}{R_{\mathrm{Exh}} T_{\mathrm{Ds}}}} \cdot ar_{\mathrm{EGR,eff}} \cdot p_{\mathrm{Ds}} \cdot \Psi\left(\frac{p_{\mathrm{Ds}}}{p_{\mathrm{Us}}}\right) \tag{1}$$

where $R_{\mathrm{Exh}}$ represents the gas constant of the exhaust gas, $T_{\mathrm{Us}}$ and $p_{\mathrm{Us}}$ the exhaust temperature and pressure upstream of the EGR valve, and $p_{\mathrm{Ds}}$ the exhaust pressure downstream of the EGR valve. In (1), except for the effective area of the EGR valve, $ar_{\mathrm{EGR,eff}}$, all signals can be measured by sensors or be provided by other

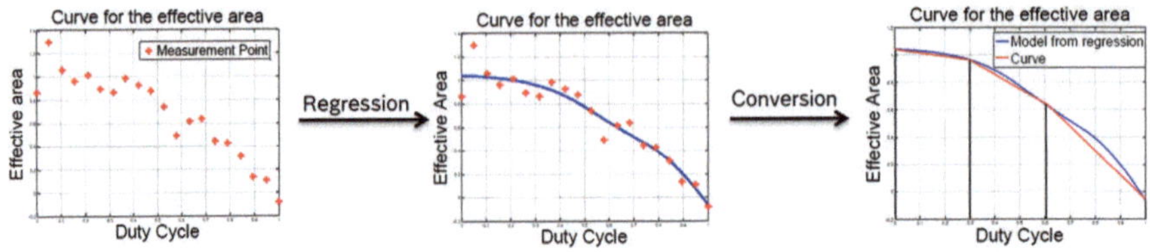


**Fig. 2** Schematic illustration of the working process for base-application of ECU-Function

submodels. Also, the effective area depends only on the duty cycle of the EGR valve, it is represented with following relation:

$$ar_{\mathrm{EGR,eff}} = f(DC_{\mathrm{EGR}}) \tag{2}$$

This relation in (2) can theoretically be depicted through a geometrical approach, which will not work out in practice because of some other nonnegligible factors, such as turbulence at the valve. Therefore, in order to model the EGR mass flow with high accuracy, the relation *f* in (2) should be experimentally indentified with a statistical model. And the process of identification is called calibration of the ECU-functions, which is displayed in Fig. 2.

There, with variation of the duty cycle, $DC_{\mathrm{EGR}}$, measurement data with the structure $\mathbf{D} := [\dot{m}_{\mathrm{EGR}}, T_{\mathrm{Us}}, p_{\mathrm{Ds}}, p_{\mathrm{Us}}]$ is collected. Based on the measurement data, **D**, measurement points for the calibration of the effective area, $\mathrm{MP} = [ar_{\mathrm{EGR,eff}}, \mathrm{Var}(ar), DC_{\mathrm{EGR}}]$, are produced through (1) (red stars). The variance, $\mathrm{Var}(ar)$, is determined by the rule of error propagation. With the measurement points, MP, a statistical model (blue curve) is fitted by the selected regression model and then converted into a curve/map. There are a number of such submodels in the ECU-functions, e.g. submodel for modeling the exhaust gas back pressure, volumetric efficiency, etc., which are connected through physical equations and should be calibrated. In this paper an automated and optimal calibration for the network of such submodels in the ECU-functions for diesel engine air system is investigated.

## 2 State of the Art Technique for Base-Calibration of ECU-Functions

In the calibration of submodels in the ECU-functions for air system, the grid experimental design has established itself as a useful approach, where the measurement points are generated by equidistant test plans made up of fuel quantity and engine speed with and without EGR. The corresponding experimental designs are demonstrated in Figs. 3 and 4 respectively.

Using the collected data, curves and maps are then directly calibrated. Meanwhile, other air actuators, for example, throttle valve, VGT, etc., are regulated

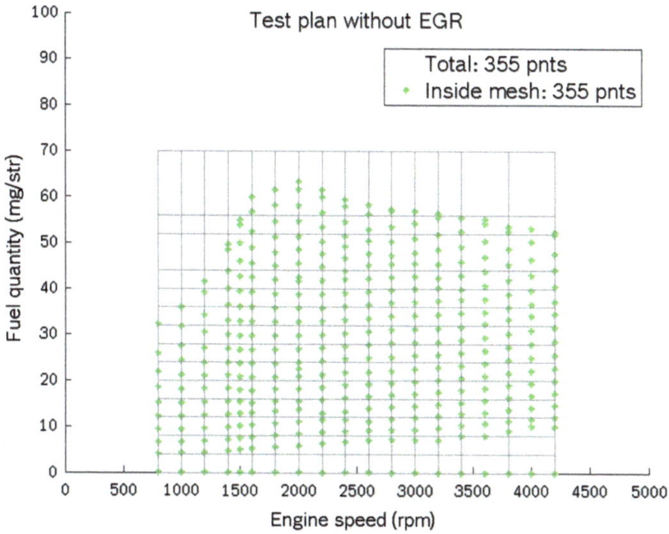


**Fig. 3** Equidistant experimental design without EGR for calibration of the air system

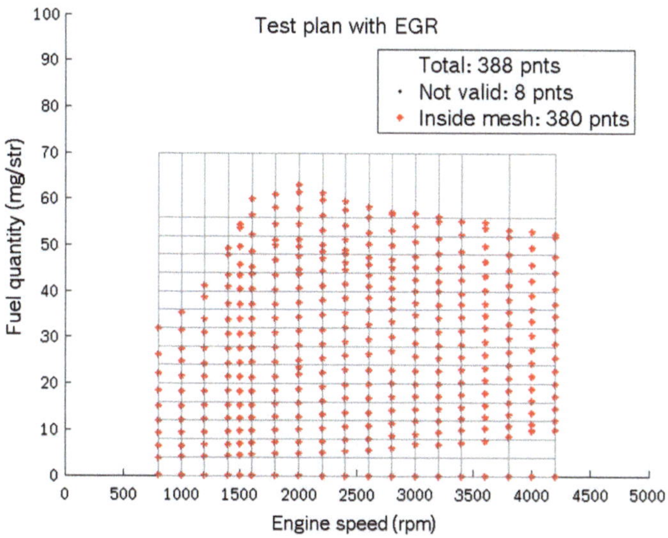


**Fig. 4** Equidistant experimental design with EGR for calibration of the air system

with a standard control system. While the equidistant experimental design is easy to implement, this simple approach does has a few drawbacks:

1. The equidistant experimental design is not "intelligent" enough, which can lead to a waste of the measurement points. It will be worse, when there are some local effects to be considered

2. In the equidistant experimental design, the physics of the system are not considered and utilized, which means that the tolerance of the measurement points are not considered or checked when the test plan is generated
3. Not all air actuators are exploited, which may mean that the coverage of the input space is not guaranteed.

# 3 Proposed Approach

First, the problem to solved is described in detail in Sect. 3.1. Then a systematic approach is presented in Sects. 3.2 and 3.3.

## 3.1 Description of the Problem

In order to provide a thorough understanding, the system to be calibrated is mathematically described in Fig. 5.

Three submodels (SM) are shown schematically, each submodel consists of physical equations, $f^{\mathrm{SM}}(\mathbf{X})$, with system variables, $\mathbf{X}$, and an empirical model, $\mathrm{EM}^{\mathrm{SM}}(u)$, with the input $u$. In addition, the three submodels are connected by physical equations. A submodel then can be represented with the following equations:

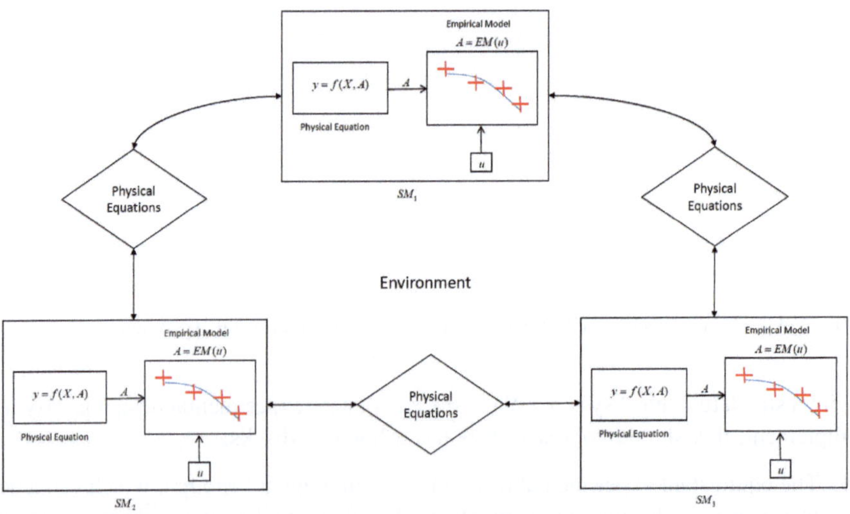


**Fig. 5** System consisting of several submodels as a network

$$\begin{aligned} y^{\mathrm{SM}} &= f^{\mathrm{SM}}(\mathbf{X}^{\mathrm{SM}}, A^{\mathrm{SM}}, Q^{\mathrm{SM}}) \\ A^{\mathrm{SM}} &= \mathrm{EM}^{\mathrm{SM}}(u^{\mathrm{SM}}) \end{aligned} \tag{3}$$

where the terms $y^{\mathrm{SM}}$ and $A^{\mathrm{SM}}$ respectively represent the output of the submodel and of the empirical model, $\mathrm{EM}^{\mathrm{SM}}$. The terms $Q^{\mathrm{SM}}$ are the environmental constants that have an impact on the corresponding submodel. The whole system can then be described with the following equation system:

$$\begin{aligned} y^{\mathrm{SM}_1} &= f^{\mathrm{SM}_1}\left(\mathbf{X}^{\mathrm{SM}_1}, \mathrm{EM}^{\mathrm{SM}_1}\left(u^{\mathrm{SM}_1}\right), Q^{\mathrm{SM}_1}\right) \\ y^{\mathrm{SM}_2} &= f^{\mathrm{SM}_2}\left(\mathbf{X}^{\mathrm{SM}_2}, \mathrm{EM}^{\mathrm{SM}_2}\left(u^{\mathrm{SM}_2}\right), Q^{\mathrm{SM}_2}\right) \\ y^{\mathrm{SM}_3} &= f^{\mathrm{SM}_3}\left(\mathbf{X}^{\mathrm{SM}_3}, \mathrm{EM}^{\mathrm{SM}_3}\left(u^{\mathrm{SM}_3}\right), Q^{\mathrm{SM}_3}\right) \\ 0 &= \mathbf{G}\left(y^{\mathrm{SM}_1}, y^{\mathrm{SM}_2}, y^{\mathrm{SM}_3}, \mathbf{X}^{\mathrm{SM}_1}, \mathbf{X}^{\mathrm{SM}_2}, \mathbf{X}^{\mathrm{SM}_3}\right) \end{aligned} \tag{4}$$

where the last set of equations, **G**, stand for the connections between the submodels. The aim of this work is to identify the empirical models of all submodels, namely automated and with little effort as possible. In summary, an optimal global test plan $U^*$, which can maximize the sum of the information contents for calibration of all submodels, should be constructed for the entire system. The corresponding optimization problem can be mathematically described with following formula:

$$U^* = \arg\max_{U} \sum_{i=1}^{m} I_{\mathrm{SM}_i}\left(U^{\mathrm{SM}_i}\right) \tag{5}$$

where $m$ is the number of submodels. The global experimental design $U$ is composed of local experimental design $U^{\mathrm{SM}_i}$ for single submodel $\mathrm{SM}_i$, namely $U = [U^{\mathrm{SM}_1}, U^{\mathrm{SM}_2}, \cdots, U^{\mathrm{SM}_m}]$. The function $I_{\mathrm{SM}_i}(U^{\mathrm{SM}_i})$ represents the information content, which is delivered by local experimental design $U^{\mathrm{SM}_i}$ for identification of single submodel $U^{\mathrm{SM}_i}$, and has yet to be defined. The information content of experimental design depends on the following two factors:

- The distribution of the measurement points of the experimental design in the input space. An experimental plan should possess more information when it covers the input space well and places relatively more measurement points, where there is a lot going on
- The variance of the generated data points, e.g. as $\mathrm{Var}(ar_{\mathrm{EGR,eff}})$ in Sect. 1.2. A measurement point should provide solid information when its variance is small.

When it comes to the calibration of a single submodel, e.g. submodel for modeling the EGR mass flow as described in Sect. 1.2, we have no control over the variance of the effective area, $\mathrm{Var}(ar_{\mathrm{EGR,eff}})$. In this case, we can achieve a calibration of good quality only by a reasonable distribution of the measurement points. However, when the whole air system, as a network of submodels, is to be calibrated, the duty cycle of the VTG can be adjusted in order to increase the pressure ratio at the EGR pipe. In this fashion, a smaller variance in the effective area of the EGR valve can be attained. This means that for the calibration of a network of submodels the information content of a local experimental design is influenced by inputs of all submodels, which requires a small change to the optimization problem to be solved in (5):

$$U^* = \arg\max_U \sum_{i=1}^{m} I_{SM_i}(U) \tag{6}$$

where $U = [U^{SM_1}, U^{SM_2}, \cdots, U^{SM_m}]$ holds. From the mentioned optimization problem in (6), the following three subtasks should be considered:

- Selection a suitable regression model for the calibration, see Sect. 3.2
- Creation of a local experimental design for the calibration of a single submodel, see Sect. 3.3
- Creation of a global experimental design for the calibration of network of submodels, see Sect. 3.3.

## 3.2 *Gaussian Process Regression for the Calibration of ECU-Functions*

In the identification of the empirical model, based on the experiment measurements, $\mathbf{D} := \{x_i, y_i | i \in \{1\ldots N\}\}$, a relationship between $x$ and $y$ should be investigated. In addition, $y = [y_1, y_2, \ldots, y_i]$ stands for the vector of the measured values and $N$ is the number of measurement points. In general, this problem can be formulated by the following formula:

$$y = f(x) + \varepsilon \tag{7}$$

where the term $\varepsilon$ is the variance of the normally distributed measurement noise. In recent years, a great interest in the *Gaussian Process Regression* (GPR) has developed, especially in the community of *Machine Learning*, [1–3]. Unlike other regression models, e.g. polynomial approach, *Feedforward Neural Network*, etc., at GPR the prior distribution is defined over the estimated function $\hat{f}$ itself, instead over some parameters. By definition in [2], a *Gaussian Process* (GP) is a stochastic process, in which every finite subset is normally distributed. With a GP as the prior distribution, the estimated function can be described as follows [2]:

$$\hat{f}(x) \sim GP(m(x), k(x, x')) \tag{8}$$

Here, the term $m(x)$ is the mean value function and $k(x, x')$ is the covariance function of the GP. Normally, a zero mean value function, $m(x) = 0$, is chosen, because the covariance function in GP is flexible enough to model any mean value function [2]. In *Kriging*, a procedure similar to GPR in the field of geography, linear function is used as a mean value function, see [4, 5] for further discussion. Now the specification of the covariance $k(x, x')$ is left to completely determine the GP. Many different covariance functions are proposed in the literature. In this work, the classic *Squared Exponential* (SE) covariance function is used:

$$k(x, x') = \sigma_{\mathrm{f}} \exp\left(\frac{||x - x'||^2}{-2l^2}\right) \tag{9}$$

because on the one hand, the hyperparameters $l$ and $\sigma_{\mathrm{f}}$ of the SE covariance function can reflect the property of the model and are easy to understand. Term $l$ is called the *Characteristic Length-Scale* and describes the horizontal strength of the correlation between two points, while $\sigma_{\mathrm{f}}$ affects the vertical behavior of the model. On another hand, the SE covariance is infinitely differentiable, so that the model can capture details of any high order from the data, which is a valuable property [6].

**Model Training and Prediction** When a normally distributed measurement noise $p(y|x) = \mathcal{N}(y|\hat{f}(x), \sigma_{\mathrm{n}})$ is assumed, the covariance of the measurement points $y$ is given by:

$$\mathrm{cov}(y) = K(x, x) + \sigma_{\mathrm{n}}^2 I \tag{10}$$

where $K(x, x)$ is the Gram matrix of the SE covariance function $k(x, x')$ with the input vector $x = [x_1, x_2, \cdots, x_{\mathrm{N}}]$. According to the definition of GP, the multivariate distribution of the measurement points $y$ and the predicted function value at the point $x^*$, $\hat{f}(x^*)$, can be written as the following [2]:

$$\hat{f}(x^*) \sim \mathcal{N}(m(x^*), \mathrm{cov}(x^*)), \tag{11}$$

where

$$m(x^*) = K(x^*, x)\left[K(x, x) + \sigma_{\mathrm{n}}^2 I\right]^{-1} y \tag{12}$$

$$\mathrm{cov}(x^*) = K(x^*, x^*) - K(x^*, x)\left[K(x, x) + \sigma_{\mathrm{n}}^2 I\right]^{-1} K(x, x^*) \tag{13}$$

The prediction can only be done if the hyperparameters $l$, $\sigma_{\mathrm{f}}$ in the SE covariance function and $\sigma_{\mathrm{n}}$ from the measurement noise are known. In order to estimate the hyperparameters, $\mathcal{H} = [l, \sigma_{\mathrm{f}}, \sigma_{\mathrm{n}}]$, the *Log-Likelihood* of it, $\log p(y|\mathcal{H})$ is minimized using the following formula:

$$\log p(y|\mathcal{H}) = -\frac{1}{2} y K_{\mathrm{y}}^{-1} y - \frac{1}{2} \log|K_{\mathrm{y}}| - \frac{N}{2} \log(2\pi) \tag{14}$$

where $K_y = K(x, x) + \sigma_n^2 I$. Another option for the model training of GPR is the use of the *Cross Validation Technique* (CV), which requires more computing time, see [2, 7] for further analysis.

### *3.3 Algorithm for Automated and Optimal Base-Calibration of ECU-Functions*

Now we are able to approximate the empirical model by statistical model (GPR). Then the question remains, where should the measurement points be positioned in the input space? In fact, a reasonable experimental design, which can maximize the information content for the approximation of the empirical models, is required.

**Calibration of a Single Submodel** Before dealing with the calibration of a network of submodels, first we focus on the creation of a local experimental design for the optimal calibration of a single submodel. Since no analytical fixed assumption is made about the relationship to be approximated in the GPR, for example, as ti is in the polynomial approach, it is impossible and meaningless to solve the optimization problem in (6) directly. Instead, the idea of sequential experimental design, which is based on the concept of "Active Sampling" (AS), [8, 9], emerges as a practical solution for this problem. During this process, the experimental design is not completely fixed from the start but is created step by step. Typically, an algorithm for sequential experimental design comprises three components: initial experimental design, active experimental design and automatic stopping criterion (SC). The mechanism of it is shown in Fig. 6.

Before further details of the algorithm are discussed, it is important to note that all calculations are carried out on the grid of the input space, $u_{\text{all}} = (u_{\min}, u_{\max})$, and repetition of the measurement points is not allowed.

In the initial experiment in Fig. 6, with a loose equidistant test plan, $U_0$, the initial GPM, $\text{GPM}_0$, is fitted, which offers an overview of the to be approximated process. Here, the variance of the $\text{GPM}_0$, $\text{Var}(\text{GPM}_0)$ is also estimated. Note the fact that the initial experimental design should cover the vertices and edges of the input space, so that no extrapolation is necessary.

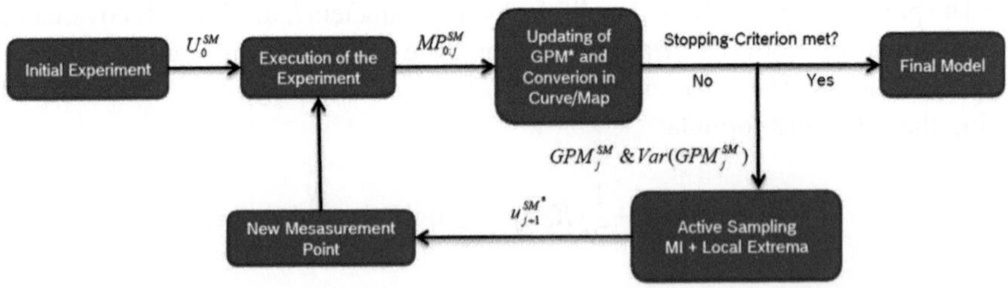


**Fig. 6** Schematic illustration of sequential experimental design for the calibration of a single submodel

After the initial experiment, the algorithm steps into a loop. In each loop, the optimal input for the next iteration is selected based on the current GPM and the developed rule for AS. For example, after j iterations, with the current GPM, $\mathrm{GPM_j}$, and its variance, $\mathrm{Var(GPM_j)}$, the input for the j + 1-th iteration, $u_{\mathrm{j}+1}$, should be selected based on the rule for AS. In this work, the following rule for AS is used:

$$u^*_{\mathrm{j}+1} = \begin{cases} \text{An unoccupied local extreme of the current GPM} \\ \arg \max\limits_{u_{\mathrm{j}+1} \subset (u_{\mathrm{all}} \backslash u_{0:\mathrm{j}})} I(u_{\mathrm{j}+1}), \quad \text{if all local extremes are occupied} \end{cases} \tag{15}$$

where the term $u_{\mathrm{all}} \backslash u_{0:\mathrm{j}}$ represents the set of all possible inputs in the input space for the j + 1-th iteration. That means, if there are still unoccupied local extremes at the current GPM, then the local extreme with the greatest curvature is chosen. The curvature of the current GPM is defined numerically by the following formula:

$$K(u) = \begin{cases} \left| \frac{\partial^2 \mathrm{GPM_j}(u)}{\partial u^2} \right| & \text{for 1 − dimensional model} \\ \left| \frac{\partial^2 \mathrm{GPM_j}(u)}{\partial u_1^2} + \frac{\partial^2 \mathrm{GPM_j}(u)}{\partial u_2^2} \right| & \text{for 2 − dimensional model} \end{cases} \tag{16}$$

where the term $u_1$ and $u_2$ stands for the two dimensions of the model. If all local extremes are already occupied, then all possible inputs are checked on a criterion, which is based on the concept of "Mutual Information" (MI). The term MI is derived from information theory and is used to describe the relationship between two random variables [6]. For example, for two continuous random variables, X and Y, the MI between them, MI (X: Y), is defined by the following formula [6]:

$$\mathrm{MI(X:Y)} = \int\limits_Y \int\limits_X p(x,y) \log\left(\frac{p(x,y)}{p(x)p(y)}\right) \mathrm{d}x\mathrm{d}y \tag{17}$$

In principle, MI(X: Y) describes how much the uncertainty about Y is reduced when X is known [10]. Active sampling, which is based on the MI, for the fitting of GPM is not new and already studied in [11]. With $\mathrm{MI(GPM_m : GPM_{um})}$, the reduction of the uncertainty of the unmeasured region can be quantified. Before we explain further details of the rule for AS, a few terms and designations are clarified by Fig. 7.

The Fig. 7 shows the basic situation of the algorithm by j-th iterations:

- Collected measurement points by j-th iteration: $\mathrm{MP}_{0:\mathrm{j}} = \left[u_{0:\mathrm{j}}, A_{0:\mathrm{j}}, \mathrm{Var}\left(A_{0:\mathrm{j}}\right)\right]$ (black stars)
- Current GPM: $\mathrm{GPM_j}$ (red curve) and error bars: $\mathrm{Var(GPM_j)}$ (green curves)
- The set of measured inputs: $u_{\mathrm{m}} = u_{0:\mathrm{j}}$, the set of unmeasured inputs: $u_{\mathrm{um}} = u_{\mathrm{all}} \backslash u_{0:\mathrm{j}}$
- The set of the estimated model values at the measured inputs, $\mathrm{GPM_m} = \mathrm{GPM}(u_{\mathrm{m}})$ and its variance $\mathrm{Var(GPM_m)}$

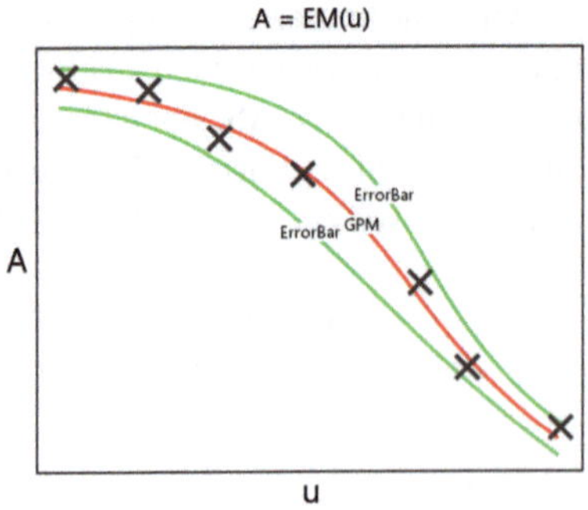


**Fig. 7** Basic situation for active experimental design by j-th iteration

- The set of the estimated model values at the unmeasured inputs, $\mathrm{GPM_{um}} = \mathrm{GPM}(u_{\mathrm{um}})$ and its variance $\mathrm{Var}(\mathrm{GPM_{um}})$

In GPR a normal distribution of the model value at the point $u$, $\mathrm{GPM}(u)$, is assumed, where the estimated mean value, $\mathrm{GPM}(u)$, and the standard deviation, $\sigma_{\mathrm{u}}$, are calculated using the formulas in (12) and (13). In addition, the estimated model values at the measured and unmeasured inputs, $\mathrm{GPM_m}$ and $\mathrm{GPM_{um}}$, can be treated as random variables with multivariate normal distribution. If a good approximation of the relation is the goal, a measurement point is considered optimal when this can lead to a maximum reduction of uncertainty of the unmeasured region. With all of the previous considerations the information content of a measurement point for the calibration of a single submodel, $\mathrm{MP_{j+1}} = \left[u_{\mathrm{j+1}}, A_{\mathrm{j+1}}, \mathrm{Var}\left(A_{\mathrm{j+1}}\right)\right]$, can be defined by the following formula:

$$I\left(u_{\mathrm{j+1}}\right) = \mathrm{MI}\left(\mathrm{GPM_m} \cup A_{\mathrm{j+1}} : \mathrm{GPM_{um}}\right) - \mathrm{MI}(\mathrm{GPM_m} : \mathrm{GPM_{um}}) \tag{18}$$

In summary the information content of the input, $I(u_{\mathrm{j+1}})$, describes the reduction of the uncertainty about the unmeasured region, if this input is selected. After derivation the following formula can be used to calculate the information content [11]:

$$I(u_{\mathrm{j+1}}) = \log\left[\frac{\mathrm{Var}(A_{\mathrm{j+1}}) - K(u_{\mathrm{j+1}}, u_{\mathrm{m}})K^{-1}(u_{\mathrm{m}}, u_{\mathrm{m}})K(u_{\mathrm{m}}, u_{\mathrm{j+1}})}{\mathrm{Var}(A_{\mathrm{j+1}}) - K(u_{\mathrm{j+1}}, u_{\mathrm{nm}})K^{-1}(u_{\mathrm{nm}}, u_{\mathrm{nm}})K(u_{\mathrm{mn}}, u_{\mathrm{j+1}})}\right] \tag{19}$$

where the term $K(u_1, u_2)$ stands for the covariance matrix using the hyperparameters from the current GPM. With the optimal input, $u^*_{\mathrm{j+1}}$, the new measurement point, $\mathrm{MP_{j+1}} = \left[u_{\mathrm{j+1}}, A_{\mathrm{j+1}}, \mathrm{Var}\left(A_{\mathrm{j+1}}\right)\right]$, is generated and added to the existing measurement points, $\mathrm{MP_{0:j}}$. Then the new GPM is fitted and is available for the j + 1-th iteration. This process will stop when the so-called automatic stopping criterion is met.

**Automatic Stopping Criterion** For the automation of the measurement process, a suitable automatic stopping criterion (SC) plays an important role. During the measurement, the GPM and the corresponding curve/map change constantly. If the

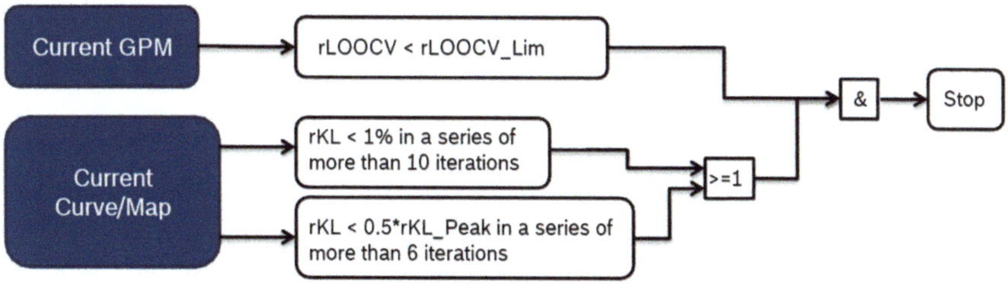


**Fig. 8** The logic in the automatic stopping criterion

algorithm is working properly, the change of the GPM and corresponding curve/map should be reduced step by step. A functioning SC should be able to recognize a stable phase of the change, where adding more measurement points no longer makes sense. Since the curve/map is the end product of the algorithm, at the SC of this work the change of the curve/map during the measurement process, rKL, is primarily investigated. As an additional source of information, the change of the GPM is used to recognize the extreme cases, e.g. overfitting of the measurement data. The SC used in this work is presented in Fig. 8.

The logic of the SC consists of two branches, which are respectively based on the current GPM and curve/map. In the first branch the relative leave-one-out cross-validation error (rLOOCV) is given by the following definition:

$$\mathrm{rLOOCV} = \frac{1}{N}\sum_{i=1}^{N}\frac{|A_\mathrm{i} - \hat{A}_{\mathrm{fit}\backslash\mathrm{i}}|}{|A_\mathrm{i}|} \tag{20}$$

Each time, a measurement point, $\mathrm{MP_i} = [u_\mathrm{i}, A_\mathrm{i}, \mathrm{Var}(A_\mathrm{i})]$, is excluded from the set of all collected $N$ measurement points and is no longer considered for the fitting of the GPM. The remaining $N - 1$ measurement points are then used to fit a new GPM, $\mathrm{GPM}_{\mathrm{fit}\backslash\mathrm{i}}$. Based on the $\mathrm{GPM}_{\mathrm{fit}\backslash\mathrm{i}}$ the model value at input $u_\mathrm{i}, \hat{A}_{\mathrm{fit}\backslash\mathrm{i}} = \mathrm{GPM}(u_\mathrm{i})$, is calculated and compared with the corresponding measured value $A_\mathrm{i}$. All measurement points will go through the same process, the resulting average relative error is then compared with a limit, which is defined with the following formula:

$$\mathrm{rLOOCV}_{\mathrm{lim}} = \frac{1}{N}\sum_{i=1}^{N}\frac{\sigma_{\mathrm{A_i}}}{|A_\mathrm{i}|} \tag{21}$$

where the term $\sigma_{\mathrm{A_i}}$ represents the standard deviation of the measured value $A_\mathrm{i}$. In the first branch it is checked whether the GPM is fitted correctly or not. If the limit is exceeded, there is a danger that overfitting or underfitting takes place. In the second branch, an attempt is made to detect a stable plateau regarding the change of the curve/map. The metric used is the relative change of the values at the support points of the curve/map in the course of measurement, rKL, as defined in the following formula:

$$rKL_j = \frac{|KL_j - KL_{j-1}|}{|KL_{j-1}|} \tag{22}$$

where the term $KL_j$ stands for the values at the support points in j-th iteration. To find a stable stage, in this work the following heuristic approach is used: A phase of the change is considered to be stable if one of following two conditions is satisfied:

- The relative change of KL, rKL, remains less than 1 % in a series of more than 10 iterations
- The relative change of KL, rKL, remains less than $0.5 * rKL_{Peak}$ in a series of more than 6 iterations, where the term $rKL_{Peak}$ represents the closest peak of the change.

**Calibration of a Network of Submodels** With the approach of sequential experimental design presented in Sect. 3.3 a single submodel can already be calibrated optimally. However, as analyzed in Sect. 3.1, the variance of the measurement points should be considered in the calibration of a network of submodels, therefore in this section, the approach for the single submodel will be expanded to accommodate this feature in the calibration for the network. Similar to the algorithm for a single submodel, the calibration of the network of submodels is also carried out in the framework of the sequential experimental design. The process of the approach consists of two phases and is shown in Fig. 9.

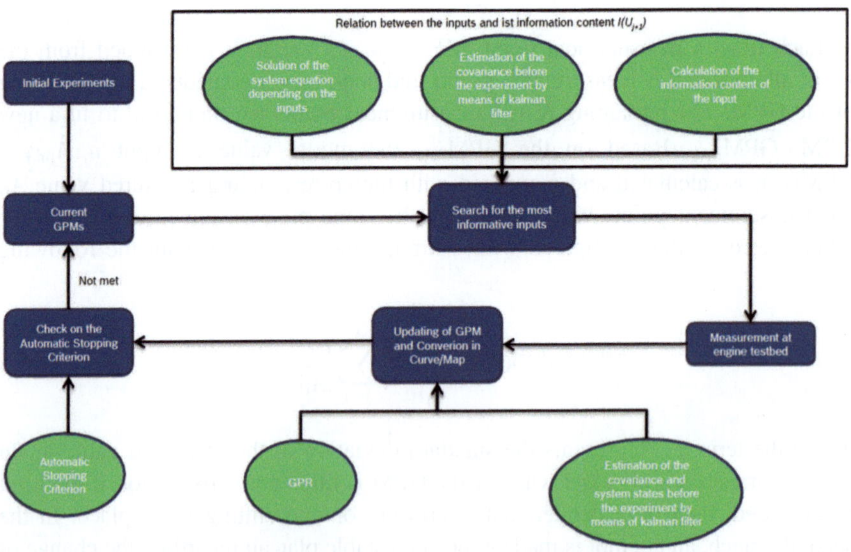


**Fig. 9** Schematic illustration of sequential experimental design for calibration of the network of submodels

- Phase 1 Initial Sampling: in this phase, the empirical models of all submodels are identified with equidistant initial test plans, $U_0 = [U_0^{\mathrm{SM}_1}, U_0^2, \cdots, U_0^{\mathrm{SM_m}}]$, where the index 0 stands for the 0-th iteration and the term $U_{\mathrm{SM}_1}^0$ represents the initial experimental plan for $\mathrm{SM}_1$. Based on experience, the equidistant experimental plan for the 1-D empirical model includes 5–10 measurement points, for 2-D empirical models a 4 × 4 grid should be a good choice. In summary, initial test plans for $m$ submodels should look like following matrix:

$$U_0 = \begin{pmatrix} u_{0,1}^{\mathrm{SM}_1} & u_{0,2}^{\mathrm{SM}_1} & \cdots & u_{0,k_0^{\mathrm{SM}_1}}^{\mathrm{SM}_1} \\ u_{0,1}^{\mathrm{SM}_2} & u_{0,2}^{\mathrm{SM}_2} & \cdots & u_{0,k_0^{\mathrm{SM}_2}}^{\mathrm{SM}_2} \\ \vdots & \vdots & \ddots & \vdots \\ u_{0,1}^{\mathrm{SM_m}} & u_{0,2}^{\mathrm{SM_m}} & \cdots & u_{0,k_0^{\mathrm{SM_m}}}^{\mathrm{SM_m}} \end{pmatrix} \tag{23}$$

Notably, the initial experimental plan should cover the entire input area so that in phase 2, no extrapolation in the empirical model is necessary. After the experiments shown in (23) are performed, all relevant system variables $\mathbf{X}$ and outputs $y$ of each submodel are collected to calculate the measurement points $A_0$ and their variances $\mathrm{Var}(A_0)$ at each empirical model. The matrice for measurement points $A_0$ and their variance $\mathrm{Var}(A_0)$ should have the same structure as the experimental design $U_0$ in (23).

$$A_0 = \begin{pmatrix} A_{0,1}^{\mathrm{SM}_1} & A_{0,2}^{\mathrm{SM}_1} & \cdots & A_{0,k_{\mathrm{SM}_1}^0}^{\mathrm{SM}_1} \\ A_{0,1}^{\mathrm{SM}_2} & A_{0,2}^{\mathrm{SM}_2} & \cdots & A_{0,k_{\mathrm{SM}_2}^0}^{\mathrm{SM}_2} \\ \vdots & \vdots & \ddots & \vdots \\ A_{0,1}^{\mathrm{SM_m}} & A_{0,2}^{\mathrm{SM_m}} & \cdots & A_{0,k_{\mathrm{SM_m}}^0}^{\mathrm{SM_m}} \end{pmatrix} \tag{24}$$

$$\mathrm{Var}(A_0) = \begin{pmatrix} \mathrm{Var}\left(A_{0,1}^{\mathrm{SM}_1}\right) & \mathrm{Var}\left(A_{0,2}^{\mathrm{SM}_1}\right) & \cdots & \mathrm{Var}\left(A_{0,k_{\mathrm{SM}_1}^0}^{\mathrm{SM}_1}\right) \\ \mathrm{Var}\left(A_{0,1}^{\mathrm{SM}_2}\right) & \mathrm{Var}\left(A_{0,2}^{\mathrm{SM}_2}\right) & \cdots & \mathrm{Var}\left(A_{0,k_{\mathrm{SM}_2}^0}^{\mathrm{SM}_2}\right) \\ \vdots & \vdots & \ddots & \vdots \\ \mathrm{Var}\left(A_{0,1}^{\mathrm{SM_m}}\right) & \mathrm{Var}\left(A_{0,2}^{\mathrm{SM_m}}\right) & \cdots & \mathrm{Var}\left(A_{0,k_{\mathrm{SM_m}}^0}^{\mathrm{SM_m}}\right) \end{pmatrix} \tag{25}$$

where the element $A_{0,1}^{\mathrm{SM}_1}$ in (24) represents the calculated measurement value for the empirical model $\mathrm{SM}_1$, when the input is $u_{0,1}^{\mathrm{SM}_1}$. Element $\mathrm{Var}\left(A_{0,1}^{\mathrm{SM}_1}\right)$ in (25) is the estimated variance. Based on this matrix for measurement points, $\mathrm{MP}_0 = [U_0, A_0, \mathrm{Var}(A_0)]$, for each empirical model a GPM, $\mathrm{GPM}_0 = \left[\mathrm{GPM}_0^{\mathrm{SM}_1}, \mathrm{GPM}_0^{\mathrm{SM}_2}, \cdots, \mathrm{GPM}_0^{\mathrm{SM}_m}\right]$, with error bars, $\mathrm{Var}(\mathrm{GPM}_0) = \left[\mathrm{Var}\left(\mathrm{GPM}_0^{\mathrm{SM}_1}\right), \mathrm{Var}\left(\mathrm{GPM}_0^{\mathrm{SM}_2}\right), \cdots, \mathrm{Var}\left(\mathrm{GPM}_0^{\mathrm{SM}_m}\right)\right]$, is fitted.

- Phase 2 Experiments with Active Sampling: This phase consists of several iterations, the number of iterations is not given a priori but is determined by the SC. In the j-th iteration a combination of the inputs from all $m$ empirical models, $U_j = \left[u_j^{\mathrm{SM}_1}, u_j^{\mathrm{SM}_2}, \cdots, u_j^{\mathrm{SM}_m}\right]$, is searched, which can maximize the sum of the information contents. The corresponding optimization problem for the j-th iteration is given by:

$$U_{j+1}^* = \arg\max_U \sum_{i=1}^{m} I_{\mathrm{SM}_i}\left(U_{j+1}\right) \tag{26}$$

To depict the relationship between the combination of the inputs of the current iteration, $u_j$, and the acquired information contents for the calibration of the sub-models $\mathrm{SM}_i$, namely the function $I_{\mathrm{SM}_i}(U_j)$, two steps are required (Fig. 10):

1. At first, based on the combination of the inputs, $U_j$, and the environment constants, $Q = [Q^{\mathrm{SM}_1}, Q^{\mathrm{SM}_2}, \cdots, Q^{\mathrm{SM}_m}]$, the system equation in (4) is solved with respect to the system variables, $\mathbf{X}_j^{\mathrm{SM}} = \left[\mathbf{X}_j^{\mathrm{SM}_i}, \mathbf{X}_j^{\mathrm{SM}_2}, \ldots, \mathbf{X}_j^{\mathrm{SM}_m}\right]$, and outputs, $y_j^{\mathrm{SM}} = \left[y_j^{\mathrm{SM}_1}, y_j^{\mathrm{SM}_2}, \cdots, y_j^{\mathrm{SM}_m}\right]$.

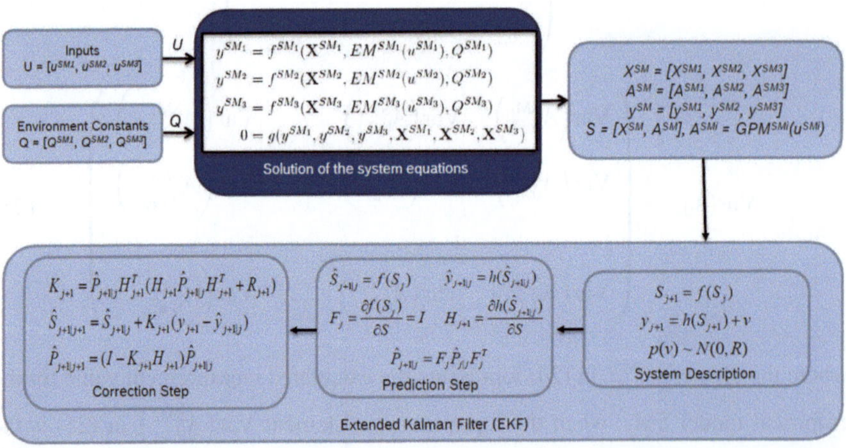


**Fig. 10** Schematic illustration of the process for determination of the relation between inputs and its information content

2. In this step, the variances of the outputs of all empirical models, $\mathrm{Var}(A_{\mathrm{j}}) = \left[\mathrm{Var}\left(A_{\mathrm{j}}^{\mathrm{SM}_1}\right), \mathrm{Var}\left(A_{\mathrm{j}}^{\mathrm{SM}_2}\right), \cdots, \mathrm{Var}\left(A_{\mathrm{j}}^{\mathrm{SM}_m}\right)\right]$, are predicted with an *Extended Kalman Filter* (EKF), before the experiments are carried out. Within the framework of EKF, the transition of the whole system from the j-th iteration to the j + 1-th iteration can be described with following formulas:

$$\mathbf{S}_{\mathrm{j}+1} = f(\mathbf{S}_{\mathrm{j}}) \tag{27}$$

$$y_{\mathrm{j}+1} = h(\mathbf{S}_{\mathrm{j}+1}) + V \tag{28}$$

$$p(\mathbf{V}) : \mathcal{N}(0, \mathbf{R}) \tag{29}$$

where the term $\mathbf{S}$ in (27) stands for the union of the system variables of all submodels, $\mathbf{X}$, and the measured values of empirical models, $A = \mathrm{GPM}(U)$, namely $\mathbf{S} = \mathbf{X} \cup A = \left[\mathbf{X}^{\mathrm{SM}_1}, A^{\mathrm{SM}_1}, \mathbf{X}^{\mathrm{SM}_2}, A^{\mathrm{SM}_2}, \cdots, \mathbf{X}^{\mathrm{SM}_m}, A^{\mathrm{SM}_m}\right]$. The Term $y = [y^{\mathrm{SM}_1}, y^{\mathrm{SM}_2}, \cdots, y^{\mathrm{SM}_m}]$ is the outputs of all submodels and (28) is the observation equation, where the term $\mathbf{R} = \mathrm{diag}[\mathrm{Var}(y^{\mathrm{SM}_1}), \mathrm{Var}(y^{\mathrm{SM}_2}), \cdots, \mathrm{Var}(y^{\mathrm{SM}_m})]$ is the covariance matrix of the measurement noise, $\mathbf{V}$. The estimation of the system variables, $\mathbf{S}$, and their covariance is carried out according to the calculation scheme in [10], with the following two steps:

(a) Prediction Step:

$$\hat{\mathbf{S}}_{\mathrm{j}+1|\mathrm{j}} = f(\mathbf{S}_{\mathrm{j}}) \tag{30}$$

$$F_{\mathrm{j}} = \frac{\partial f(\mathbf{S}_{\mathrm{j}})}{\partial \mathbf{S}} \tag{31}$$

$$\hat{y}_{\mathrm{j}+1|\mathrm{j}} = h\left(\hat{\mathbf{S}}_{\mathrm{j}+1|\mathrm{j}}\right) \tag{32}$$

$$H_{\mathrm{j}+1} = \frac{\partial h\left(\hat{\mathbf{S}}_{\mathrm{j}+1|\mathrm{j}}\right)}{\partial \mathbf{S}} \tag{33}$$

$$\hat{P}_{\mathrm{j}+1|\mathrm{j}} = F_{\mathrm{j}} \hat{P}_{\mathrm{j}|\mathrm{j}} F_{\mathrm{j}}^{T} \tag{34}$$

where $F_{\mathrm{j}}$ and $H_{\mathrm{j}+1}$ are respectively the Jacobi matrix for the motion Eq. (27) and the observation Eq. (28). Since in our application the steady-state properties of the system are studied, $F_{\mathrm{j}} = I_{|S|}$ is valid, where $|\mathbf{S}|$ is the number of system states.

(b) Correction Step:

$$K_{j+1} = \hat{P}_{j+1|j} H_{j+1}^{T} \left( H_{j+1} \hat{P}_{j+1|j} H_{j+1}^{T} + \mathbf{R}_{j+1} \right) \tag{35}$$

$$\hat{\mathbf{S}}_{j+1|j+1} = \hat{\mathbf{S}}_{j+1|j} + K_{j+1} \left( y_{j+1} - \hat{y}_{j+1|j} \right) \tag{36}$$

$$\hat{P}_{j+1|j+1} = \left( I - K_{j+1} H_{j+1} \right) \hat{P}_{j+1|j} \tag{37}$$

In the correction step, the system states $S_{j+1|j+1}$ and their covariance matrix $\hat{P}_{j+1|j+1}$ are optimally estimated, in the sense that the *mean square error* (MSE) of the estimator with following definition [12]

$$\mathrm{MSE}_{\mathrm{KF}} = E\left[ \left( \mathbf{S} - \hat{\mathbf{S}}_{j+1|j+1} \right)^{2} \right] \tag{38}$$

is minimized. Here, the term **S** represents the real system states. What must also be noted is that the calculation in (36) can yet not be performed, since the measurements, $y_{j+1}$, are still missing. However, the estimation of the covariance matrix, $\hat{P}_{j+1|j+1}$, is already feasible, which is also the key point of the entire algorithm.

From the estimated covariance matrix, $\hat{P}_{j+1|j+1}$, the estimated variances of the outputs of all empirical models, $\mathrm{Var}(A_j) = \left[ \mathrm{Var}\left( A_j^{\mathrm{SM}_1} \right), \mathrm{Var}\left( A_j^{\mathrm{SM}_2} \right), \cdots, \mathrm{Var}\left( A_j^{\mathrm{SM}_m} \right) \right]$, are extracted. If the estimated variances, $\mathrm{Var}(A_j)$, is used in the definition of the function of the information content in (19), the function, $I_{\mathrm{SM}_i}(U_j)$, in (26) results for the corresponding empirical model.

Using this function, the optimization problem of searching for the optimal combination of inputs for the next iteration, $U_{j+1}^{*}$, can be solved by a grid search or by more advanced optimization methods. Based on the optimal combination of inputs, $U_{j+1}^{*}$, the experiments can be performed and the measurement data can be collected. If we go through the calculations from (30) to (37), this time including (37), the system states, **S**, can be corrected on the basis of the measurement data. In this way, the GPM and its error bars of all empirical models can be updated and used in the next iteration. Phase 2 will continue until the predetermined SC is met at all empirical models.

## 4 Application and Results

In this section the algorithm developed in Sect. 3.3 is employed in a concrete example, namely, the calibration of the following three submodels in ECU-function for diesel engine air system:

**Table 1** Existing data set generated from the experiments with equidistant test plan

| $N_{\text{Griddata}_A}$ | $N_{\text{Griddata}_B}$ |
|---|---|
| 355 | 385 |

- Effective area of the throttle valve, $ar_{\text{TVA}}(DC_{\text{TVA}})$, for the modeling of pressure upstream of the throttle valve, $p_{\text{TVA,Us}}$
- Effective area of the EGR valve, $ar_{\text{EGR}}(DC_{\text{EGR}})$, for the modeling of EGR mass flow, $\dot{m}_{\text{EGR}}$
- Volumetric efficiency, $\lambda(q, n_{\text{Eng}})$, for the modeling of cylinder charge, $\dot{m}_{\text{Cyl}}$.

In fact, the algorithm is used to find the smallest set of measurement points in an existing data set, with which a calibration of equal or better quality for the above three submodels can be achieved. The existing data set are generated from the experiments with equidistant test plan with and without AGR, as presented in Sect. 2. In Table 1, the existing data set is described.

In data set A, the engine was measured without EGR and in data set B, with EGR. The algorithm for this example is demonstrated in Fig. 11.

In order to predict the variance of the measurement points, $\text{Var}(ar_{\text{TVA}}(DC_{\text{TVA}}))$, $\text{Var}(ar_{\text{EGR}}(DC_{\text{EGR}}))$ and $\text{Var}(\lambda(q, n_{\text{Eng}}))$, before the measurement is carried out, the system to be calibrated is described within the framework of the EKF as in the following formulas:

$$\mathbf{S}_{\text{j}+1} = f(\mathbf{S}_{\text{j}}) \tag{39}$$

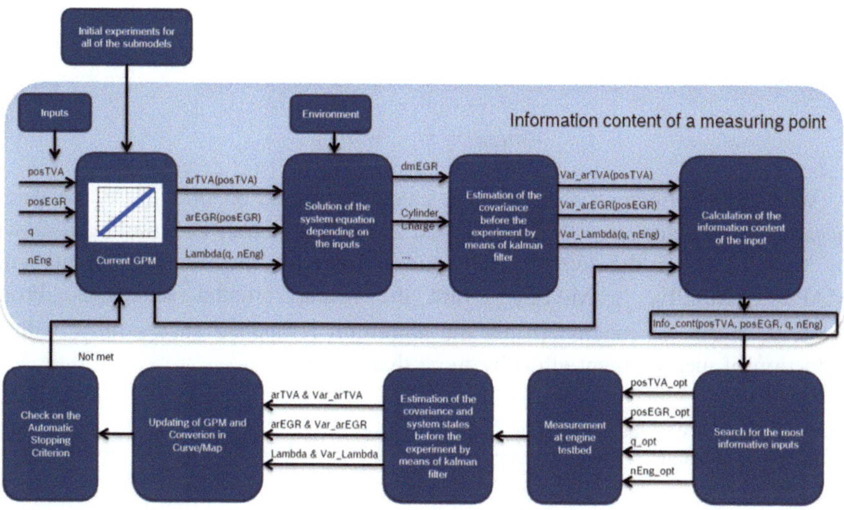


**Fig. 11** Schematic illustration of sequential experimental design for the calibration of a network of submodels

$$\dot{m}_{\mathrm{Air,j+1}} = \dot{m}_{\mathrm{Cyl,j+1}} - \dot{m}_{\mathrm{EGR,j+1}} + \mathrm{Var}(\dot{m}_{\mathrm{Air}}) \tag{40}$$

$$p_{\mathrm{TVA,Us,j+1}} = h\left(ar_{\mathrm{EGR}}\left(D_{\mathrm{EGR,j+1}}\right), \dot{m}_{\mathrm{EGR,j+1}}, p_{\mathrm{EGR,Ds,j+1}}, T_{\mathrm{TVA,Us,j+1}}\right) + \mathrm{Var}\left(p_{\mathrm{TVA,Us}}\right) \tag{41}$$

where

$$\dot{m}_{\mathrm{Cyl}} = \frac{p_{\mathrm{EGR,Ds}} \cdot f\left(\lambda\left(q, n_{\mathrm{Eng}}\right)\right)}{R_{\mathrm{Air}} \cdot T_{\mathrm{EGR,Ds}}} \tag{42}$$

and

$$\dot{m}_{\mathrm{EGR}} = \sqrt{\frac{2}{R_{\mathrm{Exh}} T_{\mathrm{EGR,Us}}}} \cdot ar_{\mathrm{EGR}}\left(DC_{\mathrm{EGR}}\right) \cdot p_{\mathrm{EGR,Us}} \cdot \Psi\left(\frac{p_{\mathrm{EGR,Ds}}}{p_{\mathrm{EGR,Us}}}\right) \tag{43}$$

The cylinder charge, EGR mass flow and pressure upstream of the throttle valve for the j + 1-th iteration, $\dot{m}_{\mathrm{Cyl}}, \dot{m}_{\mathrm{EGR}}$ and $p_{\mathrm{TVA,Us}}$, are modeled respectively by the ideal gas equation of state and throttle equation. Function $h$ in (41) represents a rearranged throttle equation.

Regarding the comparison concept, half of the existing data set is taken as a training set; the rest then serves as a validation set. Based on the training set, the desired curve/map, $ar_{\mathrm{TVA}}(DC_{\mathrm{TVA}}), ar_{\mathrm{EGR}}(DC_{\mathrm{EGR}})$ and $\lambda(q, n_{\mathrm{Eng}})$, are respectively calibrated by the standard method in Sect. 2 and the proposed approach. Then the model values, $p_{\mathrm{TVA,Us}}$, $\dot{m}_{\mathrm{EGR}}$ and $\dot{m}_{\mathrm{Cyl}}$, which are calculated based on the calibrated curve/map, are compared with the measurement values in the validation set. Notable is that, because there are no direct measurement points of the EGR mass flow in the validation set, the measured value of the EGR mass flow is generated with the following formula:

$$\dot{m}_{\mathrm{EGR}} = \dot{m}_{\mathrm{Cyl}} - \dot{m}_{\mathrm{Air}} \tag{44}$$

Where the cylinder charge, $\dot{m}_{\mathrm{Cyl}}$, is calculated based on (42). Here, the *mean absolute percentage error* (MAPE) is used as an error metric. The results are shown in Table 2.where the term $N_{\mathrm{MP}}$ is the number of measurement points and $\mathrm{MAPE}_{\dot{m}_{\mathrm{Cyl}}}, \mathrm{MAPE}_{p_{\mathrm{TVA,Us}}}, \mathrm{MAPE}_{\dot{m}_{\mathrm{EGR}}}$ are the reached model accuracies. From Table 2, it can be concluded that the algorithm is applied successfully, with a reduction of measurement effort by more than 70 %.

**Table 2** Summary of the results for calibration

| Method | Grid measurement | Proposed approach |
|---|---|---|
| $N_{\mathrm{MP}}$ | 365 | 93 |
| $\mathrm{MAPE}_{\dot{m}_{\mathrm{Cyl}}}$ (%) | 0.53 | 0.5 |
| $\mathrm{MAPE}_{p_{\mathrm{TVA,Us}}}$ (%) | 0.56 | 0.33 |
| $\mathrm{MAPE}_{\dot{m}_{\mathrm{EGR}}}$ (%) | 2.11 | 2.04 |

## 5 Conclusion

In this work, a systematic approach for an optimized and automated steady-state base-calibration of the ECU-functions in diesel engine air system is presented. Both the distribution of the measurement points in the experiment space and the uncertainty of the measurement points are taken into account, in order to create an optimal global experimental plan for the calibration of a network of submodels. While a AS rule based on MI ensures a reasonable distribution of the measurement points, the variance of the measurement points is predicted by EKF and included in the function of information content.

As an example application the algorithm is used to find the smallest set of measurement points in an existing data set, with which a calibration of equal or better quality can be achieved for the following three submodels; the submodel for the modeling of pressure upstream of the throttle valve, $p_{\mathrm{TVA,Us}}$, the submodel for the modeling of EGR mass flow, $\dot{m}_{\mathrm{EGR}}$, and the submodel for the modeling of cylinder charge, $\dot{m}_{\mathrm{Cyl}}$. As a result, with less than 30 % of the measurement points from the existing date set, a slightly better quality is achieved with the proposed approach.

## References


1. Gibbs, M.N.: Bayesian Gaussian processes for regression and classification. Dissertation, University of Cambridge (1997)
2. Rasmussen, C.E., Williams, C.K.I.: Gaussian processes for machine learning. The MIT Press, Cambridge (2006)
3. Mackay, D.J.C.: Gaussian processes: a replacement for supervised neural networks? Tutorial lecture notes for NIPS 1997, University of Cambridge (1997)
4. Ankenman, B., Nelson, B.L., Staum, J.: Stochastic Kriging for simulation metamodeling. Oper. Res. **58**(2), 371–382 (2010)
5. Costa, J.-P., Pronzato, L., Thierry, E.: A comparison between Kriging and radial basis function networks for nonlinear prediction. In: International Workshop on Nonlinear Signal and Image Processing, NSIP'99, Antalya, Turkey, Paper number: 155 (1999)
6. Bishop, C.M.: Pattern recognition and machine learning (Information science and statistics). Springer, Heidelberg (2007)
7. Chapelle, O.: Some thoughts about Gaussian processes model selection and large scale. In: Nips Workshop on Gaussian Processes. Max Planck Institute for Bioligical Cybernetics (2005)
8. Castro, R.M.: Active learning and adaptive sampling for non-parametric inference. Dissertation, Rice University (2007)
9. Chen Quin Lam, M.S.: Sequential adaptive designs in computer experiments for response surface model fit. Dissertation, The Ohio State University (2008)
10. Murphy, K.P.: Machine learning: a probabilistic perspective. The MIT Press, Cambridge (2012)
11. Krause, A., Singh, A., Guestrin, C.: Near-optimal sensor placements in Gaussian processes: theory, efficient algorithms and empirical studies. J. Mach. Learn. Res. **9**, 235–284 (2008)
12. Kalman, R.E.: A new approach to linear filtering and prediction problems. J. Basic Eng. **82**, 35–45 (1960)

# Part VIII
# Model-Based Test Benches

# Mobile Online Connectivity Test Center In-lab Emulation of Real-Word Connectivity Environments for Virtual Drive Tests

**Sondos Alaa El Din, Alexander Roy and Frank Klinkenberg**

**Abstract** In the face of rising connectivity services in the car, automotive manufactures need to perform field drive testing to ensure performance and reliability over the lifecycle of the vehicle. As the testing environment is hardly reproducible; troubleshooting of events which occur only randomly during drive testing scenarios typically requires extensive repetition of the field drive tests. These challenges intensify the need for a test solution to quickly and effectively deploy new online services. Virtual Drive Testing bridges the gap between lab and field testing through real-world emulation of network environment in the lab. The solution presented here comes with significant cost reductions to the development and deployment phases. It reduces the amount of field testing by using a repeatable lab-based testing methodology.

**Keywords** Virtual drive test · Field-to-lab testing · Connectivity · Real-world emulation

## 1 Introduction

In the mobile communications industry virtual drive testing is used in the early stage of development and deployment phases of mobile devices. This test methodology offers a common platform for sharing test environments that capture challenging radio conditions at key locations of interest in the actual operator's network, where performance is critical. By replicating all elements of the real RF propagation environment e.g. delay spread, Doppler, path loss, interference, as well

S. Alaa El Din (✉) · A. Roy · F. Klinkenberg
IAV GmbH, Rockwellstr 16, 38518 Gifhorn, Germany
e-mail: sondos.alaa.el.din@iav.de

A. Roy
e-mail: alexander.roy@iav.de

F. Klinkenberg
e-mail: frank.klinkenberg@iav.de



C. Gühmann et al. (eds.), *Simulation and Testing for Vehicle Technology*,
DOI 10.1007/978-3-319-32345-9_19

as multi-RAT and multi-cell handover, performance issues can be identified and resolved early in the development phases.

Virtual drive testing integrates actual drive-test logs to recreate a realistic mobile device performance testing in a controlled lab environment.

Accordingly, the key components of real-world emulation include a data source such as commercially available RF scanner or proprietary logging tools, a highly sophisticated parsing tool to map and filter scanner results for playback on specialized channel and network emulators, and, finally, a software tool for repeatable playback of field scenarios.

IAV has developed a processing and playback tool (IAV Connectivity Virtual Drive Test-Toolset) to test different automotive connectivity ECUs (Electronic Control Unit) across multiple geographies in a reproducible fashion as many times as required, in the lab, in a virtual environment. If an error occurs during a test drive, RF-Recorders, network loggers and DUT loggers save the real ambient conditions. The processed recording can then be used to reproduce and remedy the fault in the laboratory.

## 2 Field-to-Lab Testing Concept

A wireless signal may pass directly via line of sight (LOS) from the transmitting antennas to the receiving antennas. However, it can take multiple paths to the receiver by reflecting off buildings, vehicles, terrain, and other objects. Mobility between the transmitter and receiver causes the characteristics of these paths to be time-varying and results in rapid fluctuation of signal strength. The radio propagation effects can be characterized by fast fading, relative path delay, relative path loss, and slow shadow fading [1, 2]. Different mobile environments generate various combinations of these effects. This is standard propagation model at communications technology.

Traditional laboratory methodologies use a simulation environment consisting of channel and network emulation to verify the performance of mobile devices under harsh radio channel effects and various handover scenarios between multiple base stations in the network. However, industry-approved and standardized channel models do not capture the unique and specific conditions experienced by a device as it physically moves through an actual wireless network.

On the other hand, field testing may follow the same route each time. However, traffic as number of users in the network and—as well important—mobile transmission channel conditions as e.g. fading and Doppler effects may significantly change from drive to drive.

Virtual drive test or field-to-lab test techniques take real-world conditions into account and provide a repeatable environment for testing devices. Using this testing methodology, engineers can identify directly the faults occur in the field, even if it is in another country, and replay the scenario to evaluate the issue with sophisticated analysis tools. The same data can be used later to evaluate new devices under the same problematic conditions.

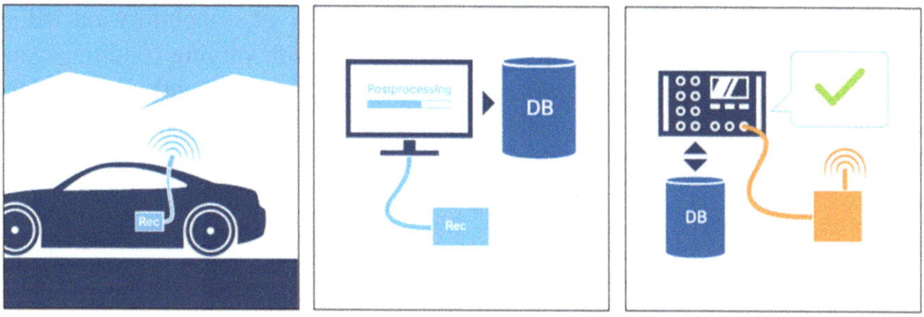


**Fig. 1** Sequence of capturing and postprocessing drive test logs for in-lab replay of ambient network and transmission channel conditions in the lab

IAV's Virtual drive test solution implements the concept of field-to-lab testing to effectively verify the performance of automotive ECUs under a controlled and repeatable environment. Figure 1 depicts the sequence of capturing and recreating real-world conditions in the lab. Once the data are captured during field test using commercial RF scanners, logging devices and diagnostic tools, postprocessing and analyzing algorithms derive the channel and network conditions to be replayed later in the lab. Scenarios with challenging network conditions are saved in a database to provide a platform of problematic conditions for further device tests.

Once a specific condition in a real-world deployment is determined a simulation environment of channel and network emulators (Fig. 2) are used for playback. As the laboratory equipment is complex, automation is a key component to accurately

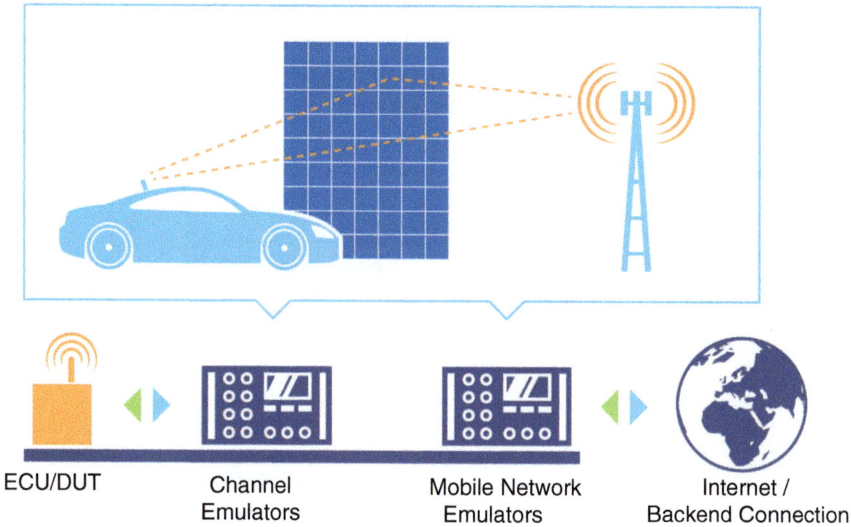


**Fig. 2** Simulation of real-world propagation effects and network conditions by highly automated channel and network emulators

replicate field mobility scenarios. Combining the effects of straightforward fading and multipath propagation with multiple base-stations with varying conditions, and real-time velocity changes—just as captured in the field—significantly increases the complexity of the test environment. And, more important, closely matches the simulation results to the real-world field scenario.

## 3 IAV Virtual Drive Test Toolset

Besides commercially available lab equipment like emulators and generators (see above) the Virtual Drive Test Toolset is the key component to enable end-to-end performance testing in the lab environment. The Toolset consists of a Processing and a Playback Tool. The Virtual Drive Test Processing Tool maps the RF data and network conditions captured from the live network to a channel emulator and extracts signaling information for configuring the network emulator. This section provides an overview of classifying radio frequency propagation environments and dynamically adapting playback and mapping of scanned data for testing automotive connectivity devices.

The general steps of the virtual drive test application can be summarized as follows (Fig. 3):

- Parsing raw drive log data: The captured raw scanner data must be parsed, processed and converted to a format which can be read by the channel and network emulators.
- Mapping data for simulation: Implementing mapping algorithm for selecting an appropriate subset of the scanned data for further analysis and simulation. Typically, a laboratory environment has fewer cells configured than were actually recorded during the drive test; therefore, some type of mapping must be used to transition that data to the simulation.
- Create output file: The output file should be a readable by the network and channel emulators for the playback functionality.

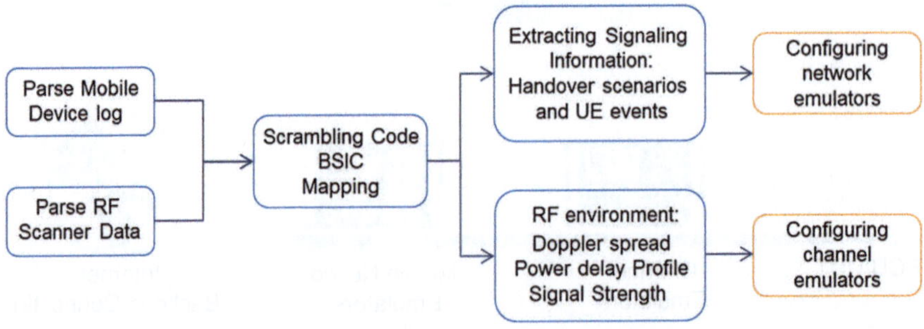


**Fig. 3** Block diagram of the IAV virtual drive test processing tool to determine parameters fed to the network and channel emulators

The captured statistical data by the scanners can be grouped based on the decoded scrambling codes (SC) and Base Station Identity Code (BSIC) which uniquely define a base station in UMTS/GSM system [3].

For each SC/BSIC a set of statistics should be parsed. A RF scanner measures signal strength, Received Signal Strength Indicator (RSSI), delay spread and a number of paths observed at that particular location and time for each scrambling code and BSIC.

The application supports the concept of a "hybrid" drive log capture, where logs of a mobile modem are used in conjunction with a RF scanner. While the scanner captures detailed information about RF environment, the mobile's modem captures signaling information recorded by a diagnostic tool such as:

- Serving cell data: that is the specific cell to which the user equipment (UE) is registered
- SIB: Information/Parameters for intra-frequency cell reselections, information on intra-frequency neighboring cells, Information on inter-frequency neighboring cells, information for reselection to GSM cells.
- Measurement Control messages
- Call events such as call setups and terminations

Due to hardware limitations (limited number of network emulators) the processing tool selects a subset of the scanned data for further analysis and simulation. Typically a subset of the acquired statistical data can provide an accurate characterization of the field test environment to evaluate performance of a wireless communication device. To play back and analyze the captured statistical data in a laboratory, the statistical data can be ranked and sorted based on the following signal properties:

- Descending power $E_c/I_0$ (the ratio of the received energy per chip to the total received power spectral density) level ratios. If the log contains more cells than simulated cells, those with the lowest power are not mapped.
- Signals from the "highest" ranked scrambling codes are mapped to the simulated cells.
- When a hybrid drive log with serving cell data is loaded, the application should ensure mapping the current serving cell to one of the simulated cells. The serving cell mapping may switch between different simulated cells in the final output. This mapping algorithm is evaluated at each state.

After analyzing and conditioning the data, the Drive Test Processing Tool generates a set of files that are ready for use with the laboratory equipment, i.e. the channel and network emulators. The abstracted data gathered from field measurements is transformed into a setting file. This file is used to feed the channel emulators to recreate the field conditions, including slow fading, fast fading, path delay and fading power spectral density. Different kind of handover scenarios [3], such as soft, softer, intra-band, hard, inter-band hard handover are simulated by the network emulators. A programming code is developed to automatically synchronize the laboratory equipment and initiate the network events in alignment to the simulated field conditions.

## 4 Measurement Campaign

A measurement campaign was conducted with a commercially available smart-phone as DUT. Logs pulled from the DUT contain lots of valuable information, from the various fading and multipath effects to how many base-stations or sectors are visible at any time. Also GNSS (GPS/GLONASS) information as location and velocity is reported. A mapping algorithm dynamically controls how the original scanner data is mapped to the number of mobile cellular cells supported in the simulation. GNSS recordings are replayed by a multi-GNSS constellation generator which reproduces genuine satellites signals. It emulates the vehicle and satellite motion, signal characteristics and atmospheric effects.

The IAV Connectivity Test Center (Fig. 4) is able to simulate up a high number of WCDMA/LTE and/or GSM/CDMA base-stations. It also supports the generation of a high number of hybrid and dynamic GNSS satellites signals. Besides Fig. 5 shows the virtual drive test playback tool which fed the channel and network emulators with parameters extracted by the virtual drive test Processing Tool. It also

**Fig. 4** IAV mobile connectivity test center

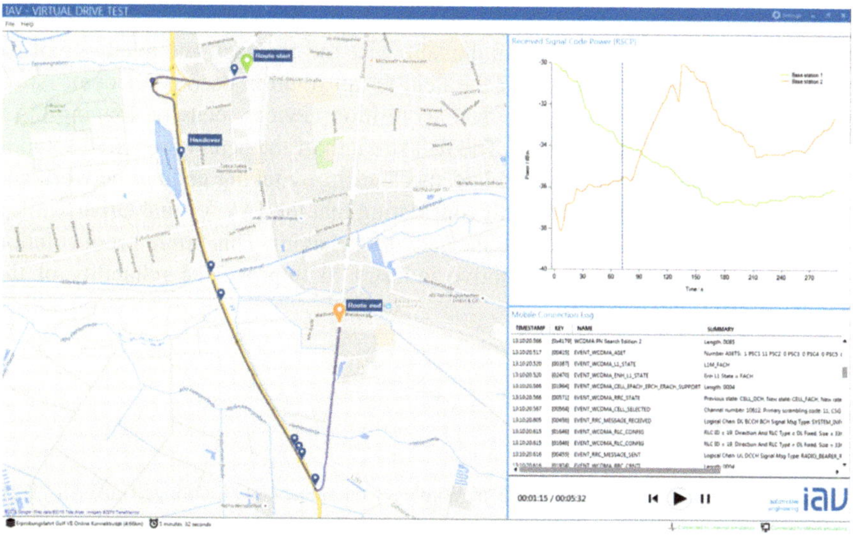


Fig. 5 IAV virtual drive test tool plays back recorded wireless scenarios and controlling channel and network emulator

depicts the drive route and received signal strength for the current simulation time. The map will also show the handover and the specific radio access technology (RAT) events if hybrid logs are processed.

The IAV Drive Test Playback Tool has the option to stop and replay the scenario. This option enables the test of the ECU's performance along a critical driving path as many times as required.

The powerful automation capability allows repetitive trialing of virtualized field test routes for different use case scenarios saved in the databases. Using a large representative set of drive logs in a field-to-lab solution makes it possible to verify the ECU's E2E performance under controlled environment.

## 5 Conclusions

Field-to-lab techniques provide a repeatable test environment for testing devices in an E2E-fashion to reproduce and verify not only sporadic field issues. The most indispensable part of the field-to-lab test methodology involves analyzing the drive test recordings and logs and conditioning the data for playback. A primary purpose of the processing section is to create a bridge between the simulations by approximating real-world conditions. Recordings of critical ambient conditions—taken on different drive routes from around the world—are collected in a scenario library that can be reused at all times.

The key advantage of this approach comes in the form of measured field results that can be repeated consistently in the lab. Once the accuracy and reliability of the system are proven, the benefits of the field-to-lab solution become evident.

Over recent years, automotive data acquisition devices focus on logging CAN messages and vehicle parameters. The real strength of the herein presented solution is expanding the conventional logging mechanism by taking cellular network and GNSS data into account. It is worth considering that the IAV Virtual Drive Toolset will contribute to more insight into the network behavior. This future-proof solution will ultimately enable the automotive industry to improve the reliability of the online connectivity services.

## References

1. Tse, D., Viswanath, P.: Fundamentals of Wireless Communication. Cambrige University Press (2005)
2. Rappaport, T.S.: Wireless Communictions: Princples and Practice, 2/E. Prentice Hall (2001)
3. Chevallier, C., Brunner, C., Garavaglia, A., Murray, K., Baker, K.: WCDMA (UMTS) Deployment Handbook Planing and Optimization Aspects. John Wiley and Sons Ltd (2006)

# High-Dynamic Accurate Railway Catenary Emulation by Real-Time Mechanical Impedance Control for Pantograph Testing

**Alexander Schirrer, Guilherme Aschauer and Stefan Jakubek**

**Abstract** The developments in modern hardware-in-the-loop testing tend towards realistically emulating the designated environment of the unit-under-test already in the laboratory. Using the example of an innovative high-speed pantograph test rig it is demonstrated how complex, distributed-parameters system dynamics (catenary), described via coupled partial differential equations can efficiently be transferred into a high-fidelity real-time-capable model. Careful modeling choices and efficient finite element codes provide these real-time capabilities. Then it is shown how the dynamics of the obtained model can be tracked by test rig actuators by using impedance control. A control concept is presented that not only leads to superior performance but also assures consistent energy and momentum exchange in hardware-in-the-loop testing, allowing for yet more realistic results. First experimental results are presented that validate the proposed concepts: impedance control results showcase the advanced catenary emulation functionality, and the efficiency of the real-time-enabled finite element code developments is analyzed. These building blocks deliver superior performance for their upcoming integration into the overall HiL control implementation.

**Keywords** Hardware-in-the-loop · Pantograph · Finite elements · Partial differential equations · Impedance control · Conserved quantities

This work was financially supported by the Austrian Research Promotion Agency (project 841331). The authors are grateful for the support of Siemens pantographs team (Siemens AG Austria, MO MLT BG PN).

A. Schirrer (✉) · G. Aschauer · S. Jakubek
Division of Control and Process Automation, Institute of Mechanics and Mechatronics, Technische Universität Wien, Getreidemarkt 9, 1060 Vienna, Austria
e-mail: alexander.schirrer@tuwien.ac.at

G. Aschauer
e-mail: guilherme.aschauer@tuwien.ac.at

S. Jakubek
e-mail: stefan.jakubek@tuwien.ac.at



C. Gühmann et al. (eds.), *Simulation and Testing for Vehicle Technology*,
DOI 10.1007/978-3-319-32345-9_20

# 1 Introduction

High ride velocities in railway operation are possible only if the complex dynamic interaction between the overhead catenary and the railway current collectors (pantographs) are carefully considered and engineered.

The coupled dynamics of these elastic systems is being excited to oscillations, and the current collector may lose contact (resulting in the formation of electric arcs and heavy wear), or high contact force peaks occur resulting in immediate damage of the pan head or the catenary.

To improve dynamic performance, modern high-speed current collectors are designed as elaborate mechatronic systems, possibly subjected to active control of the contact force, that require significant design, testing and calibration effort. Supportive tools are model-based simulations of the interaction, laboratory hardware-in-the-loop (HiL) tests at various complexity levels, and finally test rides on an actual track. Model-based simulation or HiL-testing, however, require to address the challenging distributed-parameter dynamics of the catenary: wave propagation effects, low damping, and system stiffness are the key aspects to be addressed appropriately.

In order to reduce design effort, a Siemens high dynamic pantograph test rig is being enhanced with real-time-emulation of accurate catenary dynamics in a joint development effort with the TU Wien. The goal is to minimize the expensive, actual track testing by providing realistic, reliable, and reproducible virtual test rides for pantographs in a lab environment.

To reach this goal, this work proposes two main contributions: firstly, the catenary dynamics is modeled in a detailed yet efficient way to enable real-time simulation and model-based control. Secondly, a model-based predictive test rig control concept is provided that accomplishes real-time catenary emulation at the highest possible level of accuracy and physical trustworthiness.

The catenary is a flexible structure comprised of a carrier wire supported by masts and holding a contact wire via so-called droppers. The relevant flexible components are the two wires (under axial tension) whose vertical dynamics can be well modeled by partial differential equations (PDEs) following the Euler-Bernoulli beam theory. Modeling approaches in literature range from simplified varying-stiffness mass-oscillator surrogate models over utilizing the string equation (disregarding bending stiffness) to the Euler-Bernoulli description. To solve the arising PDE problems, discretizations via finite differences [1] and via finite elements (FE) [2], as well as spectral formulations [3] have been proposed. Various studies cover high-accuracy simulations, and real-time usage in HiL testing applications has been reported with a spectral description [3]. A real-time-capable model based on finite differences (FD) for Euler-Bernoulli-based catenary representations in moving coordinates is proposed in Ref. [1]. In the present text, a novel real-time-enabled finite element description of such catenary in moving coordinates will be outlined.

As catenaries are spatially extended (essentially unbounded) and show wave propagation with low damping, truncating the computational domain leads to boundary effects that need to be considered to achieve trustworthy simulation results: to avoid spurious reflections from the (artificial) computational boundaries, special absorbing boundary formulations need to be implemented. To let waves travel out of the computational domain, absorbing boundary conditions (ABCs) have been proposed: Easy-to-use low-order ABCs for the wave equation were derived by Engquist and Majda [4]. Higdon developed ABCs in Ref. [5] for the discretized two-dimensional wave equation by starting from the discrete dispersion relation. However, both variants are only usable in their basic, low-order formulations because of the fact that their computational effort drastically increases with the order. Reference [6] gives a comprehensive review on high-order ABCs for different types of wave equations (e.g. dispersive or time-dependent). Analytic ABCs for complex equations such as the extended Euler-Bernoulli PDE treated here are not available, but a suitable, generic optimization-based ABC formulation has been proposed in Ref. [7]. In contrast to applying ABCs, an alternative approach to producing absorbing boundary behavior is to extend the computational domain by a special region with wave-absorbing properties, based on a complex change of coordinates, called perfectly matched layer (PML). Again, the analytical derivation of PMLs is so far only known for a few variants of PDEs and its use typically requires many additional equations that have to be solved at each time-step. In Ref. [8] it is demonstrated how the properties of PMLs can be efficiently emulated by the use of static state-feedback control. Formulating the catenary dynamics in moving coordinates and realizing absorbing boundary behavior allows to drastically reduce the necessary degrees of freedom for given accuracy requirements near the pantograph contact point.

The concept of emulating a dynamic behavior is called impedance control, introduced in Ref. [9] and widely used in close-to-reality HiL testing, e.g. Ref. [10]. The focus here lies on a smooth integration of complex virtual system dynamics that should be emulated. To reduce the phase lag intrinsic to most classical control concepts and to be able to naturally incorporate constraints directly in the control design, model predictive control is utilized. An optimization problem is solved in each sampling interval to determine the optimal future control moves. The first control move is applied to the system in a receding-horizon manner, i.e. the optimization is repeated at the next time step.

Besides typical tracking goals (simple impedance control would just track the outputs of the virtual system model as done in [3]), high-accuracy HiL testing results are improved if the ever-present imperfections (model errors, actuator limits) of the control system and test rig are compensated for. With a suitable mathematical formulation, the physical conservation quantities (energy and momentum) exchanged between the virtual system and the unit under test can be made consistent through control action. In Ref. [11] it was shown that physical trustworthiness (in terms of correct fuel consumption) in an automotive test rig application could be improved when the conservation of energy between the UUT and the virtual system model had been ensured by the controller. In the control concept outlined in the present paper, this multitude of goals is addressed simultaneously in the control law.

## 2 Problem Setup

In order to accomplish the realistic HiL testing of a full-sized pantograph (*unit under test*) in contact with a virtual, dynamic, high-fidelity model of a catenary (*virtual system environment*), the test rig realizing this contact must be controlled appropriately.

A high-dynamic HiL test rig setup for pantograph testing with real-time catenary emulation has been jointly developed further with Siemens (illustrated in Fig. 1). The actual pantograph is in contact with the test rig end effector. This contact force is measured, processed, and fed in an adapted form into the virtual catenary simulation model. Its response is tracked by a high-dynamic model-predictive *controller* and the pantograph contact point is actuated so as to accurately establish the dynamic coupling of the real pantograph and the virtual catenary systems. The tested pantograph literally *feels* the same dynamic response as if it would run along a physical catenary. Additional measures have been developed to ensure stability and the consistency of the conserved quantities (energy and momentum) exchanged between the pantograph and the catenary.

### 2.1 Catenary Modeling

A typical railway catenary system as depicted in Fig. 2 consists of a contact wire under axial tension which interacts with the train's pantograph and is connected via droppers to the carrier wire (also axially loaded). This carrier wire is mounted on inertia-fixed masts, and the droppers are typically modeled as one-sided springs with high tensile stiffness and zero stiffness when compressed.

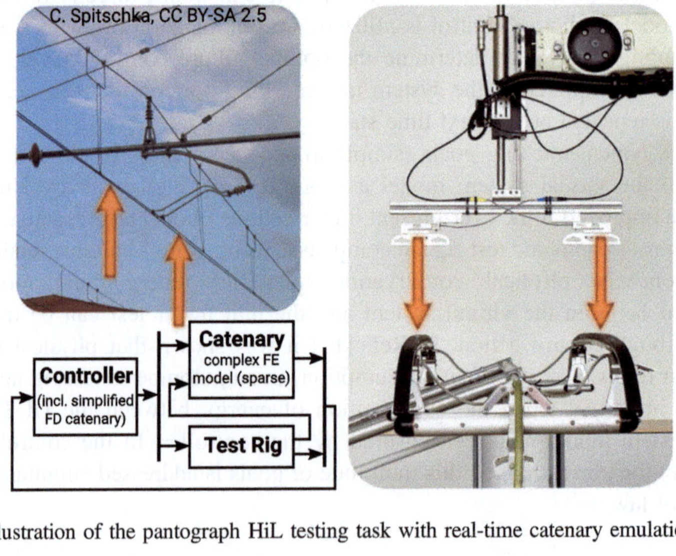


**Fig. 1** Illustration of the pantograph HiL testing task with real-time catenary emulation

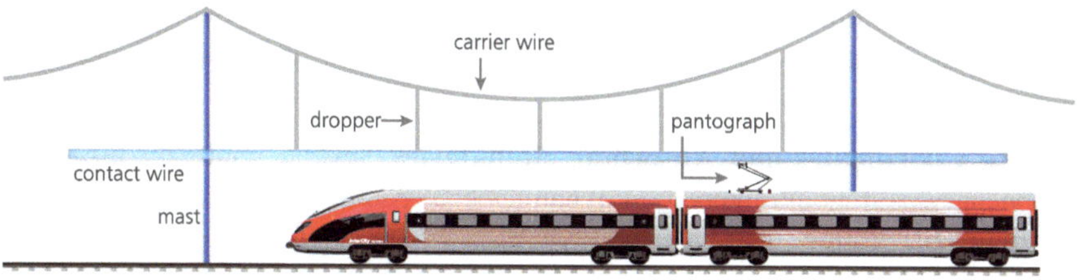


**Fig. 2** Schematics of a typical catenary configuration

The vertical dynamics of each wire can be modeled by the Euler-Bernoulli Beam (EBB) under axial pre-tension (see Ref. [12]):

$$\rho A \frac{\partial^2 \bar{w}}{\partial t^2} + \beta \frac{\partial \bar{w}}{\partial t} = -EI \frac{\partial^4 \bar{w}}{\partial x^4} + T \frac{\partial^2 \bar{w}}{\partial x^2} + \bar{f}, \tag{1}$$

where $\rho$ is the mass density, $A$ the cross section area, $\beta$ a velocity-proportional damping coefficient, $E$ Young's modulus of elasticity, $I$ the geometrical moment of inertia, and $T$ the axial pre-tension force. The vertical displacement field of the beam is denoted by $\bar{w}(x,t)$ and the distributed vertical force per unit length acting on the wire is denoted by $\bar{f}(x,t)$ (bars denote catenary-fixed coordinates). The spatial coordinate $x$ is counted horizontally on the undeformed wire (admissible for small vertical displacements), and $t$ denotes time.

The dropper coupling of these two wires can be realized via vertical force terms collected into $\bar{f}(x,t)$: with the deformed dropper length $l_d(t) = w_{ca}(x_d,t) - w_{co}(x_d,t)$ of the dropper at position $x_d$ (co and ca indicate contact and carrier wire, respectively), the one-sided dropper force is typically modeled as

$$F_d(t) = \begin{cases} k_d\left(l_d(t) - l_{d,0}\right) & \text{if } (l_d(t) - l_{d,0}) > 0 \\ 0 & \text{otherwise} \end{cases}, \tag{2}$$

where $k_d$ is the tensile dropper stiffness and $l_{d,0}$ denotes the undeformed dropper length. Then, the corresponding force density coupling terms are

$$\bar{f}_{d,co}(x,t) = -\bar{f}_{d,ca}(x,t) = \delta(x - x_d)F_d(t), \tag{3}$$

with the Dirac delta distribution $\delta$ or a suitable regularization function [12]. Likewise, the force acting at the pantograph's position is considered via a contribution to the contact wire's forcing:

$$\bar{f}_{panto,co}(x,t) = \delta(x - x_{panto,0} - vt)F_{panto}(t). \tag{4}$$

As the catenary shows distributed-parameter dynamics represented by coupled PDEs, appropriate approximation methods are needed to numerically solve the equations of motion over time to perform simulation or model-based predictive control. Finite difference and finite element methods will be employed (and extended with respect to their real-time-capabilities) to arrive at the necessary models for the HiL testing task.

**Moving Coordinates and Absorbing Boundaries** Several important aspects in the catenary/pantograph interaction dynamics give rise to significant challenges for real-time HiL testing. The catenary dynamics is low-damped, shows high wave propagation velocities, and also high train velocities should be tested. The simple choice of catenary-fixed coordinates is problematic in this setting: Long segments of the catenary need to be modeled, and in order to allow realistic or unlimited testing durations, special formulations are required (for example, to repeatedly shift the computational domain to keep the pantograph near the domain's center [3]). Additionally, when truncating the essentially unbounded catenary to obtain a manageable computational domain size, simple (clamped) boundary conditions produce spurious reflections of wave components at the truncation points (these wave components would otherwise leave the truncated domain). Because of the small system damping, these reflected waves do not die out and eventually destroy physical trustworthiness of the simulated contact point interaction.

As a novel solution to these modeling challenges, a novel catenary representation in *moving coordinates* (i.e. pantograph-fixed), combined with efficient *absorbing boundary* formulations turn out to produce highly efficient catenary models. The modified catenary model needs to consider only a piece of the catenary moderately extended forward and backward around the pantograph position. This approach has the potential to maximize model quality for given computational real-time requirements. The authors have developed such novel representation for the catenary/pantograph interaction problem [1] by introducing the moving spatial coordinate $\xi = x + v\,t$ where $v$ denotes the (constant) train velocity. The PDEs are rewritten into the new coordinates $(\xi, t)$ which gives rise to mixed-derivative terms and direction-dependent wave propagation speeds. Now, the dropper and mast positions move over time (producing a time-varying model structure), and the pantograph position lies at a constant coordinate $\xi_{\text{panto}}$.

The corresponding beam equation, written in the moving coordinates $(\xi, t)$ for the displacement field $w = w(\xi, t)$ subject to the vertical force density $f(\xi, t)$ reads

$$\rho A \frac{\partial^2 w}{\partial t^2} + \beta \frac{\partial w}{\partial t} = -EI \frac{\partial^4 w}{\partial \xi^4} + (T - \rho A v^2) \frac{\partial^2 w}{\partial \xi^2} + \beta\, v \frac{\partial w}{\partial \xi} + 2\, v \rho A \frac{\partial^2 w}{\partial t\, \partial \xi} + f \quad (5)$$

in the domain

$$-L_{\text{PML}} - \frac{L}{2} \leq \xi \leq \frac{L}{2} + L_{\text{PML}}, 0 \leq t. \quad (6)$$

Thereby, the domain of interest is of length $L$, and boundary layers of thickness $L_{\mathrm{PML}}$ have been added. At the boundaries of this extended domain, simple clamped boundary conditions are defined:

$$w(\xi = -L_{\mathrm{PML}} - \frac{L}{2}, t) = w(\xi = L_{\mathrm{PML}} + \frac{L}{2}, t) = 0, \tag{7}$$

$$\frac{\partial w}{\partial \xi}(\xi = -L_{\mathrm{PML}} - \frac{L}{2}, t) = \frac{\partial w}{\partial \xi}(\xi = L_{\mathrm{PML}} + \frac{L}{2}, t) = 0. \tag{8}$$

The boundary layers are steered by a control law as developed in [8] to absorb outgoing waves.

## 2.2 Test Rig Modeling

A high-dynamic HiL test rig has been developed as depicted in Fig. 3 (joint development of the authors with Siemens AG Austria, MO MLT BG PN). It consists of a six-degree-of-freedom industrial robot with a large operating range and a linear drive attached to the robot's end effector that can impose high-frequency displacement trajectories. With this setup and a proper division of the movements into low-frequency 3D displacements (tracked by the robot) and high-frequency small-scale vertical movements (executed by the linear drive) both advantages can be combined, resulting in a large operation range with high dynamic maneuvers.

At the linear motor slider's end two force sensors are attached that are in contact with the pantograph (the UUT). In this work, the robot is assumed rigid and the catenary behavior is emulated with the linear drive only. It is being controlled by a

**Fig. 3** Pantograph HiL test rig (Siemens AG Austria, MO MLT BG PN)

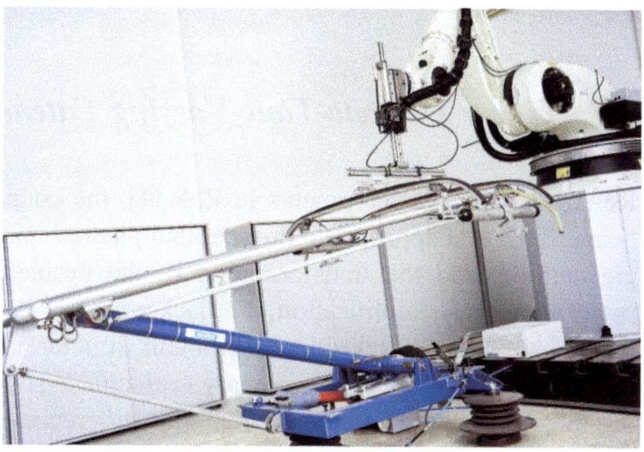

fast position tracking controller (including, for example, compensation of friction and disturbance forces) and accessed by the high-level test rig controller via a demanded position signal $x_{\mathrm{tr,dmd}}$. Its dynamics is assumed to be given as discrete-time state space model

$$\boldsymbol{x}_{\mathrm{tr},n+1} = \mathbf{A}_{\mathrm{tr}}\boldsymbol{x}_{\mathrm{tr},n} + \mathbf{B}_{\mathrm{tr}}x_{\mathrm{tr,dmd},n}, \tag{9}$$

$$\begin{bmatrix} x_{\mathrm{tr},n} \\ \dot{x}_{\mathrm{tr},n} \end{bmatrix} = \mathbf{C}_{\mathrm{tr}}\boldsymbol{x}_{\mathrm{tr},n}. \tag{10}$$

# 3 Methodology

Two main challenges are identified in the HiL testing task outlined in Sect. 2:

- The catenary dynamics must be simulated at a high detail level in real time.
- Appropriate control must act on the test rig and possibly on the catenary model to establish a physically correct, stable, high-accuracy coupling of the two systems, (virtual) catenary and (real) pantograph.

This chapter first focuses on providing methods to construct real-time-enabled catenary models: disregarding the nonlinear dropper slackening effects, a linear time-varying state space model is obtained via the finite difference method. This model will be suitable as design model for linear model predictive control (MPC). However, for maximum model accuracy, the actually simulated virtual catenary model is proposed to be modeled by finite elements with specific optimizations for real-time computation. Subsequently, prior and current proceedings on the test rig control task are summarized and a novel test rig system architecture for high-fidelity HiL test rig control is proposed.

## *3.1 Linear Discrete-Time-Varying Catenary Model via FD*

Referring to the developments in Ref. [1], the coupled system of Euler-Bernoulli beam PDEs and appropriate dropper and pantograph coupling terms (see Sect. 2.1) are defined, and the transformation of the problem into moving coordinates is performed. As a simplification, the slackening of the droppers is disregarded for this model, so that the considered problem structure becomes linear. Additionally, geometric periodicity of the catenary is assumed. The resulting PDE system is being discretized by the method of finite differences in space and time (using appropriately chosen equidistant step sizes $\Delta\xi$ and $\Delta t$), yielding a linear, periodically time-varying, time-discrete system of the form

$$\boldsymbol{x}_{n+1} = \mathbf{A}_n \boldsymbol{x}_n + \boldsymbol{B}_n F_{\text{panto},n} \tag{11}$$

$$\begin{bmatrix} w_{\text{panto},n} \\ \dot{w}_{\text{panto},n} \end{bmatrix} = \mathbf{C}\boldsymbol{x}_n. \tag{12}$$

where the state vector is $\boldsymbol{x}_n = \begin{bmatrix} \boldsymbol{w}_n^{\mathrm{T}} & \boldsymbol{w}_{n-1}^{\mathrm{T}} \end{bmatrix}^{\mathrm{T}}$, and $\boldsymbol{w}$ approximates the displacement field at the spatial grid points at time step $n$. The coefficients are time-varying but periodic due to the catenary's periodicity. The input force $F_{\text{panto},n}$ acts at the pantograph position.

Utilizing a linear model eases MPC implementation and, in particular, reduces computational load of the on-line MPC computations, but it introduces a model error between the high-fidelity catenary model and the MPC's design model that needs to be considered in controller tuning.

## *3.2 Real-Time-Capable Finite Element Model*

The finite element (FE) method [13] is well known and widely used as a flexible and versatile modeling tool to formulate, approximate, and solve complex distributed-parameter system dynamics. Here, in the context of real-time-capable test rig impedance control with distributed-parameter virtual system dynamics, the dynamics will be modeled by a real-time-specific FE approach.

**General Approach** The FE method approximates a distributed-parameter problem by dividing the problem domain into numerous subdomains, and in each of these finite elements, the governing PDEs are considered in a weak form only. The dependent field variables are approximated by a set of local Ansatz-functions, and their linear combination constitutes the approximated solution. For problems in structural mechanics, typically the spatial domain is being discretized and the Ansatz-functions are functions of the spatial coordinates. For the present problem of modeling catenaries (in one spatial dimension), extensions of the well-known Euler-Bernoulli (Hermitian) beam element have been derived by the method of weighted residuals.

After spatial discretization and re-assembly of all element equations, a set of ordinary differential equations (ODEs) is obtained for the generalized displacements $\boldsymbol{q}$ of the FE representation:

$$\mathbf{M}(t)\ddot{\boldsymbol{q}}(t) + \mathbf{C}(t)\dot{\boldsymbol{q}}(t) + \mathbf{K}_q(t)\dot{\boldsymbol{q}}(t) = \boldsymbol{f}(t). \tag{13}$$

Thereby, all coefficient matrices (mass matrix $\mathbf{M}$, damping matrix $\mathbf{C}$, and stiffness matrix $\mathbf{K}_q$) are generally time-varying (due to shifting elastic dropper couplings moving through the domain with velocity $-v$). The stiffness matrix $\mathbf{K}_q$ additionally depends on the dropper elongations $l_{\mathrm{d},i}(t) - l_{\mathrm{d},0,i}$ (hence on $\boldsymbol{q}$),

$$\mathbf{K}_q(t) = \mathbf{K}_0(t) + \sum_i \mathbf{K}_i(t)\sigma(l_{\mathrm{d},i}(t) - l_{\mathrm{d},0,i}), \tag{14}$$

where $\sigma$ denotes the Heaviside step function. Typically, the catenaries are considered periodic, so the coefficient matrices are also periodic. The methods presented here, however, do not require periodicity and could be implemented with increased data management effort.

**Real-Time Capability by Sparse Implicit Newmark Integration** To solve the system dynamics (13) forward in time, numerical integration procedures have to be utilized. It turns out that the ODE system has several key features that need to be addressed to efficiently solve it under real-time requirements:

- the ODE system is piecewise linear (one-sided elastic dropper couplings),
- it is of high order, and its coefficient matrices are only sparsely populated,
- these coefficients are periodically time-varying, and
- the ODE system is moderately stiff due to high mechanical stiffness of the dropper constraints.

**Explicit integration schemes** can directly exploit problem sparsity, and explicit single-step methods directly produce a set of explicit discrete-time difference equations that can directly be incorporated into model predictive control design. Moreover, time-varying coefficients can easily be incorporated. However, these methods are typically unsuitable for stiff problems because their stable time step sizes quickly become too small. Expressing the state-space dynamics in a slower time base (time step aggregation) destroys sparsity and becomes cumbersome when the system matrices have to be updated (e.g., when dropper couplings turn inactive). It is noted that for $n$ equations in (13), an explicit sparse time step costs $\mathcal{O}(n)$.

**Implicit integration schemes**, on the contrary, allow larger stable time steps especially for stiff systems, and some schemes are even unconditionally stable so that the choice of time step size needs only be based on the accuracy vs. computational speed trade-off. In implicit schemes, a system of equations needs to be solved in each time step, which is typically more time-consuming than explicit computations. If, however, sparsity is exploited in all computation steps, linear runtime order $\mathcal{O}(n)$ per time step is attainable for certain sparsity patterns. In combination with an arbitrary time step size, this represents a significant advantage for real-time computation.

One suitable and widely used family of numeric integration methods is the so-called (family of) **Newmark integration** method(s). It reliably solves dynamic problems of the specific form (13), even if they are moderately stiff, and sparsity can be exploited to improve computational efficiency. The method is parametrized by the real-valued parameters $(\gamma, \beta)$ with $0 \leq \gamma \leq 1$, $0 \leq \beta \leq 1$. For $\gamma \geq 0.5$ and $\beta \geq \frac{(2\gamma+1)^2}{16}$, the scheme becomes **unconditionally stable**. [13]

First, it is assumed that the displacements and velocities at the next time step $t_{n+1} = t_n + \Delta t$ satisfy

$$\boldsymbol{q}(t_{n+1}) = \boldsymbol{q}(t_n) + \Delta t \dot{\boldsymbol{q}}(t_n) + (\Delta t)^2 \left[ \left( \frac{1}{2} - \beta \right) \ddot{\boldsymbol{q}}(t_n) + \beta \ddot{\boldsymbol{q}}(t_{n+1}) \right], \tag{15}$$

$$\dot{\boldsymbol{q}}(t_{n+1}) = \dot{\boldsymbol{q}}(t_n) + \Delta t [(1-\gamma) \ddot{\boldsymbol{q}}(t_n) + \gamma \ddot{\boldsymbol{q}}(t_{n+1})], \tag{16}$$

which can be interpreted physically, for certain parameter choices, for example as constant-average-acceleration method or as linear-acceleration method. [13]

Inserting into the ODE system Eq. (13), the new accelerations can be computed as

$$\ddot{\boldsymbol{q}}(t_{n+1}) = \mathbf{K}_{cm}^{-1} \tilde{\boldsymbol{f}}(t_{n+1}), \tag{17}$$

with

$$\mathbf{K}_{cm} = \mathbf{K}_q \beta (\Delta t)^2 + \mathbf{C} \gamma \Delta t + \mathbf{M}, \tag{18}$$

$$\begin{aligned} \tilde{\boldsymbol{f}}(t_{n+1}) = & \boldsymbol{f}(t_{n+1}) - \mathbf{C}[\dot{\boldsymbol{q}}(t_n) + \Delta t (1-\gamma) \ddot{\boldsymbol{q}}(t_n)] \\ & - \mathbf{K}_q \left[ \boldsymbol{q}(t_n) + \Delta t \dot{\boldsymbol{q}}(t_n) + (\Delta t)^2 \left( \frac{1}{2} - \beta \right) \ddot{\boldsymbol{q}}(t_n) \right]. \end{aligned} \tag{19}$$

Then, (15) and (16) yield the new displacements and velocities. Note that with sparse problem data $(\mathbf{M}, \mathbf{C}, \mathbf{K}_q)$, matrix $\mathbf{K}_{cm}$ remains sparsely populated and all operations to construct $\tilde{\boldsymbol{f}}(t_{n+1})$ are multiplications of sparse matrices with vectors. Consequently, the runtime complexity of this method is determined by the attainable complexity in solving (17), typically between $\mathcal{O}(n)$ and $\mathcal{O}(n^2)$. Time-dependency and piecewise-state-dependency of the problem data is accounted for by constructing (or efficiently updating) the problem matrices in each time step.

### 3.3 *Impedance Control*

Impedance control was introduced in 1984 by Hogan [9] as the successor to active stiffness control in robotics, where a position-dependent force was formulated as a control reference (or vice versa). The goal of impedance control is to mimic a dynamic behavior on an arbitrary actuator by tracking a dynamic relationship rather than a pre-defined reference trajectory by controlling the position and velocity difference between the test rig and the impedance model to zero:

$$e_{\text{pos}} = x_{\text{im}} - x_{\text{tr}}, \quad e_{\text{vel}} = \dot{x}_{\text{im}} - \dot{x}_{\text{tr}}. \tag{20}$$

Typically, two different implementations are possible, called cascaded impedance control (CIC) and integrated impedance control (IIC), both depicted in Fig. 4, also see Ref. [10].

Both structures differ in the way the impedance model is treated in the control design. In CIC a reference tracking control is realized independent of the utilized impedance model. It offers great flexibility in terms of modularity since one control design can be reused with different impedance models and even on-line changes are possible. However, this method suffers from an intrinsic phase-lag since plain trajectory tracking control is used.

In IIC the test rig as well as the impedance model are already contained in the design plant of the controller and there is no longer a reference trajectory to be tracked, but the output difference should be controlled to zero. The controller is thus aware of the behavior of the impedance model, leading to a superior control performance. This is especially true if the dynamics of the UUT and its influence on the impedance model can be modeled too, making it possible to reliably predict the future behavior of the design plant (described in the next section). However, the design of the controller is strongly coupled to the current impedance model used and may require a new control design for each impedance model used.

## 3.4 Model Predictive Control

As was demonstrated in the last section any type of control can be used to realize impedance control. However, in this work MPC will be used and it is assumed that

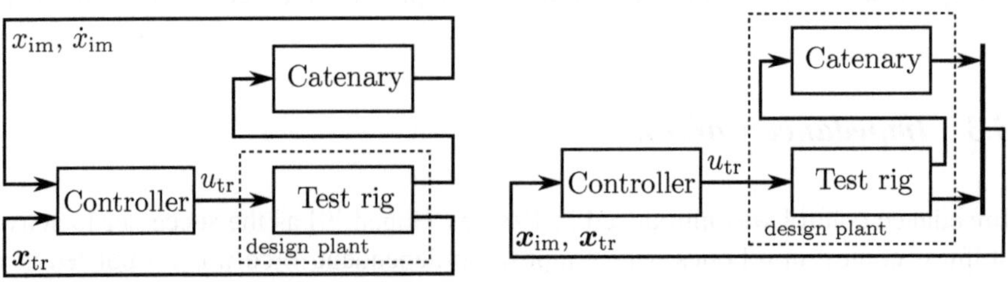


**Fig. 4** Classical structure of cascaded impedance control (CIC, *left*) and integrated impedance control (IIC, *right*)

the plant to be controlled is given in linear, periodic, time-varying, discrete-time state-space form:

$$\begin{aligned} \boldsymbol{x}_{n+1} &= \mathbf{A}_n \boldsymbol{x}_n + \mathbf{B}_n \boldsymbol{u}_n + \mathbf{E}_n \boldsymbol{z}_n, \\ \boldsymbol{y}_n &= \mathbf{C} \boldsymbol{x}_n, \end{aligned} \tag{21}$$

and that at sampling instant $n$ the current state vector $\boldsymbol{x}_n$ can either be measured or constructed from measurements. A control horizon $n_c$ is defined and the vector of future control moves is given by:

$$\boldsymbol{U}_n = [\boldsymbol{u}_n^{\mathrm{T}}, \boldsymbol{u}_{n+1}^{\mathrm{T}}, \ldots, \boldsymbol{u}_{n+n_{c-1}}^{\mathrm{T}}]^{\mathrm{T}}. \tag{22}$$

Then, the vector of the next $n_c$ outputs

$$\boldsymbol{Y}_n = [\boldsymbol{y}_{n+1}^{\mathrm{T}}, \boldsymbol{y}_{n+2}^{\mathrm{T}}, \ldots, \boldsymbol{y}_{n+n_c}^{\mathrm{T}}]^{\mathrm{T}}. \tag{23}$$

can be predicted, e.g.

$$y_{n+2|n} = \mathbf{C}\mathbf{A}_{n+1}\mathbf{A}_n \boldsymbol{x}_n + \mathbf{C}\mathbf{A}_{n+1}\mathbf{B}_n \boldsymbol{u}_n + \mathbf{C}\mathbf{B}_{n+1}\boldsymbol{u}_{n+1}, \tag{24}$$

where for space reasons the influences of the disturbance have been omitted, however, their derivation is straightforward by copying terms including $\boldsymbol{u}$ and making the following substitutions: $\boldsymbol{u} \rightarrow z$ and $\mathbf{B} \rightarrow \mathbf{E}$.

The upcoming control moves (22) are obtained by minimizing a quadratic cost function:

$$\begin{aligned} \text{minimize} \quad & (\boldsymbol{Y}_{\text{ref}} - \boldsymbol{Y})^{\mathrm{T}} \mathbf{Q} (\boldsymbol{Y}_{\text{ref}} - \boldsymbol{Y}) + \boldsymbol{U}^{\mathrm{T}} \mathbf{R} \boldsymbol{U} \\ \text{subject to} \quad & \mathbf{M}_{\mathrm{U}} \boldsymbol{U} \leq \gamma, and \\ & \mathbf{M}_{\mathbf{x}} \boldsymbol{X} \leq \gamma_x, \end{aligned} \tag{25}$$

where the matrices $\mathbf{Q}$ and $\mathbf{R}$ are weighting matrices and are tuning parameters for the designer. Iterative solvers have to be used for obtaining a solution to (25) but at the advantage of incorporating state, input and output constraints directly in the controller. Traditionally, MPC was first introduced in the process industry but with increasing computational power and more sophisticated algorithms, see e.g. Ref. [14] for a comprehensive review, real-time MPCs can now be applied at sampling rates of several kHz [15].

To realize impedance control with the IIC structure (see Fig. 4) the state-space design system from (21) is chosen as:

$$\mathbf{A} = \begin{bmatrix} \mathbf{A}_{\text{tr}} & 0 \\ 0 & \mathbf{A}_{\text{im}} \end{bmatrix}, \quad \mathbf{B} = \begin{bmatrix} \mathbf{B}_{\text{tr}} & 0 \\ 0 & \mathbf{B}_{\text{im}} \end{bmatrix}, \quad \mathbf{C} = [\,\mathbf{C}_{\text{tr}} \quad -\mathbf{C}_{\text{im}}\,], \tag{26}$$

where the matrices in (10) and the FD discretized state-space catenary model are used.

This way the controller knows about both, the test rig dynamics as well as the impedance model dynamics and tries to track their difference to zero (the reference vector is chosen as $\boldsymbol{Y}_{\text{ref}} = \mathbf{0}$).

## 3.5 Conserved Quantities

In order to obtain close-to-reality results in HiL testing it is necessary that the UUT and the impedance model exchange the correct amount of certain conserved quantities. This was for example demonstrated in Ref. [11] on an automotive test rig.

Figure 5 shows a bond-graph representation of a generic HiL test rig. The exchanged quantities in an effort and flow sense are force and velocity. The relevant conserved quantities derived therefrom are energy and momentum.

Just considering the mechanical part of energy (called work) the difference between the energy transferred into the UUT and the impedance model is defined as:

$$\Delta E(t) = \int_0^t (F_{\text{tr}}(\tau)\dot{x}_{\text{tr}}(\tau) - F_{\text{im}}(\tau)\dot{x}_{\text{im}}(\tau))\mathrm{d}\tau, \tag{27}$$

and the error in transferred momentum reads:

$$\Delta p(t) = \int_0^t (F_{\text{tr}}(\tau) - F_{\text{im}}(\tau))\mathrm{d}\tau. \tag{28}$$

If it is assured that the force at the UUT /test rig interface is directly used as an input to the impedance model, and perfect impedance matching would be possible, i.e. (20) is zero at all times, then energy and momentum consistency is of no concern because (27) and (28) are zero at all times.

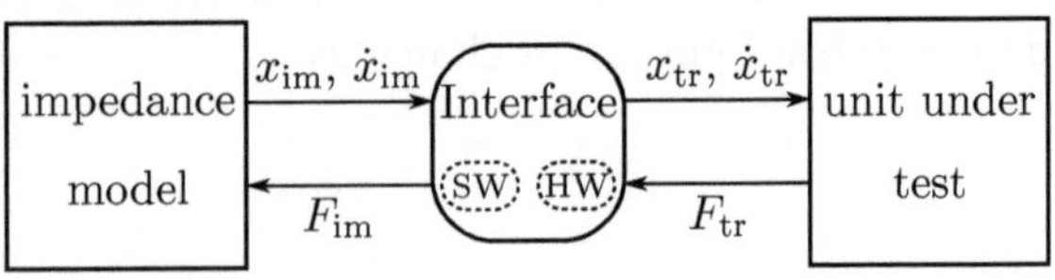


**Fig. 5** Illustration of impedance coupling in pantograph HiL testing task

To ensure long-term consistency even under the presence of ubiquitous position and velocity errors in (20), a virtual correction force $F_{\text{cor}}$ is introduced, that additionally acts on the impedance model. In the simplest case $F_{\text{im}}$ is then chosen to $F_{\text{im}} = F_{\text{tr}} + F_{\text{cor}}$ and the momentum error in (28) just reduces to the integral over $F_{\text{cor}}$, that can directly be addressed by linear control systems theory.

The value of $F_{\text{cor}}$ can be chosen by the test rig control because it is treated as an additional input in the optimization process (25). Because of the quadratic nature of the terms in (27) it is not possible to treat the energy error in classical linear theory. However, by requiring a decay in energy error $(|\Delta E_{n+1}| < |\Delta E_n|)$ bounds on $F_{\text{cor},n+1}$ can be derived that assures long-term consistency in energy transferred between the UUT and the impedance model.

## 4 Real-Time Implementation

As a real-time platform for implementing the control with the developed methods a dSPACE ds1006 was chosen. It consists of a 2.8 GHz quad-core processor and additionally I/O extension cards were installed for test rig communication and monitoring tasks. So far, three of these four processor cores are used for the following tasks:

Core A ($f_s = 5000\,\text{Hz}$) : All monitoring tasks as well as measurement acquisition run on this core. Also a basic reference tracking controller and friction compensation task is running

Core B ($f_s = 200\,\text{Hz}$) : The impedance control realized via MPC is running on the second core. The test rig model used for this purpose is already a controlled system, namely the one realized by the control in core A (5000 Hz), and the control input is a reference trajectory for the linear drive. The impedance model used is an FD-discretized model of the catenary system with absorbing layers presented in Ref. [8]

Core C ($f_s = 200\,\text{Hz}$) : A complex finite element model of the catenary is simulated here to maximize virtual system model fidelity, from which the FD-discretized model in core B is being initialized. This model also captures complex phenomena like multi-traction and dropper slackening

Most of the development was done in MATLAB/Simulink. The dSPACE integration into these programs was used for easily building executable files for the real-time computer.

Unfortunately the functionality for sparse matrices computation as well as the quadratic programming capabilities are not available for code generation and thus external libraries have to be chosen for these tasks.

As a quadratic problem solver *qpOASES* [16] was chosen, a freely available open-source C++ implementation of an active-set algorithm proposed in [17]. It has several features that makes it particularly useful for MPC tasks and has also easy integration in MATLAB/Simulink.

*Eigen* is a free (open-source) C++ template library for sparse and dense linear algebra, including matrices, vectors and a variety of numerical solvers, see Ref. [18]. It integrates into MATLAB by the use of C++ S-functions and can thus easily be used in the existing MATLAB/Simulink project. *Eigen* supports fixed-size as well as dynamically-sized matrices and is very fast when using explicit vectorization by CPU extensions such as SSE.

The integration of both libraries into the build-chain for real-time compilation on dSPACE targets is straightforward.

## 5 Results

First experimental results are shown indicating the performance of the proposed setup. The control concept is validated in actual test rig operation where a simplified catenary model is being emulated. As the real-time FE model code is still under development, this first set of results is based on a linear time-varying explicit catenary representation obtained by a finite difference discretization. This model is utilized both as control design as well as catenary simulation model. Subsequently, first real-time FE implementation results in terms of runtime analysis are stated, showing the feasibility of the final concept.

### *5.1 Controller with Finite Difference Model*

Figure 6 shows an experimental result of catenary emulation on the test rig. Thereby, a catenary with parameters as listed in Table 1 is modeled and approximated via the method of finite differences. A moving-coordinate representation and absorbing boundaries as developed in [8] are utilized, and a linear, periodically time-varying, explicit, discrete-time state space system structure is obtained. Dropper slackening is not considered (thus the catenary model remains linear), and this model is utilized both as MPC design model and as catenary simulation model.

The experiment shows the controlled test rig under manual one-sided excitation. Initially, the control only aims at tracking and momentum error rejection (up until time $t = 50$ s). The virtual correction force remains zero and thus momentum is always consistent, but a significant energy error accumulates. Then, around time $t = 50$ s, the controller is configured to also enforce energy consistency, and subsequently the energy error can successfully be eliminated by utilization of the virtual correction force. Even though (exact) momentum consistency is violated now, the momentum error is being stabilized and rejected successfully.

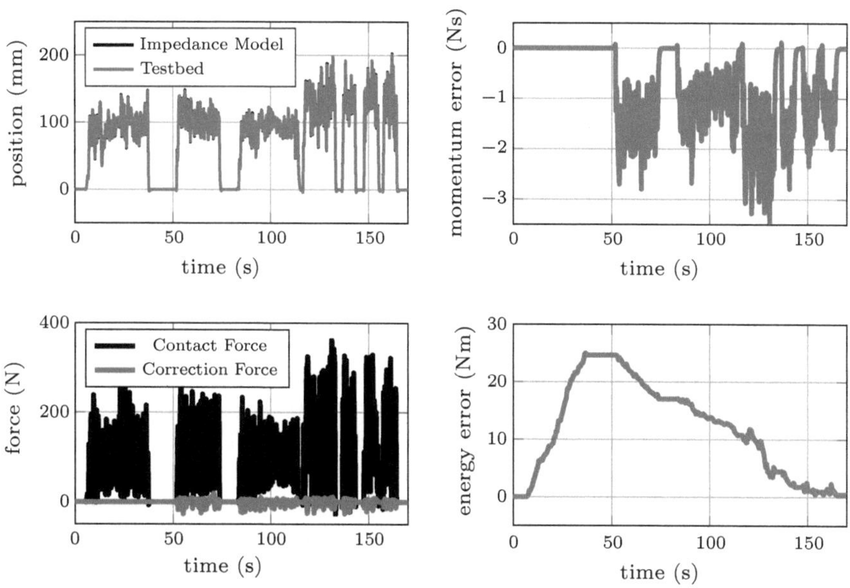


**Fig. 6** Experimental results of virtual catenary emulation on the test rig (LTV catenary model via FD method): Initially, only tracking and momentum goals are followed, leading to energy error accumulation. After $t = 50$ s, the controller is switched to enforce energy consistency and successfully recreates consistency

**Table 1** Catenary configuration for experimental test case

| Catenary Parameters | | | Contact Wire | | | Carrier Wire | | |
|---|---|---|---|---|---|---|---|---|
| $L$ | 60 | m | $\rho A$ | 1.35 | kg/m | $\rho A$ | 1.35 | kg/m |
| $\Delta x$ | 1.5385 | m | $EI$ | 150 | $Nm^2$ | $EI$ | 150 | $Nm^2$ |
| $\Delta t$ | 5.8824 | ms | $\beta$ | 0 | kg/s/m | $\beta$ | 0 | kg/s/m |
| $n_{droppers}$ | 4 | [] | $T$ | 20000 | N | $T$ | 20000 | N |

## 5.2 *First Results Towards Real-Time Finite Elements*

The full real-time implementation of a moving-coordinates catenary finite element model are under development and will be published in the future, but first results on scalability and runtime feasibility are shown in the following. The finite element for a moving-coordinate Euler-Bernoulli beam equation with axial pre-tension has been obtained by the Galerkin-FE-method [13]. For a test beam, Fig. 7 shows measurements of the time required to compute one Newmark integration step over the number of modeled degrees of freedom for various implementations: MATLAB results indicate results obtained on a desktop PC under MATLAB, either directly through m-code, or via a C++ S-function implementation of the Newmark algorithm. The dSPACE results have been obtained on the respective real-time platform.

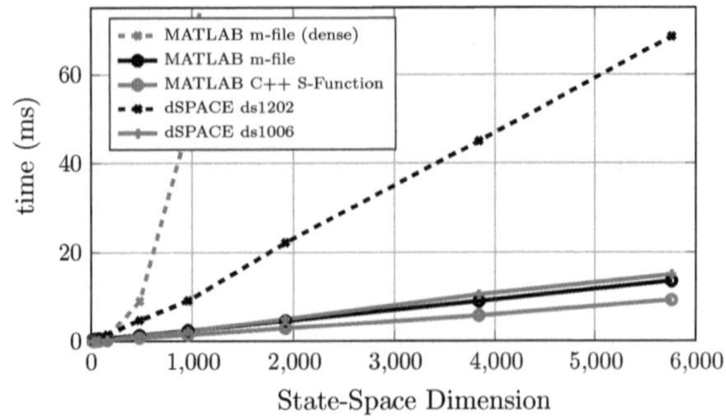


**Fig. 7** Run-time analysis for solving one time step of the Newmark-algorithm for different state-space dimensions (sparse)

The MATLAB/*dense* variant carries out all computations with dense matrices and scales prohibitively badly. It is unusable already for 1000 degrees of freedom up. In turn, all other implementations exploit sparse algebra (MATLAB's internal sparse functionality, respectively sparse algorithms implemented via *eigen*) and scale favorably, essentially with $\mathcal{O}(n)$ for the tested problems. The high run-times of the ds1202 system (MicroLabBox) are attributed both to the slower clock speed of the processor therein, as well as the missing support for CPU vectorization extensions. It is evident that exploiting sparsity enables computation times compatible with reasonable real-time requirements.

## 6 Conclusions

A novel concept for high-fidelity impedance control for hardware-in-the-loop testbeds has been shown, with the focus on high physical accuracy and trustworthiness. Model fidelity is maximized by a real-time-enabled finite-element formulation of the virtual system dynamics.

First experimental results for a pantograph testbed hardware-in-the-loop impedance control are shown and FE real-time studies are shown. The developments remain to be integrated which will be subject of future publications, but the main building blocks could be successfully validated in this work.

**Acknowledgments** ZIM – Zentrales Innovations programm Mittelstand.

## References

1. Ritzberger, D., Talic, E., Schirrer, A.: Efficient simulation of railway pantograph/ catenary interaction using pantograph-fixed coordinates. Vienna Int. Conf. Math. Model. **8**, 61–66 (2015)
2. Jung, S.P., Kim, Y.G., Paik, J.S., Park, T.W.: Estimation of dynamic contact force between a pantograph and catenary using the finite element method. J. Comput. Nonlinear Dyn. **7**(4), 041006 (2012)

3. Facchinetti, A., Bruni, S.: Hardware-in-the-loop hybrid simulation of pantograph–catenary interaction. J. Sound Vib. **331**(12), 2783–2797 (2012)
4. Engquist, B., Majda, A.: Absorbing boundary conditions for numerical simulation of waves. Proc. Natl. Acad. Sci. **74**(5), 1765–1766 (1977)
5. Higdon, R.L.: Absorbing boundary conditions for difference approximations to the multidimensional wave equation. Math. comput. **47**(176), 437–459 (1986)
6. Givoli, D.: High-order local non-reflecting boundary conditions: a review. Wave Motion **39** (4), 319–326 (2004)
7. Schirrer, A., Talic, E., Aschauer, G., Kozek, M., Jakubek, S.: Optimization based determination of highly absorbing boundary conditions for linear finite difference schemes. J. Sound Vib. **365**, 45–69 (2016)
8. Ritzberger, D., Schirrer, A., Jakubek, S.: Emulating the properties of a perfectly matched layer with an optimal feedback controller. J. Comput. Phys. (in review) (2015)
9. Hogan, N.: Impedance control: an approach to manipulation: Part I: Theory. J. Dyn. Syst. Meas. Contr. **107**(1), 1–7 (1985)
10. Priesner, R., Jakubek, S.: Mechanical impedance control of rotatory test beds. IEEE Trans. Industr. Electron. **61**(11), 6264–6274 (2014)
11. Jakubek, S., Luchini, E., Oberhummer, A., Pfister, F.: A model-based interfacing concept for accurate power hardware-in-the-loop systems. Math. Comput. Model. Dyn. Syst. **22**(1), 1–20 (2016)
12. Arnold, M., Simeon, B.: Pantograph and catenary dynamics: a benchmark problem and its numerical solution. Appl. Numer. Math. **34**(4), 345–362 (2000)
13. Liu, G.R., Quek, S.S.: The Finite Element Method: A Practical Course. Butterworth-Heinemann, 2nd edn. (2013)
14. Yu-Geng, X., De-Wei, L., Shu, L.: Model predictive control—status and challenges. Acta Automatica Sinica **39**(3), 222–236 (2013)
15. König, O., Gregorčič, G., Jakubek, S.: Model predictive control of a dc–dc converter for battery emulation. Control Eng. Pract. **21**(4), 428–440 (2013)
16. Ferreau, H., Kirches, C., Potschka, A., Bock, H., Diehl, M.: qpOASES: a parametric active-set algorithm for quadratic programming. Math. Program. Comput. **6**(4), 327–363 (2014)
17. Ferreau, H., Bock, H., Diehl, M.: An online active set strategy to overcome the limitations of explicit mpc. Int. J. Robust Nonlinear Control **18**(8), 816–830 (2008)
18. Guennebaud, G., Jacob, B.: Eigen v3. http://eigen.tuxfamily.org (2010)

# Part IX
# Parameter Identification

# Efficient In-Vehicle Calibration by the Usage of Automation and Enhanced Online DoE Approaches

**Silja Thewes, Matthias Krause, Christoph Reuber, Markus Lange-Hegermann, Rafael Dziadek and Martin Rebbert**

**Abstract** Data-driven modeling techniques in combination with efficient test planning procedures are widely used for calibration tasks. Such modeling techniques will also be required for a successful transfer of vehicle calibration activities to the test bench and finally to the office in a so called "Road to rig to desktop" approach. A reduction in development time and cost for prototypes are the major drivers behind this development. However, specific aspects of vehicle calibration like drivability and final approval will still be conducted mainly in the vehicle for the next years. Well established approaches and structured processes are carried over from test bench calibration to the vehicle to ensure that all aspects of vehicle calibration can be managed within an appropriate time and cost frame. An effective drivability calibration requires the objectification of the subjective impressions, but also the appropriate modeling and optimization of the vehicle behavior. Gaussian Process models and their extensions are applied to this task. The typically manual iteration loops are replaced by an automated procedure, containing: online test planning, maneuver driving, evaluation, and modeling. The approach is demonstrated on selected use cases of the drivability calibration domain.

S. Thewes (✉) · M. Krause · C. Reuber · M. Lange-Hegermann · R. Dziadek · M. Rebbert
FEV GmbH, Aachen, Germany
e-mail: thewes_s@FEV.COM

M. Krause
e-mail: krause_m@FEV.COM

C. Reuber
e-mail: reuber@FEV.COM

M. Lange-Hegermann
e-mail: hegermann@FEV.COM

R. Dziadek
e-mail: dziadek@fev.com

M. Rebbert
e-mail: Rebbert@FEV.com



C. Gühmann et al. (eds.), *Simulation and Testing for Vehicle Technology*,
DOI 10.1007/978-3-319-32345-9_21

# 1 Introduction

Engine calibration is always faced with an increasing number of calibration tasks and requirements on the one hand and with the demand for high efficiency in terms of numbers of measurements on the other hand. At the test bench this has encouraged the use of automated testing procedures and model based approaches like the Design of Experiments (DoE) [1, 2] already for a long time.

In contrast to the test bench, in-vehicle calibration has not yet seen this stringent application of efficient methods and procedures. However, specific aspects of vehicle calibration like drivability and final approval will still be conducted mainly in the vehicle for the next years. So there is significant potential of reduction in time and cost for the in-vehicle calibration by usage of automation and model based DoE approaches. First remarkable benefits are achieved in taking over the standard concept of fully automated campaigns from the test bench to the vehicle. In this way the reproducibility and quality of measurement data can be achieved in a shorter time and thus build the foundation of model based approaches. Providing respective automation frameworks that support an easy description of the designated task, preferably in terms of a graphical language, is the backbone of such an approach.

The methodology of Design of Experiments has become an indispensable tool to support the work of the calibration engineer [3]. Classical—offline—DoE methods are characterized by four strictly separated successive steps: the generation of a set of design points that are going to be tested, the measurement of the design points, the generation of mathematical models for the engine behavior represented by the measurements, and finally the optimization with respect to these models. This chronological order implies that for any step no information is required from the subsequent step.

One typical counter-example where information from a subsequent step is relevant is the generation of a test plan in case that the valid or appropriate design space is not known. An inappropriate design space contains engine settings that are inapplicable for calibration, pollutes the measurement system, or risks engine damage. Such engine settings should be avoided in the measurement phase. It is not possible to exactly determine which design points lead to these undesired engine settings beforehand, since the exact engine behavior is not known. However, estimations can be made by pre-screening of the design space taking into account engineering knowledge. Alternatively estimations on valid engine settings can be made if information from the measured data or modeling techniques are taken into account.

For this reason, online DoE combines the steps of test planning, measurement acquisition and modeling [4–11]. The test plan is created iteratively while already doing the first measurements. This procedure allows to include information from previous measurements. With increasing number of measurements more and more information is gathered and it is possible to create increasingly better data-driven models of the engine behavior and its limits. In return such a model can be used to determine next measurement points that most probably are valid. Hence, unnecessary or undesired measurements at invalid points can be avoided.

One important component of this online DoE methodology is the modeling part. Gaussian processes have proved to be an adequate tool for this task as they provide enough flexibility to model even complex dependencies, but at the same time can deal with small numbers of measurements [12, 13]. An enhancement of the Gaussian processes are the so-called warped Gaussian processes that can provide an improved model quality by virtue of intrinsic output transformations that are automatically found during the training of the model [14].

These warped Gaussian processes allow to model even complex behavior with large changes in the modeled output data [15]. The additional flexibility of the warped Gaussian processes involves an increased complexity of the model description reflected by an increased number of underlying parameters. So a careful decision has to be made whether warped or standard Gaussian process models are better suited to the specific modeling task. Model selection criteria like the Akaike information criterion (*AIC*) [16] can be used to evaluate the benefit of one model class over the other and allow for implementation of an automatic selection feature.

To present the benefits of automation and online DoE approaches for an in-vehicle calibration task, the calibration of a tip-in is chosen. It is naturally conducted in the vehicle and typically performed by highly specialized personal with years of experience in this domain. To replace the subjective impressions of the driver, the transient vehicle behavior is aggregated to objectified characteristic values [17] and can thus be used for subsequent modeling.

The paper is structured as follows: Sect. 2 contains the automation of maneuver driving for in-vehicle calibration. Section 3 describes the online methods applied during the automated testing. Section 4 introduces the modeling and model selection approach used by the online methods. In Sect. 5 a use case from the domain of drivability calibration for the described approach is presented. Finally, Sect. 6 summarizes the results.

## 2 Automated Maneuver Driving

In contrast to testing procedures at the test bench, one major challenge of in-vehicle calibration is the reproducibility of the testing procedure, namely the execution of the driving maneuver. The importance of a repeatable execution of the maneuvers becomes indispensable when the analysis of the vehicle behavior is based on measurement signals only. For this reason, it has been for a long time of great interest for FEV to ensure the acquisition of high quality measurements in the vehicle and to provide the respective test automation possibilities. As a consequence, the Vehicle Test Automator tool (VTA) has been developed.

The main goal of the VTA is to assist the driver in semi- or fully-automated conduction of maneuvers and calibration campaigns [18]. Hence, input from the driver can be manipulated by the VTA, e.g. pedal positions. Additionally, the VTA can process well-timed manipulations of calibration parameters in the vehicle, record measurements, perform online calculations, and analyze data.

The user defines the tasks by means of a user-friendly graphical workflow in the VTA. This workflow supports sequence diagrams, flow charts, and state machines. The top-layer of the work flow used for the subsequently described online DoE procedure contains a flow chart with three main steps, cf. Fig. 1: retrieving new target points, performing the measurement, and training the model. These steps are executed iteratively until a defined abortion criterion is met.

The workflow engine provides a variety of standard actions and calculations, but is easily extendable by domain specific activities. Of special interest for the current application is the possibility to integrate user defined algorithms via a standardized interface. In this way, external algorithms can be used to execute the integral steps of the online DoE procedure of retrieving new target points and of training models, cf. Fig. 1.

The step of measuring the next target point comprises the manipulation of the calibration parameters, and the measurement acquisition and processing. Practically this approach is applied by the connection to calibration tools like ETAS Inca, Vector CANape, dSPACE ControlDesk NG or ATI VISION utilizing standard calibration automation interfaces like ASAM MCD3-MC [19] or specific tool APIs.

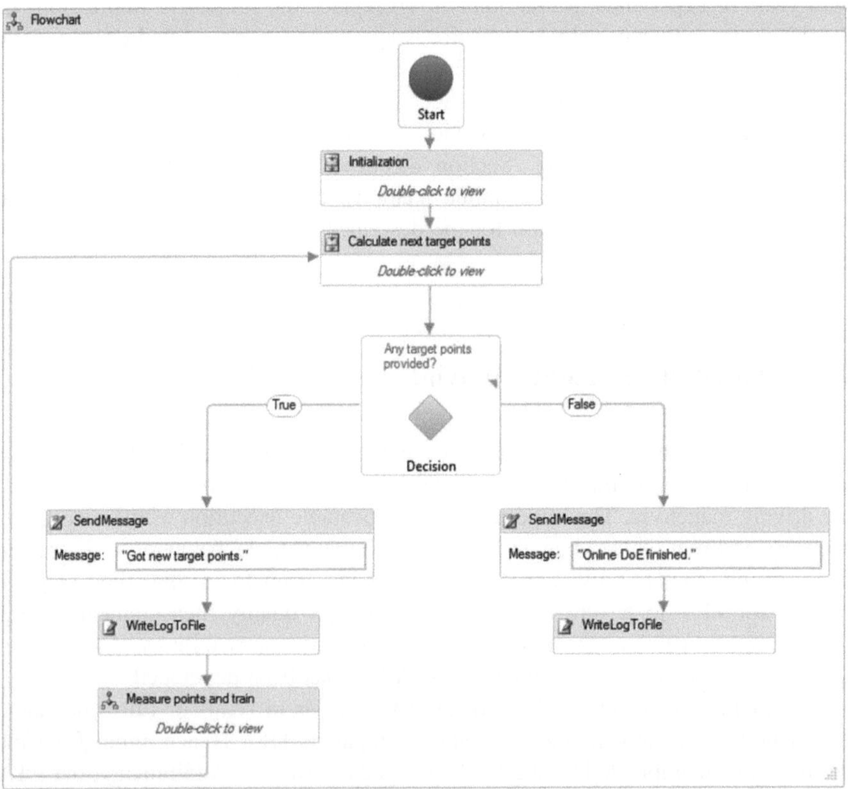


**Fig. 1** Top-layer of user defined online DoE procedure flow chart

Moreover, the use of the automation protocol ASAM ACI facilitates the connection to test benches or chassis dynamometers and thus extends the herewith shown approaches. All measurement acquisition and label manipulation is realized through this interface.

In the subsequently shown example of tip-in maneuvers especially the correct realization of throttle requests is of major interest and has to be solved individually depending on the software structure provided by the concerning ECU supplier. It turned out to be best to apply those manipulations as early as possible, e.g. directly after the pedal value sensor, to make all subsequent and relying software parts work based on the desired values.

The description of the driving maneuver itself is realized in the tool by means of state machines. State machines turned out to be most convenient because of their high understandability and usability by calibration engineers. Figure 2 visualizes the state machine description of the realized tip-in maneuver. After checking for the correct gear—modeling requires the same gear as different gear sets have different oscillation behavior, hence fundamentally different characteristics—the vehicle is accelerated to a dedicated engine speed from which the initially required coast down phase can be performed. This step is essential as a tip-in is fundamentally characterized by an instantaneous load alteration that is typically caused by a pedal step and its respective torque request. After coast down, as soon as the target tip-in engine speed is reached, the pedal step is applied via a manipulation of the respective calibration labels. The engine mounting loads and driveline clearances

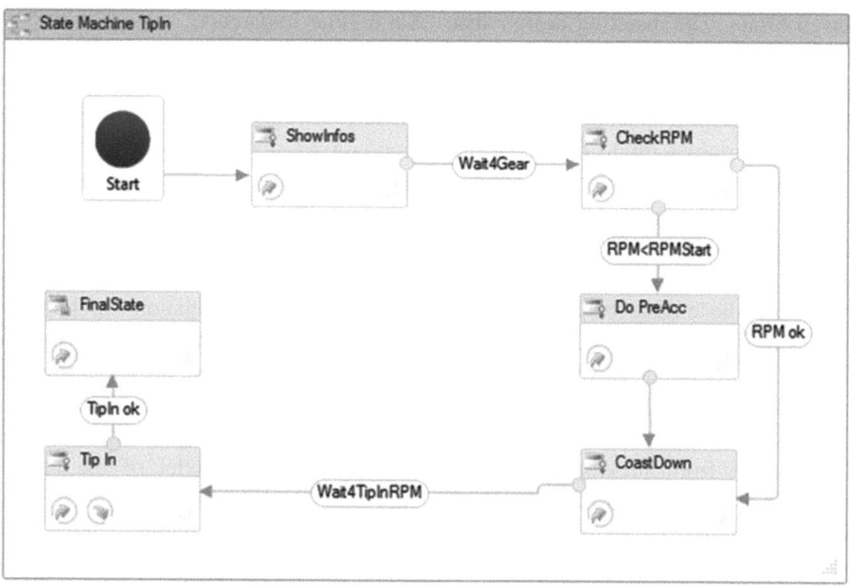


**Fig. 2** State machine description of user-defined tip-in maneuver

are reversed, typically being notice-able as a respective jerk, and the vehicle is accelerated. The conducted maneuver ends after a certain time of engine speed difference, depending on the current operation point.

## 3 Online DoE Approach

Online DoE features or adaptive DoE techniques are used in a wide range of applications, for example in generating global models [6] or determining borders [4]. For powertrain calibration, a complete automated modeling has been discussed [2], also different strategies have been introduced [7–11], and changes between strategies have been evaluated [6].

In typical calibration campaigns the parameters that are decisive for the description of a certain engine behavior are known. In contrast, it is often unclear how to define a suitable value range for these engine parameters that allow valid measurements. For the purpose of this paper, a valid measurement is defined as a measurement with engine settings that are relevant for and applicable to calibration. Invalid engine settings are avoided by defining two kinds of limits: On the one hand limits on the actuated variables, given by the test plan, and more importantly on the other hand by limits on observed variables, i.e. variables that are going to be measured at each point.

Limits or bounds on the actuated variables can easily be taken into account during the planning of the design points, simply by allowing only values within the given limits. The resulting design space is a regular hypercube.

In contrast, it is much harder to consider bounds on the observed variables, as they cannot easily be translated to bounds on the actuated variables. Even if the valid design space is a simple hypercube for the observed variables, its form for the actuated variables might have a complex non-convex shape, holes of any kind, or might even not be connected. The relationship between actuated and observed variables that is needed to translate bounds from the one space to the other is determined by the a priori unknown engine behavior.

To approximate the relationship between actuated and observed variables, a model of the engine behavior can be used to predict the relationship between actuated input variables of the model and observed output variable of the model. The more precise the prediction of this model, the more accurate the description of the border of the valid design space in input space of the actuated variables.

This modeling task is an integral part of the online DoE algorithm, as it uses information from previous measurements to generate a model of the engine behavior that is then utilized to intelligently plan the position of the next measurement point. The quality of the engine behavior models increases in the course of the online DoE procedure by successively adding new measurement data to it.

The complete online DoE approach is composed of three phases:

1. In a first initial phase, some initial valid measurements are gathered to generate preliminary models.
2. The second phase explores the borders of the valid design space by iteratively adding new measurement points.
3. The third phase enhances model quality by adding points from an initially defined test plan taking into account the border estimations from phase two.

The transition between the three phases can be triggered fully automatically, e.g. by defining the required number of valid measurements in each phase.

In the first phase, a small number of initial points are measured. These initial measurement points can originate from a test plan automatically generated around a presumably safe valid point, e.g. in the center of the original design space. Since the measurement points are chosen near the safe point, limit violations during this initial phase may occur but are unlikely. However, at the end of the initial phase, a small number of valid measurements is able to generate the models of the engine behavior which are needed in the subsequent second phase.

The second phase iteratively adds measurements to the test plan such that the known valid design space grows. Each new measurement point is allowed only within a certain probing region. This probing region is computed from a trade-off that includes for example the risk of violating a border. This risk increases with two factors: first, leaving the relative proximity to the already known measurements in input space, and, second, predicting an output that approaches or even exceeds a limit.

The extent of the probing region in phase two is a result of an automatic trade-off. If the probing region is larger, then the known design space grows faster, but there is also a higher risk of violating a limit. The applied algorithm is designed such that the probing region automatically adapts to the behavior requested by the calibration engineer. For each output limit, the behavior can be anything between the following two:

1. Try to avoid any limit violation. A lot of measurements are required to explore the borders of the valid design space.
2. Try to explore the borders of the valid design space as fast as possible with the risk of limit violations.

In case that a limit violation is detected and the point is stable, the algorithm automatically takes this limit violation into account and adapts the models and the probing region accordingly. Apart from a limit violation, an invalid measurement can also result from insufficient stability of the measurement. Then, this point is not included in the model data and the algorithm adapts the probing region by introducing additional constraints that exclude this instable measurement from the design space.

The goal of choosing the next point is a rapid expansion of the probing region. This is achieved using strongly scattered input values, i.e., out of the probing region the one point is selected that increases the scattering of the input values the most.

Here, scattering is defined using the well-known $D$-optimal designs for homogeneous linear models [20–25], which was suggested for Online DoE by [11]. Such designs are well-known to have many points near the border of the design space and, hence, new measurement points are preferably chosen at the border of the probing region. Note that extending a $D$-optimal design by a point can be computed efficiently, since adding a point $x_1$ to an existing design $X$ yields a better $D$-optimal design than adding a point $x_2$ if and only if $x_1(X^{\mathrm{T}}X)^{-1}x_1^{\mathrm{T}} > x_2(X^{\mathrm{T}}X)^{-1}x_2^{\mathrm{T}}$.

Once the borders of the valid design space are known with a sufficient accuracy, the third phase starts with filling up this valid design space according to a given design pattern, e.g. a space filling design. Measurements from the previous phase one and two can be incorporated here, such that the information from these measurements is not lost, but in contrast leads to a higher quality of the resulting models.

## 4 Data-Driven Modeling

An integral part of the online DoE approach is the modeling of the engine behavior based on measurement data. A class of well suited data-driven models is the one of Gaussian process models [12], as they offer high flexibility to model even complex dependencies. A Gaussian process model provides the most likely model description for the training data by means of Bayesian inference. This approach avoids overfitting by automatically trading off the complexity of the model with the goodness of the fit to the data. Thus, Gaussian process models can also deal with a low number of measurement points that occur during the initial phase of an online DoE approach.

A Gaussian process model is characterized by its hyper-parameters. For each set of hyper-parameters $\theta$ the likelihood $p$ is given that the model describes the relationship between input data $X$ and output data $y$ best

$$\log p(y|X,\theta) = -\frac{1}{2}y^T K y - \frac{1}{2}\log \det K - \frac{n}{2}\log 2\pi,$$

where $K$ is the covariance matrix and $n$ is the number of data points.

Standard Gaussian processes assume that the data that the model tries to interpret has a Gaussian distribution. However, this might often not be the case, e.g. if data values change by orders of magnitude from one part of the input space to another [15]. An enhanced version of Gaussian process models, so-called warped Gaussian process models [14], addresses this problem by transforming the output data, hence warping the distribution of the data. This transformation is automatically determined while training the model, such that the optimal transformation with respect to the likelihood of the model is found.

Warped Gaussian process models might not be highly beneficial for all uses cases. In some cases, e.g. where the data is very homogeneous, transformations do not necessarily lead to a strong increase in model accuracy and instead result in overfitting; in such cases, standard Gaussian process models are preferable. Moreover, the intrinsic transformation in warped Gaussian process models entails an increased number of hyper-parameters that need to be optimized. This impedes the determination of an appropriate model for warped Gaussian processes, especially for a very low number of data points. A model selection criterion can be applied to select the best suited of the two modeling classes. An example for such a criterion is the Akaike information criterion AIC [16]. It penalizes model classes with a complex model structure, as the selection term depends on the number of hyper-parameters h:

$$-2 \log p(y|X, \theta) + 2 \cdot h.$$

Hence, it favors simpler model classes when an additional complexity in the model structure does not lead to an equivalently increased likelihood of the model to represent the data and as such automatically avoids overfitting.

## 5 Use Case from Drivability Calibration

The following use case is taken from the domain of drivability calibration. The drivability for the standardized driving maneuver of a tip-in [26] is investigated. The drivability is evaluated by means of two objective criteria that characterize the jerk and the response: the vibrational dose value of the first jerk $VDV_{FJ}$ [27] and the time to 80 % of the maximum acceleration $t_{80\ \%}$. These two objective criteria can be measured in the vehicle, as they are based on physical principles. These two objective criteria are anti-correlated: while for the $VDV_{FJ}$ a high value corresponds to a dynamic behavior of the vehicle, a high value for $t_{80\ \%}$ is associated with a slow response and vice versa. So in drivability calibration, one usually has to find a trade-off between both values.

For the investigated vehicle and driving maneuver, there are mainly two ECU variables that can be used to influence the drivability characteristics: the time constant *T1* and the amplification factor *Kd* for the torque demand shaping at positive gradient. Not all of the possible values for *T1* and *Kd* are applicable to calibration, as e.g. a very large jerk may be experienced as uncomfortable, while a very slow response may be perceived as not dynamic enough. For this reason, bounds are defined for the two objective criteria:

1. A constant upper bound for $VDV_{FJ}$ independent of the pedal position
2. A lower bound for $VDV_{FJ}$ that increases with increasing pedal position
3. An upper bound for $t_{80\ \%}$ that decreases with increasing pedal position

The latter two bounds are defined as characteristic maps over pedal position and engine speed, cf. Fig. 3. This means that the bounds for $VDV_{FJ}$ and $t_{80\ \%}$ depend on the operating point.

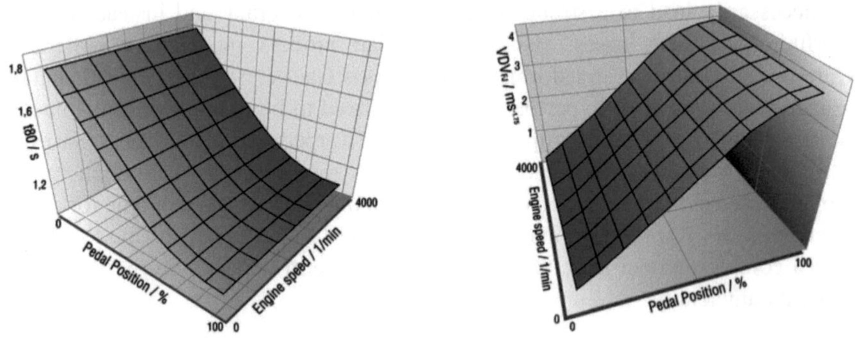


**Fig. 3** Characteristic maps for the upper bound of $t_{80\ \%}$ (*left*) and the lower bound of $VDV_{FJ}$ (*right*)

For all of the bounds on $VDV_{FJ}$ and $t_{80\ \%}$ the shape is defined and known in the output space. However, it is unknown how these bounds yield non-linear constraints of the input space of *T1* and *Kd*, and how these constraints change with the operating point.

The combined approach of automated maneuver driving and online DoE is applied to automatically measure points which are highly likely to be applicable to calibration of the dynamic of the vehicle, i.e. points that fulfill the restrictions on the objective criteria as defined above.

The principles of the online DoE approach are first demonstrated graphically in an illustrative example, which fixes the operating point defined by pedal position and engine speed. The second example is then a real calibration example; it includes pedal position and engine speed in the parameter variations to derive global models that can be used for the calibration of the complete operating range.

This first, illustrative example focusses on the local behavior for 50 % pedal position and 2000 $\text{min}^{-1}$ engine speed. The allowed range for the parameters *T1* and *Kd* is 0.05–0.5 s and 0.1–0.5 respectively. Assuming that the points in the center of these input bounds are valid, the first points are measured close to *T1* = 0.275 s and *Kd* = 0.3. After a number of ten initial measurements, the algorithm starts to iteratively add further measurement points by intelligently deciding on the input values for the next measurement point. Depending on the algorithm settings, this decision can be made rather conservative or rather aggressive.

A conservative decision selects a point that has a low risk of violating an output limit, but that does not increase the scattering of the input values so much, cf. Fig. 4. Hence, only two measurements out of 32 violate an output limit. So almost all measurements are applicable to calibration. However, the extent of the known design space does not yet cover all allowed value combinations for *T1* and *Kd*. Further measurements would be needed such that the algorithm also selects points closer to the output limits, e.g. points with high values for *T1* and *Kd*. The choice of

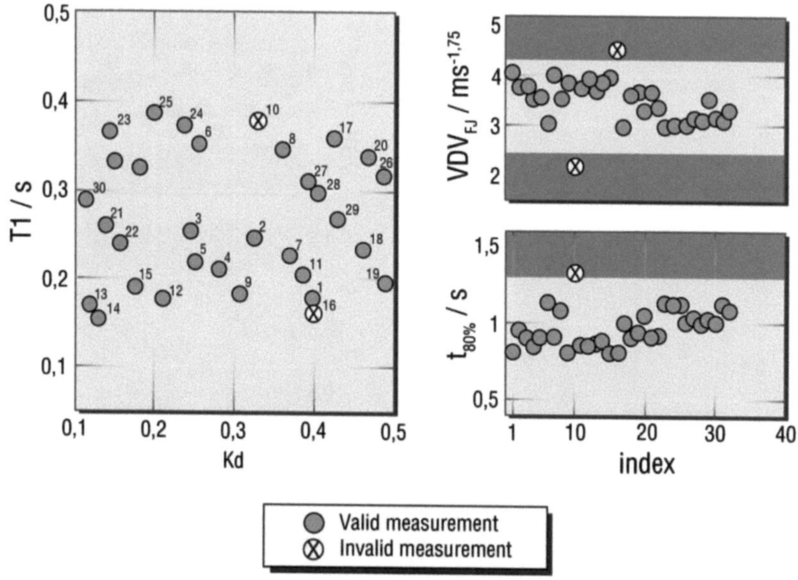


**Fig. 4** Online DoE approach with conservative behavior. *Left* Distribution of valid and invalid points in the input space. *Right* Distribution of valid and invalid points in the output space

measurement point of the online DoE algorithm automatically leads to a higher scattering in *Kd* than in *T1*, as the influence of *T1* on the engine behavior is much higher than the one of *Kd*.

In contrast, an aggressive decision selects a point that increases the scattering more at the expense of a higher risk of violating a limit, cf. Fig. 5. In the example, the settings lead to quick coverage of the complete valid design space. However, 8 out of 38 measurements violate an output limit. These limit violations can be found in two invalid regions of the design space. For high values of *T1* and small values of *Kd* the response during the tip-in is too slow, such that either the upper limit of $t_{80\ \%}$ is violated or the lower limit of $VDV_{FJ}$. For short *T1* timings, the first jerk is above the upper limit for the $VDV_{FJ}$.

For both conservative and aggressive settings of the algorithm, the typical trade-off between $VDV_{FJ}$ and $t_{80\ \%}$ is obtained, cf. Fig. 6. This trade-off can subsequently be used for calibration to control the dynamic of the driving behavior.

For the measurements of $t_{80\ \%}$ and $VDV_{FJ}$ there is no risk of damage to the engine. Hence, an overly conservative setting is not needed. To prevent too many measurements which are not applicable for calibration, these overly aggressive settings are also not profitable. In the next example, an appropriate setting in between the illustrative extreme settings above is chosen.

Usually there is no interest in the calibration of a single operating point, but of the complete operating range. Hence, the previous example is extended to four input dimensions including engine speed and load. Here, the operating area ranges

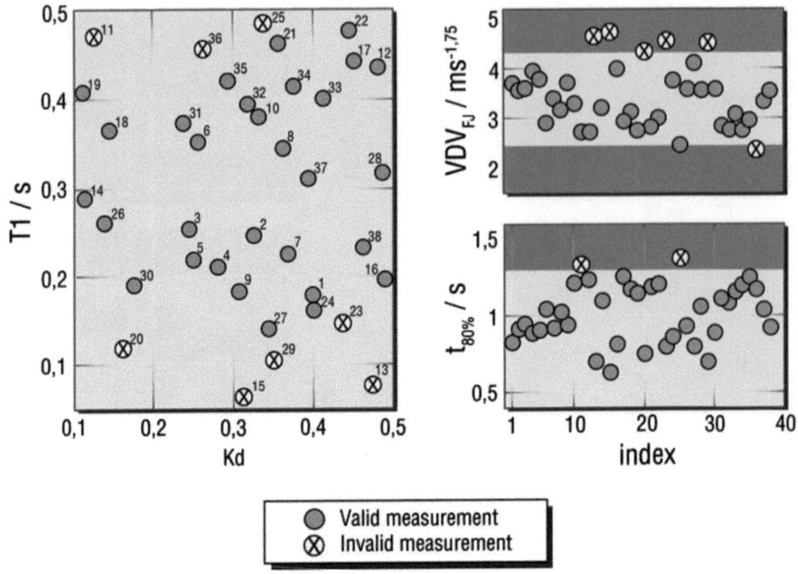


**Fig. 5** Online DoE approach with aggressive behavior. *Left* Distribution of valid and invalid points in the input space. *Right* Distribution of valid and invalid points in the output space

**Fig. 6** Trade-off between $VDV_{FJ}$ and $t_{80\ \%}$ for a tip-in at 2000 min$^{-1}$ and 50 % pedal position. *Black points* represent measured data, *gray points* represent predictions obtained from Gaussian process models for a set of 10,000 randomly distributed points

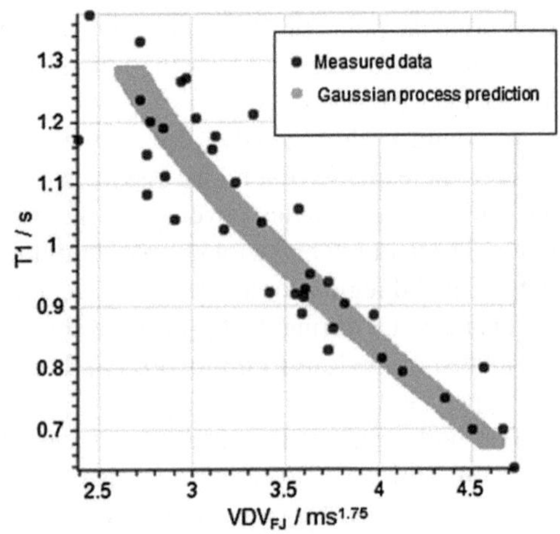


from 30 to 80 % pedal position (pedal positions above 80 % are considered as full load with the same ECU settings as for 80 %) and from 1000 to 4000 min$^{-1}$ engine speed.

In a first step, the extent of the valid design space is explored in the four input dimensions by fully automatically applying phases one and two of the online DoE approach in the vehicle. In phase one, a few initial points are measured in a central region of the design space. In phase two, points are subsequently added such that

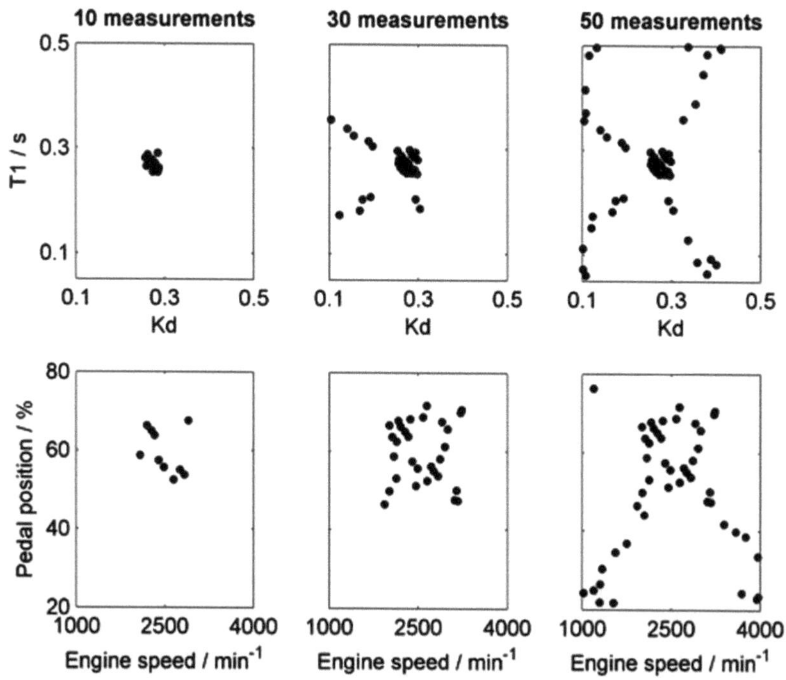


**Fig. 7** Distribution of the points in the input space after 10, 30, and 50 measurements as projections on the speed-load and *Kd*-*T1* plane

the scattering of the input values increases. One way of increasing the scattering of the input values is to add points at the border of the known design space. So in this phase, the algorithm tends to select points that increase the size of the known input space, cf. Fig. 7. At the same time, the algorithm takes into account that the valid design space is confined by the output limits, so not all points that increase the scattering can be selected as they might violate an output limit. The exact position of the output limits in the input space is not known a priori. However, their position is estimated on the basis of predictions from the Gaussian process models. With increasing number of measurements these model predictions become more precise, such that the shape of the valid design space is also known with higher accuracy. This increase in accuracy means that the volume percentage of falsely classified design space becomes smaller, cf. Fig. 8, i.e. a lower percentage is classified as invalid although it is part of the valid design space or classified as valid although an output limit is violated. The determination of the valid design space works independent of its shape, since it is based on model predictions in the output space. Hence, even complex shapes of the valid design space in the input space are identified, such as holes and non-convex boundaries, cf. Fig. 9.

The first two phases aim at quickly exploring the boundaries of the valid design space. In a subsequent step, phase three of the online DoE approach is applied to

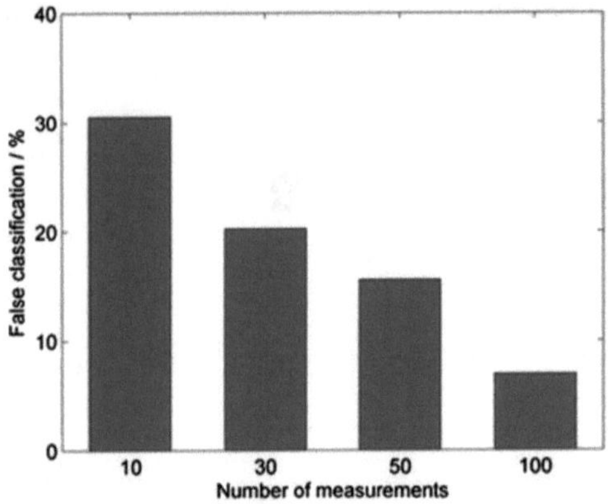


**Fig. 8** Percentage of false classified volume (valid as invalid or invalid as valid) in the original input hyper cube after 10, 30, 50, and 100 measurements. The reference volume is based on 300 measurements

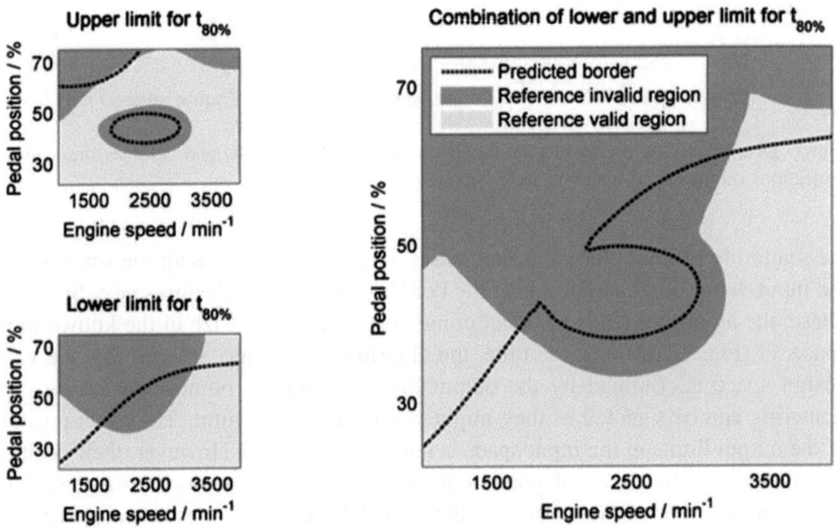


**Fig. 9** Prediction of the border of the valid design space at $Kd$ = 0.1 and $T1$ = 0.5 s after 46 measurements. Reference boundaries are based on 300 measurements

automatically fill up this known valid design space with points. The points are taken from an automatically generated test plan that contains 150 points distributed in a space-filling manner within the currently known borders of the valid design space. Points are picked in the order given by the test plan, measured, and added to the models. These models already take into account the points from phases one and two, but are now successively fed with additional measurements from phase three.

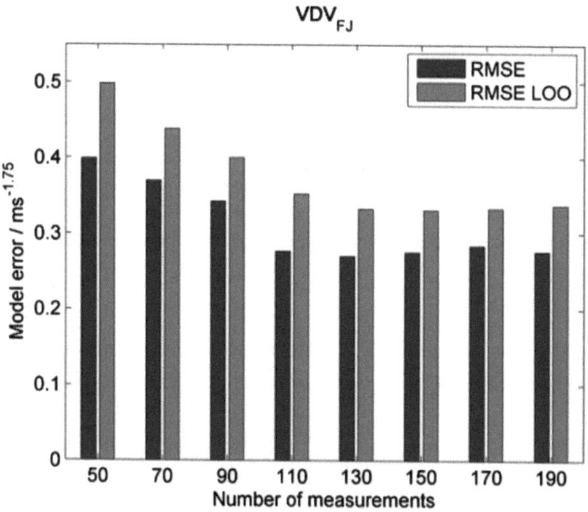


**Fig. 10** Model error in terms of root mean square error (RMSE) and leave-one-out root mean square error (RMSE LOO) for $VDV_{FJ}$ after 50, 70, 90, 110, 130, 150, 170, 190 measurements

With increasing number of measured points the quality of the models increases and the predictive power of the Gaussian process models improves, cf. Fig. 10.

In some cases, a transformation of the output data as included in warped Gaussian process models can provide models with an even better predictive power than standard Gaussian process models. The AIC is applied to compare standard and warped Gaussian process models.

As the distribution of the output data for $t_{80\ \%}$ is very homogeneous and close to a normal distribution, the model does not gain from a transformation of the output data for few measurements. The difference between standard and warped Gaussian process model for $t_{80\ \%}$ is insignificant for $\lesssim$160 measurements and hence a standard Gaussian process is preferred, cf. Fig. 11. For more measurements, the AIC prefers warped Gaussian processes.

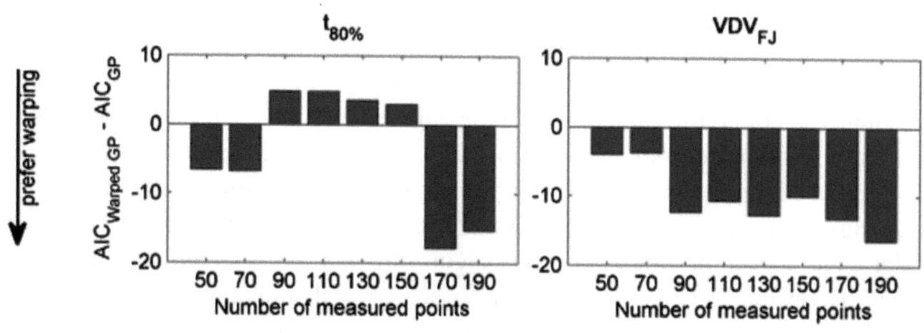


**Fig. 11** Comparison of $AIC$ criterion for standard and warped GP models of $t_{80\ \%}$ and $VDV_{FJ}$ as a function of the number of measured points

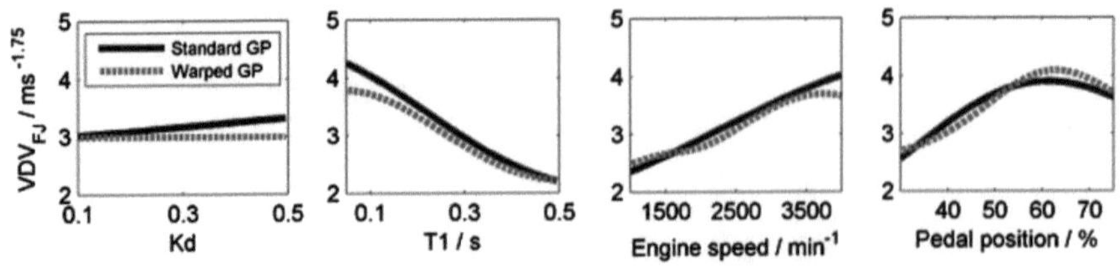


**Fig. 12** Intersection plots of standard and warped Gaussian process model in the middle of the design space with 100 measured points

For $VDV_{FJ}$, the warped Gaussian process model exhibits a significantly better *AIC* value than the standard Gaussian process model starting at 80 measurement points.

Still, the differences between the AIC values of these models is rather small. For example the models for $VDV_{FJ}$ are very similar as shown in Fig. 12. However, warped Gaussian process models could be favorable for different problems where the data requires a more complex model description, e.g. models for $NO_x$, HC, or particulates emissions.

## 6 Summary

Since vehicle calibration steps like drivability and final approval will still be conducted mainly in the vehicle for the next years it is important to apply efficient model-based approaches for this task. The target is to reduce development times on the one hand and on the other hand to manage the increasing complexity of the calibration tasks. The aim of this paper has been to explore the possibility to take over procedures and modeling techniques from test benches into the vehicle and evaluate the applicability. Therefore the combination of the automation of driving maneuvers with online DoE methods has been studied in view of an efficient in-vehicle calibration. Manual iteration loops are replaced by an automated enhanced online test planning, maneuver driving, evaluation, and modeling procedure. The presented online DoE approach consists of three phases. The first two phases focus on quickly determining the valid design space and the third phase on the modeling part. An important part is the modeling algorithm, since it is used to determine the boundaries and to model the engine behavior. Gaussian processes and their extensions turned out to fulfill this task and have the capability to determine non-convex design spaces. The developed algorithm allows for a fast exploration of the design space and takes into account a description of the risk of violating a border. The different phases of the algorithm can be adapted and combined depending on the use case. The approach has been demonstrated on selected use cases of the drivability calibration domain using a tip-in as standardized driving maneuver.

**Acknowledgements** The authors would like to thank Christoph Donker, Alexander Dornheim and Jessica Plock for their support.

## References

1. Baumann, W., Dreher, T., Röpke, K., Stelzer, S.: DoE for series production calibration. In: Design of Experiments (DoE) in Engine Development. Expert Verlag (2013)
2. Klar, H., Klages, B., Gundel, D.: Automation of model-based calibration in engine development. In: Design of Experiments (DoE) in Engine Development. Expert Verlag (2011)
3. Myers, R.H., Montgomery, D.C.: Response Surface Methodology. Wiley, New York (1995)
4. Picheny, V., Ginsbourger, D., Roustant, O., Haftka, R.T., Kim, N.: Adaptive designs of experiments for accurate approximation of a target region. J. Mech. Des. **132**(7)
5. Raidt, B.: A strategy to employ criteria for online location selection when using Gaussian processes. In: Design of Experiments (DoE) in Engine Development. Expert Verlag (2015)
6. Jones, D.R., Schonlau, M., Welch, W.J.: Efficient global optimization of expensive black-box functions. J Global Optim. **13**(4), 455–492
7. Poland, J., Knödler, K., Fleischhauer, T., Mitterer, A., Ullmann, S., Zell, A.: Modellbasierte Online-Optimierung moderner Verbrennungsmotoren Teil 1: Aktives Lernen. Motorentechnische Zeitschrift **64**(5), 432–437
8. Poland, J., Knödler, K., Fleischhauer, T., Mitterer, A., Ullmann, S., Zell, A.: Modellbasierte Online-Optimierung moderner Verbrennungsmotoren Teil 2: Grenzen des fahrbaren Suchraums. In: Motorentechnische Zeitschrift **64**(6), 520–526
9. Deregnaucourt, M., Didcock, N., Hametner, C., Jakubek, S., Lanschützer, H., Rainer, A., Stadlbauer, M., Zettel, K.: Methode zur Ermittlung eines Modells einer Ausgangsgröße eines technischen Systems. Austrian patent application AT000000512977, filed 22 Mai 2012
10. Angermaier, S., Diener, R., Huber, T., Imhof, V., Kloppenburg, E., Kruse, T., Markert, H., Streichert, F., Ulmer, H.: Verfahren zum Erstellen einer Funktion für ein Steuergerät. German patent application DE102011081346, filed 22 Aug 2011
11. Poland, J.: Modellgestützte und Evolutionäre Optimierungsverfahren für die Motorentwicklung. Dissertation Universität Tübingen (2002)
12. Rasmussen, C., Williams, C.: Gaussian Processes for Machine Learning. MIT Press, Cambridge (2006)
13. Tietze, N., Konigorski, U., Nguyen-Tuong, D.: Local Gaussian process regression for model-based calibration of engine control units. In: Simulation and Testing for Automotive Electronics V. Expert Verlag (2014)
14. Snelson, E., Rasmussen, C., Ghahramani, Z.: Warped gaussian processes. Adv. Neural Inf. Process. Syst. **16**, 337–344 (2004)
15. Thewes, S., Lange-Hegermann, M., Reuber, C., Beck, R.: Advanced Gaussian process modeling techniques. In: Design of Experiments (DoE) in Engine Development. Expert Verlag (2015)
16. Akaike, H.: Information theory and an extension of the maximum likelihood principle. In: Petrov, B.N., Csáki, F. (eds.) 2nd International Symposium on Information Theory, Tsahkadsor, Armenia, USSR, 2–8 Sept 1971. Akadémiai Kiadó, Budapest, pp. 267–281
17. Wolff, K., Kraaijeveld, R., Hoppermanns, J.: Objektivierung der Fahrbarkeit/Objective Parameters for Vehicle Driveability. In: Subjektive Fahreindrücke sichtbar machen IV. Expert Verlag (2009)
18. Beck, R., Kinoo, B., Haverkort, G., Schloßer, A., Krause, M.: Toolunterstützte Fahrbarkeitsapplikation/tool supported drivability calibration. In: Proc.: 4. Internationales Symposium für Entwicklungsmethodik (2011)

19. Association for Standardisation of Automation and Measuring Systems; ASAM MCD-3 MC standard. Retrieved from https://wiki.asam.net/display/STANDARDS/ASAM+MCD-3+MC on 29th Feb 2016
20. Nguyen, N., Miller, A.J.: A review of some exchange algorithms for constructing discrete D-optimal designs. In: Comput. Stat. Data Anal. **14**(4), 489–498
21. Welch, W.J.: Computer-aided design of experiments for response estimation. In: Technometrics **26**(3), 217–224
22. Smith, K.: On the standard deviations of adjusted and interpolated values of an observed polynomial function and its constants and the guidance they give towards a proper choice of the distribution of observations. Biometrika **12**(1), 1–85
23. Fedorov, V.V.: Theory of optimal experiments. Academic Press, London (1972)
24. Zaglauer, S., Deflorian, M.: Bayesian D-optimal design. In: Design of experiments (DoE) in engine development. Expert Verlag (2011)
25. Zaglauer, S.: D-optimal DoE with covariance matrix adaptation. In: Design of Experiments (DoE) in Engine Development. Expert Verlag (2013)
26. Pillas, J., Kirschbaum, F., Jakobi, R., Gebhardt, A., Uphaus, F.: Model-based load change reaction optimization using vehicle drivetrain test beds. In: Proc.: 14. Internationales Stuttgarter Symposium. Springer, Berlin (2014)
27. Griffin, M.J.: Handbook of Human Vibration. Academic Press, London (1996)

# Part X
# Simulation/Testing/Optimization/E/E Systems

# Average Models for Hardware-in-the-Loop Simulation of Power Electronic Circuits

**Axel Kiffe and Thomas Schulte**

**Abstract** Power electronic devices are growing in importance in automotive applications. Power converters are used in hybrid electric vehicles, but also in other vehicle applications like electric steering systems for example. For testing electronic equipment, hardware-in-the-loop (HIL) simulation is a today's standard method in the automotive industry and other fields, as well. This contribution provides an overview of different average modeling methods for the real-time simulation of power electronics, which are necessary for hardware-in-the-loop simulation. The average models are described and evaluated by the criteria computation effort, algorithmizability, generalizability and platform usability. The results are summarized to give hints to development engineers for choosing a suitable modeling approach for hardware-in-the-loop simulation of their specific applications.

**Keywords** Hardware-in-the-loop · Real-time · Simulation · Power electronic

## 1 Introduction

Power electronic devices are important in almost all industrial sectors. Power electronics is used for electric drives and power converters in many fields like automation and automotive industry. For testing such electronic equipment hardware-in-the-loop (HIL) simulation is almost a standard method today, especially in the automotive industry. HIL test benches simulate the real environment of the control unit by a real-time simulation of the plant incorporating an interface for connecting the actuator

A. Kiffe (✉) · T. Schulte
Hochschule Ostwestfalen-Lippe University of Applied Sciences, Liebigstr. 87, 32657 Lemgo, Germany
e-mail: axel.kiffe@hs-owl.de
URL: http://www.hs-owl.de; http://simts.de

T. Schulte
e-mail: thomas.schulte@hs-owl.de
URL: http://www.hs-owl.de; http://simts.de

 
C. Gühmann et al. (eds.), *Simulation and Testing for Vehicle Technology*,
DOI 10.1007/978-3-319-32345-9_22

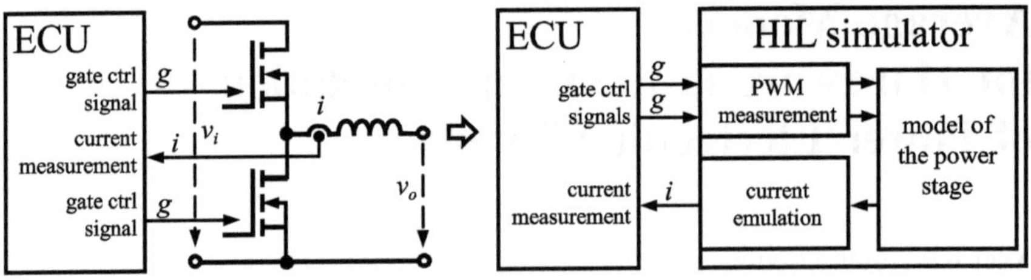


**Fig. 1** Real system (*left*) and HIL simulation (*right*) [1]

and sensor lines. This reduces the costs and improves the test efficiency by enabling automated testing in a laboratory with repeatable conditions.

A control unit, which controls power electronics, can be integrated into a HIL simulation by replacing the power stage and connecting the gate control and measurement lines to the HIL test bench, Fig. 1. The gate control signals for the semiconductor switches are captured by appropriate input channels, while the measurement signals for voltages and currents are generated by the test bench.

Since the real power stage is removed, this kind of HIL simulation can easily be adapted and heavy equipment, like electronic loads or mechanical test benches, is not required [2]. However, the development of the real-time capable models is still an ambitious challenge, due to switching events, the limited performance of the computation platform and the general properties of power electronics:

- Natural and forced switching events:
  Power electronic circuits comprising semiconductors, operated as switches. Like transistors, they are either controlled by an electronic control system via control signals or like diodes by internal values of the power electronic circuit or like thyristors even by both. Neglecting any detailed behavior, e.g. parasitic capacitances or the switching transients, the semiconductors can be considered as ideal switches, changing their states and causing switching events in the simulation. According to their control mechanism, switching events can be distinguished between internal (natural) switching events, caused by internal values, and external (forced) switching events, caused by control signals, [4]. When simulating power electronics, switching events cause changes of the system's structure. For natural switching events, the changes depend on internal values of the system, yielding implicit relations in the simulation algorithms.
- Fixed simulation step-size:
  Variable-step integration methods are common for electric circuit simulations for considering the switching events accurately. However, the time-constraints of real-time systems requires fixed-step integration [3]. For the appropriate consideration of switching events in between two simulation sample steps, special simulation algorithms are mandatory. One type of these methods are average models, which are described in Sect. 3. Generally, the system dynamics of power electronic circuits is high compared to mechanical systems. Despite the high eigenvalues, which can also be caused by the modeling approach, suitable

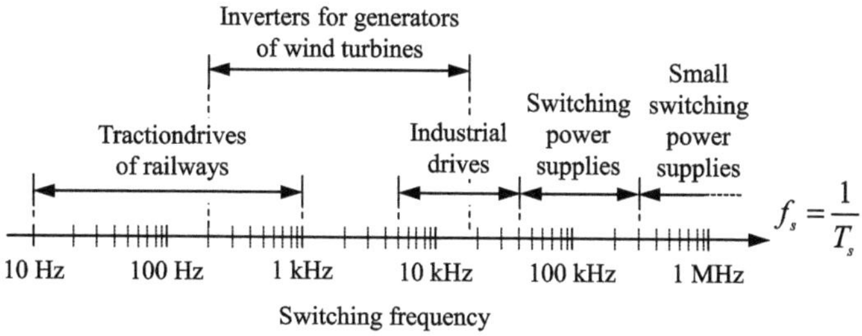


**Fig. 2** Applications and their typical switching frequencies

discretization methods like implicit A-stable or L-stable methods are required. However, the high dynamic is primarily caused by the switching of the semiconductors. In Fig. 2 some typical applications and their switching frequencies are shown. For a fundamental classification, the oversampling factor [4]

$$\kappa = T_s/T, \quad T \geq T_{calc} \tag{1}$$

can be used, where $T_s$ is the switching period, $T$ is the sampling period and $T_{calc}$ is the calculation time for one simulation step.

- Step size and simulation accuracy:
  A general issue in real-time simulation is the determination of an adequate compromise regarding model accuracy and computation effort. On one hand, the accuracy of the simulation and the modeling depth must meet the requirements, but on the other hand, the computation time must be short enough in HIL simulation for ensuring real-time capability.

Different modeling approaches for power electronic circuits have been developed in the last couple of years, which can generally be used for real-time simulation. The approaches can be classified by different aspects, but the sampling strategy and the handling of discontinuities are the most significant criteria which define the final performance. In the following section, different sampling strategies are described. Afterwards a detailed overview of modeling methods for average models is given. The methods are evaluated by considering the following criteria:

- *Computation effort*, which basically depends on the computation time per simulation step and the minimum oversampling factor that meets the requirements regarding the application.
- *Algorithmizability*, which means the applicability and systematization of the method for integration in simulation tools.
- *Generalizability*, which means the applicability of the simulation method to arbitrary circuits and large topologies.
- *Platform usability*, assessed the meaningful applicability to FPGA or microprocessors.

After describing the average methods, a classification and discussion is presented in the final part of this contribution.

## 2 Sampling Strategies

Models for real-time simulation of power electronics can be classified by their sampling strategies. For illustration, Fig. 3 shows the inductor current of a buck converter in continuous conduction mode for the low-rate synchronous and the high-rate asynchronous sampling. It is assumed, that the electronic control unit (ECU) samples synchronously to the PWM signal. For ensuring a correct sampling of the simulated signals by the ECU, two sampling strategies are suitable:

- Low-rate synchronous sampling:
  The sampling and output of the simulator is done once a switching period or with a small multiple of the sampling frequency of the ECU. Therefore it is more time for calculating the model. For this sampling strategy it is necessary to synchronize ECU and simulator to avoid beat effects, hence, synchronization is necessary. In the case of a fixed switching frequency of the circuit, synchronization can be realized by appropriate input devices, measuring the PWM signal and trigger the model.
- High-rate asynchronous sampling:
  The sampling and output of the simulator is done with large multiple of the sampling frequency of the ECU. Compared to the low-rate synchronous sampling, the time for calculating the model is significantly smaller. However, when the oversampling is high enough, synchronization can be neglected.

As a matter of principle, the dead time $T_t$ of both sampling strategies is at least one simulation step-size $0 < T_t < T$. Thus, the dead times of models for low-rate synchronous sampling are larger than for models requiring high-rate asynchronous sampling. Table 1 compares the properties of both sampling strategies. As stated in the last row of the table, different types of models are suitable for HIL-simulation. This contribution focuses on average models.

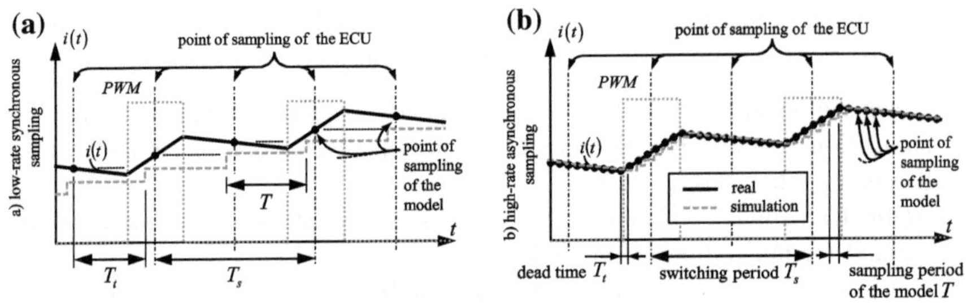


**Fig. 3** Sampling strategies of models: **a** low-rate synchronous sampling, **b** high-rate asynchronous sampling

**Table 1** Comparison of sampling strategies

| Properties | Low-rate synchronous sampling | High-rate asynchronous sampling |
|---|---|---|
| Synchronization | Necessary | Not necessary |
| Input-/output-time | Low | High |
| Calculation time | More | Less |
| Dead-time $T_t$ | High | Low |
| Models | Average models, correction methods | Oversampling models |

# 3 Average Models

Average models are well known in the field of controller design and real-time simulation. For derivation of an average model, a sequence of switch states must be already known. The sequence for the buck converter, shown in Fig. 4a, can be noticed by the inductor current in discontinuous conduction mode—see Fig. 4b.

The length of the first interval $d_1 T_s$ is directly given by the duty-cycle of the PWM-signal, which controls the transistor and can be measured by appropriate input channels of the HIL bench. The length of the second interval depends on internal values of the buck converter, because a natural switching event determine the time, when the diode changes its conduction state.

A lot of average modeling approaches can be found in the literature. Primarily, they were developed for describing the continuous conduction mode, assuming that switching events are affected by external signals, only. Therefore, this contribution distinguishes between conventional models/methods, which are restricted to the continuous conduction mode, and extended models/methods, which consider the discontinuous conduction mode.

The types of average models can be roughly divided into in-circuit averaging methods (ICAM) and space averaging methods (SPAM and dSPAM), which are described in the following subsections.

## 3.1 In-Circuit Averaging Method (ICAM)

The basic concept of the ICAM is the separation of the power electronic circuit in a fast and a slow subsystem, [5, 6]. While the values of the slow subsystem $\mathbf{x}_s = \mathbf{f}_s(\bar{\mathbf{x}}_f)$ are assumed to be constant, i.e. they already represent the average values $\bar{\mathbf{x}}_s$, the values of the fast part $\mathbf{f}_f(\bar{\mathbf{x}}_s, d_i)$ have to be averaged:

$$\bar{\mathbf{x}}_f = \frac{1}{T_s}\int_0^{T_s} \mathbf{f}_f(\bar{\mathbf{x}}_s, d_i)d\tau, \quad \bar{\mathbf{x}}_s \approx \mathbf{x}_s = \mathbf{f}_s(\bar{\mathbf{x}}_f). \tag{2}$$

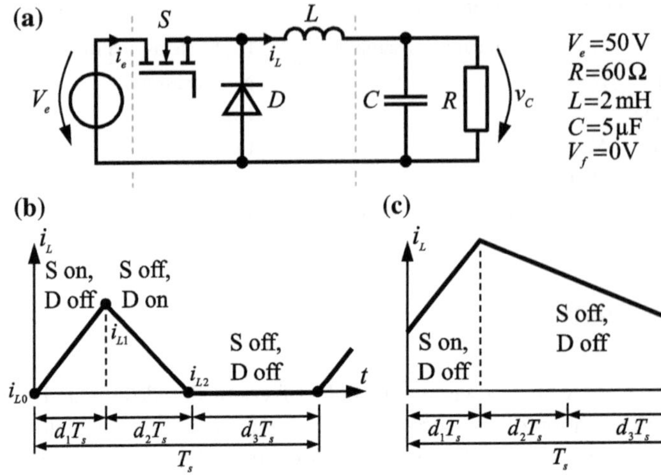


**Fig. 4** **a** Schematic of a buck converter, **b** inductor current in discontinuous conduction mode, **c** inductor current in continuous conduction mode

Thus, the modeling procedure can be divided in the four steps

- separation,
- describing the slow subsystem,
- averaging the fast subsystem and
- determination ratios.

These are described in the next subsections for the above application example. Afterwards, the simulation algorithm is described and properties of models based on the ICAM are concluded.

**Separation**

The circuit is separated in a slow and fast subsystem. For the continuous conduction mode it is usually sufficient, that the fast subsystem consists of the ideal switches, only. Nevertheless, when the discontinuous conduction mode should be considered, the fast subsystem must contain further devices. For instance, when a buck converter in discontinuous conduction mode should be simulated, the inductor must be added to the fast subsystem as well [7]. Thus, the dashed line in Fig. 4a represents the line between the fast and slow subsystem of the circuit.

**Describing the Slow Subsystem**

To determine a mathematical description of the slow subsystem, the differential equations for all linearly independent energy storage devices (inductors and capacitors) of the slow subsystem must be known. The capacitor voltage of the buck converter, depicted in Fig. 4a, is given by:

$$dv_C/dt = C^{-1}\left(i_L - R^{-1}v_C\right) \tag{3}$$

Because of the superposition principle for averaging

$$\overline{\alpha x(t) + \beta y(t)} = \alpha\bar{x}(t) + \beta\bar{y}(\tau) \tag{4}$$

the average of the capacitor voltage (3) can be written:

$$d\bar{v}_C/dt = C^{-1}\left(\bar{i}_L - R^{-1}\bar{v}_C\right) \tag{5}$$

**Averaging the Fast Subsystem**

The inductor current in (5) is described by its average $\bar{i}_L$, which is determined in this subsection. As mentioned above, the values of the slow subsystem are assumed to be constant for the fast subsystem and already represent average values [5, 6]. For the buck converter example the capacitor voltage and the input voltage assumed to be constant:

$$\bar{v}_C \approx v_C, \quad \bar{V}_e \approx V_e \tag{6}$$

Consequently, the derivative of the fast subsystem's inductor current $i_L$ and the input current $i_e$ of the buck converter can be described by:

$$\frac{di_L}{dt} = \begin{cases} (\bar{V}_e - \bar{v}_C)L^{-1} & 0 < t \leq d_1T_s \\ -L^{-1}\bar{v}_C & d_1T_s < t \leq (d_1 + d_2)T_s \\ 0 & (d_1 + d_2)T_s < t \leq T_s \end{cases}, \quad i_e = \begin{cases} i_L & 0 < t \leq d_1T_s \\ 0 & d_1T_s < t \leq T_s \end{cases}. \tag{7}$$

These equations are used to calculate the average of the derivative. It can be shown, that the average of the derivative is equal to derivative of the average and yield

$$d\bar{i}_L/dt = L^{-1}(d_1\bar{V}_e - (d_1 + d_2)\bar{v}_C). \tag{8}$$

The average of the input current is given by

$$\bar{i}_e = \frac{1}{T_s}\int\limits_0^{d_1T_s} i_L(\tau)d\tau. \tag{9}$$

In (9), the integral of the first interval of the switching period is necessary. Therefor the progression of the inductor current $i_L$ is necessary and can be determined by solving the differential Eq. (7):

$$i_L(t, i_{L0}) = \begin{cases} (\bar{V}_e - \bar{v}_C)L^{-1}t + i_{L0} & 0 < t \leq d_1T_s \\ -(t - d_1T_s)L^{-1}\bar{v}_C + i_{L1} & d_1T_s < t \leq (d_1 + d_2)T_s \\ i_{L2} & (d_1 + d_2)T_s < t \leq T_s \end{cases} \tag{10}$$

The currents $i_{L1}$ and $i_{L2}$ are given by:

$$i_{L1} = i_L(d_1 T_s, i_{L0}), \quad i_{L2} = i_L((d_1 + d_2)T_s, i_{L0}). \tag{11}$$

For a steady state time behavior ($i_{L0} = i_L(T_s, i_{L0}) = i_{L3} = i_{L2}$) with an arbitrary initial value $i_{L0}$, the input current is given by:

$$\bar{i}_e = (d_1 + d_2)^{-1} d_1 \bar{i}_L \tag{12}$$

Another method for the integral of (9) is given by determining the complete time behavior of a state variable in the switching period. Due to (10) and (11), the overall time behavior of the inductor current $i_L$ during the switching period $T_s$ is completely described, when the initial value $i_{L0}$ is known. Therefor the average of the inductor current can be calculated by using (10):

$$\begin{aligned} \bar{i}_L &= \frac{1}{T_s}\left( \int_0^{d_1 T_s} (\bar{V}_e - \bar{v}_C)L^{-1}t + i_{L0} d\tau + \int_{d_1 T_s}^{(d_1+d_2)T_s} -(t - d_1 T_s)\bar{v}_C L^{-1} + i_{L1} d\tau + \int_{(d_1+d_2)T_s}^{T_s} i_{L2} d\tau \right) \\ \Rightarrow i_{L0} &= \frac{1}{d_1 + d_2}\bar{i}_L + \frac{(d_1 + d_2)T_s}{2L}\bar{v}_C - \frac{(d_1 + 2d_2)d_1 T_s}{(d_1 + d_2)2L}\bar{V}_e - \underbrace{\frac{1 - d_1 - d_2}{d_1 + d_2} i_{L2}}_{=0} \end{aligned} \tag{13}$$

Therein, the current $i_{L2}$, which also depends on $i_{L0}$, was not replaced by $i_L((d_1 + d_2)T_s, i_{L0})$, since the last summand is still zero, which can be seen in Fig. 4b, c. When the buck converter operates in continuous conduction mode, the ratio $d_2$ is $1 - d_1$, which yield $1 - d_1 - d_2 = 0$. When the buck converter operates in discontinuous conduction mode, the inductor current and consequently the current $i_{L2}$ gets zero and the ideal diode is getting in off-state.

Finally, the initial value $i_{L0}$ can be used to calculate the input current by analytical integration or, due to the linear time behavior and the known currents $i_{L0}$, $i_{L1}$ and $i_{L2}$, directly by the trapezoidal integration rule:

$$\begin{aligned} \bar{i}_e &= \frac{1}{T_s}\int_0^{d_1 T_s} (\bar{V}_e - \bar{v}_C)L^{-1}\tau + i_{L0} d\tau = (\bar{V}_e - \bar{v}_C)L^{-1} d_1 + i_{L0} d_1 \\ &= \frac{d_1}{d_1 + d_2}\bar{i}_L + \frac{d_1 d_2 T_s}{2L}\bar{v}_C - \frac{T_s d_1^2 d_2}{2L(d_1 + d_2)}\bar{V}_e. \end{aligned} \tag{14}$$

**Determine Ratios**

In the previous steps, the switching times are assumed to be known. In the case of the continuous conduction mode, this assumption is fulfilled, because natural

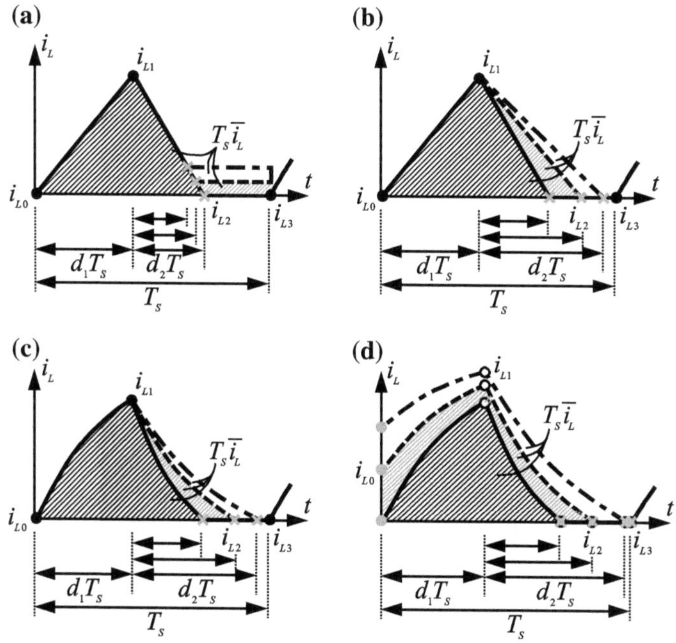

**Fig. 5** Effect of varying average of the inductor current on the ratio $d_2$

switching does not occur and all ratios can be measured by appropriate input devices of the real-time system [2]. In the discontinuous conduction mode, natural switching events occur and some ratios are depending on internal values of the circuit. In this case, the time behavior of the fast subsystem has to be evaluated. For the above example, the ratio $d_1$ can be determined by measuring the PWM-signal, which controls the transistor's switch state, whereas the ratio $d_2$ needs further investigations. In the literature, e.g. Refs. [7, 8], an inductor current for the discontinuous conduction mode in steady state ($i_{L0} = 0$) is used for calculating the ratios, like $d_2$. In Fig. 5a the current waveform according (10) is shown with an initial value $i_{L0} = 0$.

An increasing average or area ($T_s\bar{i}_L$) can only be achieved by decreasing the ratio $d_2$, which yields an area unequal to zero in the third interval. Depending on the ratio $d_2$, the current $i_{L2}$ is unequal to zero. Consequently, the ratio does not characterize the switching time of the diode, when the current becomes zero. To ensure, that the time $(d_1 + d_2)T_s$ conforms to a zero current, the slope of the second interval can be determined by:

$$i_L(t, i_{L0}) = \left(i_{L1} - \frac{i_{L2}-i_{L1}}{d_2T_s} d_1T_s\right) + \frac{i_{L2}-i_{L1}}{d_2T_s} t \quad d_1T_s < t \leq (d_1 + d_2)T_s \qquad (15)$$

The result is shown in Fig. 5b. Assuming a fixed initial value $i_{L0} = 0$, an increasing average can only reached, by increasing the ratio $d_2$, while

$i_L((d_1+d_2)T_s)=0$ is ensured. Determining the average of this waveform and reorganize to $d_2$ yields (16)

$$\bar{i}_L=\frac{(d_1+d_2)i_{L1}}{2}\Leftrightarrow d_2=\frac{2\bar{i}_L}{i_{L1}}-d_1,\quad i_{L1}=i_L(d_1T_s,0) \tag{16}$$

For exponential time behavior as depicted in Fig. 5c, Ref. [8] proposes the Equal-Area-Method for determining the equivalent linear time behaviors to apply (16). However, a similar method as for the linear time behavior can be transferred to the exponential time behavior, by determining the time constant of the second interval by the demand $i_{L2}=0$. Consequently, the known slope of the second interval, according to (10) or the time constant for the exponential behavior of the second interval is not considered. This is used by a more general method, which can be applied to current waveforms without assumptions on the initial value $i_{L0}$, like depicted in Fig. 5d. Compared to the procedure described in the previous subsection to determine the average for arbitrary intervals, the current $i_{L0}$ is determined by $i_L((d_1+d_2)T_s)=0$, which characterizes the switching event, when the diode getting in off state. For determining $i_{L0}$ (10) is applied to solve $i_{L2}=i_L((d_1+d_2)T_s,i_{L0})=0$ for the peak current $i_{L1}$ and afterwards $i_L(d_1T_s,i_{L0})=i_{L1}$ for $i_{L0}$. With $i_{L0}$ the ratio $d_2$ is the only unknown variable in (10). Therefore, the average $\bar{i}_L$ can be used to calculate the ratio $d_2$ by

$$d_2=\sqrt{\frac{T_s\bar{V}_e d_1^2+2L\bar{i}_L}{T_s\bar{v}_C}}-d_1\,. \tag{17}$$

**Realization of the Simulation Algorithm**

The model consists of a state space representation whose system and input matrix depends on the ratios and the equations for determine ratios, which refer to natural switching events. For the above application example (5), (8) and (12) or (5), (8) and (14) can be transferred to a state space representation. Equation (16) or (17) are describing the relation between average values and ratio $d_2$. Figure 6a depicts the simulation algorithm for ICAM models. In the first step, the continuous time state space representation is discretized, e.g. by trapezoidal integration rule. This is necessary during simulation, because the matrices of the state space representation depend on the ratios. In step two, the prospective state vector $\bar{\mathbf{x}}_{k+1}$ is calculated. Afterwards, the ratios (e.g. (16) or (17)), referring to natural switching events, are calculated. They must be calculated by using the prospective state vector, to ensure the correct ratios for each simulation step. This yields an implicit relation, since the ratios are already needed for calculating the prospective state vector. Therefore appropriate numerical methods are required to solve this nonlinear implicit equation. For the application example, with a scalar implicit equation, methods like the bisection or regula falsi method are suitable because convergence is guaranteed,

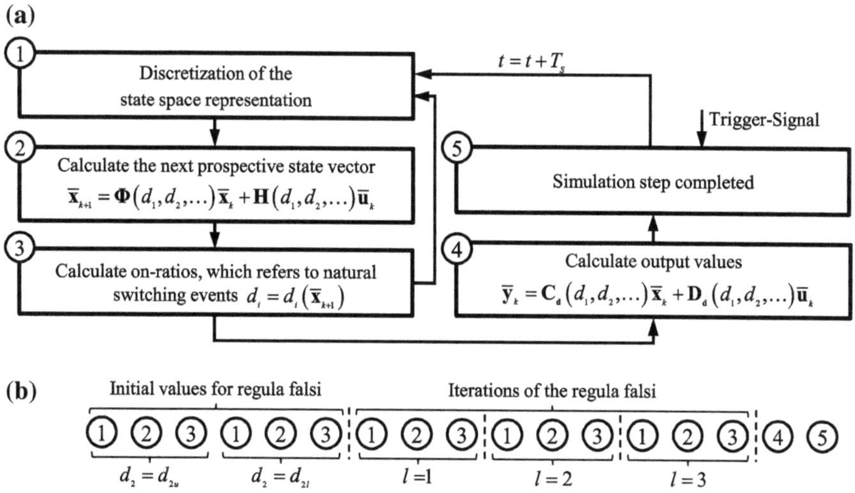


**Fig. 6** Simulation algorithm for models determined by the ICAM

when the solution is between their initial values. The initial values can be chosen to $d_{2l} = 0$ and $d_{2u} = 1 - d_1$ for the lower and upper initial value, when the signs of the corresponding evaluation of the implicit function are different for the initial values. In the other case, $d_2$ is equal to $1 - d_1$ and no iteration is necessary. The iterative procedure of the numerical methods is indicated by the arrow from step three to step one. In step four, the output values are calculated, which completes the simulation step. The following sequence of the algorithm is started by a trigger signal, which is generated, e.g., by measurement of PWM signals to synchronize ECU and HIL simulator. The sequence of the simulation algorithm for the above application example is shown in Fig. 6b. The circled numbers corresponds to the previously described steps shown in Fig. 6a. The simulation step starts with the evaluation of the implicit equation for the initial values $d_{2l}$ and $d_{2u}$. Afterwards three iterations of the regula falsi are performed ($l = 1, 2, 3$). At the end, the output equation is calculated. For comparison, the number of iteration was chosen experimentally and yields an accuracy, which is similar to the model determined by the discrete space averaging method, described in Sect. 3.3.

**ICAM's Properties**

Some fundamental properties of the ICAM are concluded from the above discussion:

1. For an appropriate balance in between accuracy and computation effort, the separation of the fast and small subsystem requires knowledge of the circuit's function and the modeling approach itself.

2. Based on the differential equation and its solution, like (10), as well as the continuity conditions of state variables, it is possible to determine the time behavior of the state variables by using average values, only.
3. By determining the state variables enables the calculation of any value and every average. However for large fast subsystems, the complexity increases significantly.
4. The complexity can be reduced by the assumption of steady state behavior for determining integrals, e.g., for output value $\bar{i}_e$. However the accuracy is reduced as well.
5. Depending on the separation of the fast and slow subsystem, the method to determine the ratios and further assumptions, a large variety of different models can be obtained.
6. For ratios referring to natural switching events, nonlinear implicit equations are obtained, which must be solved numerically. This increases the computation effort significantly and circuit specific investigations are necessary to choose the necessary number of iterations for appropriate accuracy and real-time capability.

## 3.2 State Space Averaging Method (SPAM)

The State Space Averaging Method (SPAM) differs from the ICAM in such way, that it is based on averaging matrices for the different switch states instead of signal averaging. This is done by weighting them by their ratios:

$$\dot{\bar{\mathbf{x}}} = \underbrace{\sum_{i=1}^{m} d_i \mathbf{A}_i}_{\bar{\mathbf{A}}} \cdot \bar{\mathbf{x}} + \underbrace{\sum_{i=1}^{m} d_i \mathbf{B}_i}_{\bar{\mathbf{B}}} \cdot \bar{\mathbf{u}}, \quad \bar{\mathbf{y}} = \underbrace{\sum_{i=1}^{m} d_i \mathbf{C}_i}_{\bar{\mathbf{C}}} \cdot \bar{\mathbf{x}} + \underbrace{\sum_{i=1}^{m} d_i \mathbf{D}_i}_{\bar{\mathbf{D}}} \cdot \bar{\mathbf{u}}, \quad \sum_{i=1}^{m} d_i = 1 \quad (18)$$

The matrices $\mathbf{A}_i$, $\mathbf{B}_i$, $\mathbf{C}_i$ and $\mathbf{D}_i$ are the system, input, output and feedthrough matrix of the continuous state space representation for the switch state of the $i$th interval. Additionally, $d_i$ is the corresponding ratio, the same as for the ICAM.

In Ref. [9] it is assumed, that the ratios refer to forced switching events, only, and can be measured as known. An extension to the discontinuous conduction mode, incorporating natural switching events, is given in Ref. [7]. As for the ICAM this requires ratios, which need to be calculated since they depend on internal values. Moreover, a correction matrix was used in Ref. [7] to extend (18) for getting appropriate results for the discontinuous conduction mode. A general approach was presented in Ref. [8] by adding further correction matrices for each switch state:

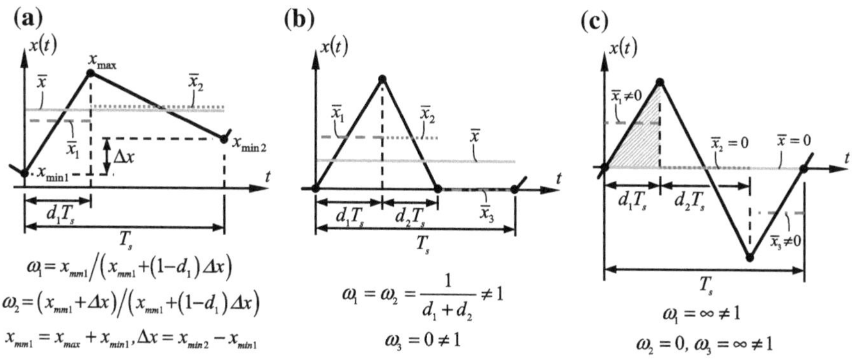


**Fig. 7** Different operation modes and their entries of the correction matrices

$$\dot{\bar{\mathbf{x}}} = \underbrace{\sum_{i=1}^{m} d_i \mathbf{A}_i \mathbf{W}_i}_{\bar{\mathbf{A}}} \cdot \bar{\mathbf{x}} + \underbrace{\sum_{i=1}^{m} d_i \mathbf{B}_i}_{\bar{\mathbf{B}}} \cdot \bar{\mathbf{u}}, \quad \bar{\mathbf{y}} = \underbrace{\sum_{i=1}^{m} d_i \mathbf{C}_i \mathbf{W}_i}_{\bar{\mathbf{C}}} \cdot \bar{\mathbf{x}} + \underbrace{\sum_{i=1}^{m} d_i \mathbf{D}_i}_{\bar{\mathbf{D}}} \cdot \bar{\mathbf{u}} \tag{19}$$

where $\mathbf{W}_i$ are diagonal correction matrices

$$\mathbf{W}_i = diag(\omega_{1,i}, \ldots, \omega_{n,i}) \tag{20}$$

with

$$\omega_{j,i} = \bar{x}_{j,i}/\bar{x}_j = \frac{1}{d_i T_s} \int_{t_{i-1}}^{t_i} x_j(\tau) d\tau \cdot \left( \frac{1}{T_s} \int_0^{T_s} x_j(\tau) d\tau \right)^{-1}, \quad t_n = T_s \sum_{l=1}^{n} d_l \tag{21}$$

Here, $j$ indicates the $j$th state variable and $i$ indicates the $i$th switch state. The calculation of the correction matrix is based on the idea of describing the average values over a subsystem of the switching period by factor $\omega_{j,i}$ and the average value over the complete switching period [8]. Figure 7 depicts different time behaviors and the corresponding entries of the correction matrices $\omega_i$ are given for discussing the SPAM.

In Fig. 7a a piece-wise linear time behavior of a buck converter's inductor current in continuous conduction mode is depicted. It can be found, that $\omega_1$ and $\omega_2$ are one and (18) is equal to (19), only, when $\Delta x$ is zero, which would be a steady-state operation. However, in several applications, the entries are nearly one and the deviations are small enough for neglecting the correction in transient operation. Figure 7b shows the steady-state operation of a buck converter in discontinuous conduction mode. Here, the correction is necessary, anyway, especially when $d_1$ and consequently $d_1 + d_2$ is getting very small. Figure 7c depicts a time behavior, whose average value is zero. Integrals, like the shaded area as well as the corresponding average over this period cannot be described by a multiplication of a factor and the overall average. The state space representation of the above application example (see Fig. 4a) is given by

$$\frac{d}{dt}\begin{bmatrix}\bar{i}_L\\ \bar{v}_C\end{bmatrix}=\left(d_1\underbrace{\begin{bmatrix}0 & \frac{-1}{L}\\ \frac{1}{C} & \frac{-1}{RC}\end{bmatrix}}_{\mathbf{A}_1}\begin{bmatrix}\omega_{1,1} & 0\\ 0 & \omega_{2,1}\end{bmatrix}+d_2\underbrace{\begin{bmatrix}0 & \frac{-1}{L}\\ \frac{1}{C} & \frac{-1}{RC}\end{bmatrix}}_{\mathbf{A}_2}\begin{bmatrix}\omega_{1,2} & 0\\ 0 & \omega_{2,2}\end{bmatrix}+d_3\underbrace{\begin{bmatrix}0 & 0\\ 0 & \frac{-1}{RC}\end{bmatrix}}_{\mathbf{A}_3}\begin{bmatrix}\omega_{1,3} & 0\\ 0 & \omega_{2,3}\end{bmatrix}\right)$$
$$\times\begin{bmatrix}\bar{i}_L\\ \bar{v}_C\end{bmatrix}+\left(d_1\underbrace{\begin{bmatrix}\frac{1}{L}\\ 0\end{bmatrix}}_{\mathbf{B}_1}+d_2\underbrace{\begin{bmatrix}0\\ 0\end{bmatrix}}_{\mathbf{B}_2}+\underbrace{d_3}_{1-d_1-d_2}\underbrace{\begin{bmatrix}0\\ 0\end{bmatrix}}_{\mathbf{B}_3}\right)\bar{V}_e=\begin{bmatrix}0 & \frac{-d_1-d_2}{L}\\ \frac{1}{C} & \frac{-1}{RC}\end{bmatrix}\begin{bmatrix}\bar{i}_L\\ \bar{v}_C\end{bmatrix}+\begin{bmatrix}\frac{d_1}{L}\\ 0\end{bmatrix}\bar{V}_e \tag{22}$$

with the correction coefficients $\omega_{1,1}, \omega_{1,2}$ and $\omega_{1,3}$ for the inductor current as well as $\omega_{2,1}, \omega_{2,2}$ and $\omega_{2,3}$ for the capacitor voltage. They can be calculated by several approaches and under diverse assumptions, like steady-state operation, linear approximations (when e.g. exponential would exist) or separation in fast and slow state variables similar to the ICAM. The correction coefficients for the capacitor voltage are often assumed to be constantly one [8], while the coefficients for the inductor current are already given in Fig. 7b. For the application example, the extended SPAM yields to the same state space representation as given by the ICAM. This can be explained, by the influence of the correction coefficient on the derivative of the average of the capacitor's voltage

$$C^{-1}d_1\omega_{1,1}+C^{-1}d_2\omega_{1,2}=C^{-1}\int\limits_0^{(d_1+d_2)T_s} i_L(\tau)d\tau\left(\int\limits_0^{T_s} i_L(\tau)d\tau\right)^{-1}. \tag{23}$$

Both integrals in (23) are equal, because in continuous conduction mode $d_2$ is $1-d_1$ and in the discontinuous conduction mode the current is zero for the interval $(d_1+d_2)T_s \leq t \leq T_s$. The input current $\bar{i}_e$ can also be expressed by a fraction of the complete average of the inductor current by correction coefficients according to (19):

$$\bar{i}_e = d_1\underbrace{\frac{\left(\frac{T_s\bar{V}_e d_1^2+2L\bar{i}_L}{(d_1+d_2)}-(d_1\bar{V}_e-d_2\bar{v}_C)T_s\right)\frac{1}{2L}}{\bar{i}_L}}_{\omega_{1,1}}\bar{i}_L. \tag{24}$$

Here, the integrals for the numerator and denominator according to (21) are determined by the approach described for the ICAM—calculating the initial value $i_{L0}$ and the integrals afterwards. It can be noticed, that the representation according to (24) is not advantageous for $\bar{i}_L = 0$, because a division by zero occurs. For the application example, the input current $\bar{i}_e$ is zero as soon as $\bar{i}_L$ becomes zero (ideal switches), which can be considered in the simulation. However, it is more advantageous to reduce the fraction in (24), which yields (14). This representation needs less computation effort and enables a description by an output equation of a linear state space representation, whose output and feedthrough matrix is depending on the ratios.

Finally the ratios referring to natural switching events are required. They can be determined by the approaches described for the ICAM. Comparing (5) and (8) with (22), the equations are the same as well as the ratios. Therefore, the sequence of the algorithm is the same as described in Sect. 3.1 and Fig. 6.

**SPAM's Properties**

Here some fundamental properties of the SPAM can be concluded from the above discussion:

1. The modeling result of SPAM and ICAM can be identical. This depends on the assumptions and simplifications.
2. For calculating the correction coefficients, assumptions on the state variables in fast and slow can be suitable to reduce the computation effort, but requires expert knowledge of the circuit's function as for the ICAM.
3. Suitable simplifications, like the assumption of steady state behavior for determining integrals, e.g., the output value $\bar{i}_e$, requires expert knowledge.
4. According to the ICAM, ratios referring to natural switching events yield nonlinear implicit relations, which must be solved numerically, increasing the computation effort significantly. Moreover, circuit specific investigations are necessary to determine an appropriate number of iterations for a balance of accuracy and calculation effort.

## 3.3 discrete State Space Averaging Method (dSPAM)

The idea of the dSPAM is the interval-wise calculation of the time behavior during the switching period, which is used to calculate the average values, afterwards [10]. Figure 8a–c illustrate the function of the dSPAM, which basis on the step-wise transition through the intervals. The matrices $\mathbf{\Phi}_i = \mathbf{\Phi}_i(d_i T_s)$, $\mathbf{H}_i = \mathbf{H}_i(d_i T_s)$, $\mathbf{C}_{di}$

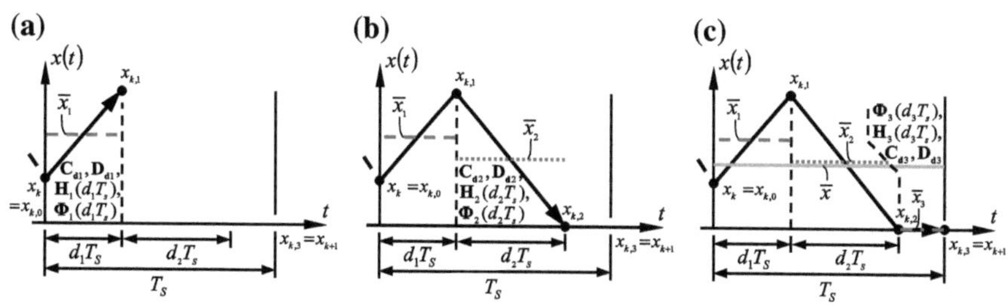


**Fig. 8** Illustration of the dSPAM's function principle

and $\mathbf{D_{d}}_i$ represent the discrete system, input, output and feedthrough matrix of the switch state according the $i$.th interval. The continuous system and input matrices ($\mathbf{A}_i$,$\mathbf{B}_i$) can be taken from (22). In Fig. 8, $\bar{x}$ denotes the average output value of the inductor current $x$.

For the application example above three intervals have to be considered:

$$\mathbf{x}_{k,1} = \mathbf{\Phi}_1\mathbf{x}_k + \mathbf{H}_1\mathbf{u}_k, \quad \mathbf{x}_{k,2} = \mathbf{\Phi}_2\mathbf{x}_{k,1} + \mathbf{H}_2\mathbf{u}_k, \quad \mathbf{x}_{k,3} = \mathbf{x}_{k+1} = \mathbf{\Phi}_3\mathbf{x}_{k,2} + \mathbf{H}_3\mathbf{u}_k. \tag{25}$$

Based on the intermediate values, the average of the output can be determined by

$$\begin{aligned}\bar{\mathbf{y}}_k &= \frac{1}{T_s}\int\limits_{t_k}^{t_k+T_s} \mathbf{y}(\tau)d\tau = \sum_{i=1}^{m} d_i\left(\mathbf{C_{d}}_i\frac{1}{d_iT_s}\int\limits_0^{d_iT_s} \mathbf{\Phi}_i(\tau)\mathbf{x}_{k,i-1} + \mathbf{H}_i(\tau)\mathbf{u}_k d\tau + \frac{1}{d_iT_s}\int\limits_0^{d_iT_s} \mathbf{D_{d}}_i\mathbf{u}_k d\tau\right)\\ &= \sum_{i=1}^{m} d_i\left(\mathbf{C_{d}}_i\overline{\mathbf{\Phi}}_i\mathbf{x}_{k,i-1} + (\mathbf{C_{d}}_i\bar{\mathbf{H}}_i + \mathbf{D_{d}}_i)\bar{\mathbf{u}}_k\right)\end{aligned} \tag{26}$$

with

$$\overline{\mathbf{\Phi}}_i = \frac{1}{d_iT_s}\int\limits_0^{d_iT_s} \mathbf{\Phi}_i(\tau)d\tau, \quad \bar{\mathbf{H}}_i = \frac{1}{d_iT_s}\int\limits_0^{d_iT_s} \mathbf{H}_i(\tau)d\tau. \tag{27}$$

For avoiding the necessity of inverse matrices of the continuous system matrices $\mathbf{A}^{-1}$, the discretization method described in Ref. [1] is suitable, which applies a zero order hold for the input $\mathbf{u} = \mathbf{u}_k = \mathbf{u}(t_k)$. Elsewise, the output matrices $\mathbf{C_{d}}_i$ and $\mathbf{D_{d}}_i$ would depend on the discretization time $d_iT_s$. Moreover, the discrete system matrix $\mathbf{\Phi}_3(d_3T_s)$ could not be determined in discontinuous conduction mode, because $\det(\mathbf{A}_3) = 0$. Accordingly, the averaged discrete matrices in (27) can be approximated by trapezoidal integration rule and yields [11, 12]:

$$\begin{aligned}\overline{\mathbf{\Phi}}_i\mathbf{x}_{k,i-1} &= \frac{1}{d_iT_s}\int\limits_0^{d_iT_s} \mathbf{\Phi}_i(\tau)d\tau\,\mathbf{x}_{k,i-1} \approx \frac{1}{2}(\mathbf{\Phi}_i(d_iT_s) + \mathbf{I})\mathbf{x}_{k,i-1} = \frac{1}{2}\left(\mathbf{x}_{k,i} + \mathbf{x}_{k,i-1}\right),\\ \bar{\mathbf{H}}_i &= \frac{1}{d_iT_s}\int\limits_0^{d_iT_s} \mathbf{H}_i(\tau)d\tau \approx \frac{1}{2}\mathbf{H}_i(d_iT_s).\end{aligned} \tag{28}$$

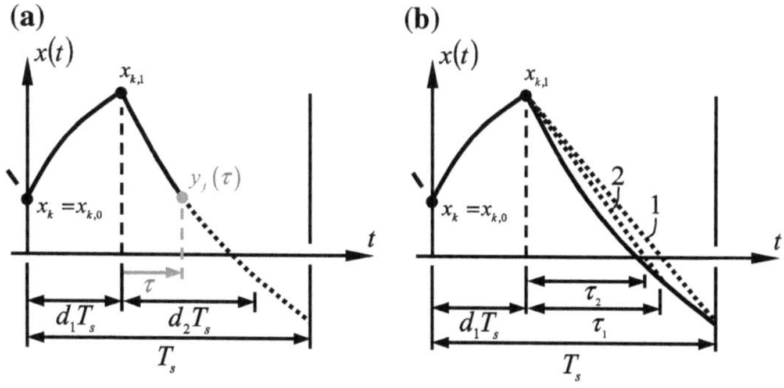


**Fig. 9** Illustration of the ratio determination for dSPAM models

This simplification is suitable for a many circuits, because the deviation is quite small for approximately linear time behavior and it is not fed back into the model.

So far, the ratios were assumed to be known. Compared to the SPAM and ICAM, the ratios can be determined much easier, because they need not be calculated from average values. Assuming that the characteristic voltage or current, e.g. the inductor current of the buck converter, is described by the *j*th row of the output vector **y**, the time behavior of the *i*th interval can be described by:

$$y_j(\tau) = \mathbf{C}_{\mathbf{d}i,j}\left(\mathbf{\Phi}_i(\tau)\mathbf{x}_{k,i-1} + \mathbf{H}_i(\tau)\mathbf{u}_k\right) + \mathbf{D}_{\mathbf{d}i,j}\mathbf{u}_k \tag{29}$$

with

$$\mathbf{C}_{\mathbf{d}i} = \begin{bmatrix} \mathbf{C}_{\mathbf{d}i,1} \\ \mathbf{C}_{\mathbf{d}i,2} \\ \vdots \end{bmatrix}, \quad \mathbf{D}_{\mathbf{d}i} = \begin{bmatrix} \mathbf{D}_{\mathbf{d}i,1} \\ \mathbf{D}_{\mathbf{d}i,2} \\ \vdots \end{bmatrix}. \tag{30}$$

In Fig. 9a the relation between $\tau$ and the *j*th output value of the state space representation of the buck converter's inductor current is depicted. For our application example, a zero crossing of the inductor current needs to be detected, to indicate the changing conduction state of the diode. Again, numerical methods, like the regula falsi method [11], are suitable, as depicted in Fig. 9b for two iterations.

For linear time behavior, one iteration yields the exact solution, already, whereas for exponential behavior two and more iterations might be sufficient.

**Realization of the Simulation Algorithm**

The overall simulation algorithm for dSPAM is depicted in Fig. 10a. In the first step, the matrices of the continuous state space representation of the *i*th interval are discretized. For small circuits it is possible to store discretized matrices for different discretization times, but for larger circuits the memory usage is usually too high. In some applications, like the buck converter, linear interpolation can replace discretization, due to linear or nearly linear time behavior [12]. In such cases, the

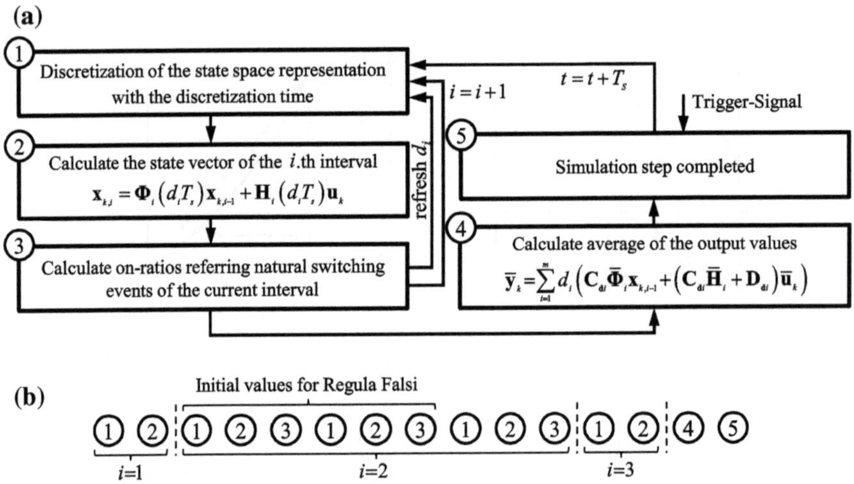


**Fig. 10** Simulation algorithm for models determined by the dSPAM

discretization is done once with the discretization time chosen as the switching period $T_S$. Afterwards the state vector at the end of the interval is calculated and used to detect switching events. To determine the switching time numerically, the discretization and the calculation of the state vector have to be repeated. This is indicated by the arrow "refresh $d_i$".

When the defined number of iteration is executed or the ratio was a measured duty-ratio of a PWM-signal, the next interval is considered and index $i$ is incremented. After finishing the last interval with the ratio

$$d_i = d_m = 1 - \sum_{l=1}^{m} d_l, \tag{31}$$

the averages of the output values are calculated in the fourth step. The overall sequence of the simulation algorithm for the application example is shown in Fig. 10b. Successively, the values for all intervals $i$ are determined. Since a natural switching event must be considered in the second interval ($i$ = 2) the regula falsi method with one iteration is used here, which represents a linear interpolation.

## dSPAM's Properties

Some fundamental properties of the dSPAM are concluded from the above discussion:

1. The computation effort for discretization depends on the implementation of the algorithm. When the discretization is done off-line before simulation start and linear interpolation is applied or matrices are stored in memory for different discretization times the effort can be reduced significantly. These advantages

have to be paid by higher memory usage or inaccuracies for exponential time behavior.

2. The computation effort increases for distinctive exponential time behavior, since the number of iterations for determining the ratios has to be increased.
3. The number of iterations can be determined easier compared to the extended SPAM or ICAM, because it is easier to understand the corresponding influence on the state variables.
4. dSPAM can be applied to arbitrary circuits. However the computation effort determines the real-time capability.

## 4 Simulation Results, Discussion and Evaluation

In the previous sections, average models for the buck converter, described at the beginning of Sect. 3, are derived. Simulation results are presented in Fig. 11, which shows the inductor's current $i_L$ and the capacitor's voltage $v_C$.

A variable-step simulation by SimPowerSystems is used as a reference. The extended SPAM and ICAM show identical results, since the methods yield the same models, as discussed in Sect. 3.2. All extended methods describe the averaged behavior quite well and the differences in between SPAM respectively ICAM and

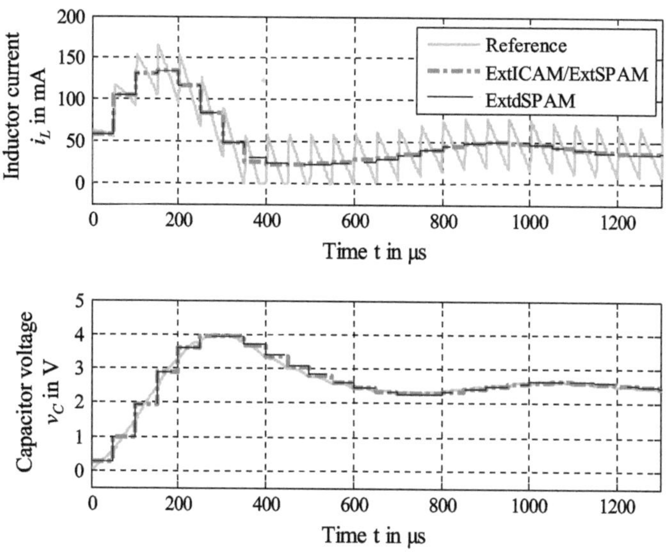


**Fig. 11** Simulation results of the buck converter

**Table 2** Conclusion of the evaluation

| Method | Computation effort per simulation step | Algorithmizable | Generalizable | Platform (Processor, FPGA) | References | Comments |
|---|---|---|---|---|---|---|
| **Conventional synchronous sampled average models—κ ≥ 1** | | | | | | |
| dSPAM | + | + | – – | P(F) | [10] | – Natural SWE not considered<br>– Online discretization necessary—computational efficiency obtained by look-up-tables for the system matrices |
| SPAM | + | + | – – | P(F) | [9, 13] | – Natural SWE not considered<br>– Average over one interval must be average over the whole period |
| ICAM | + | ? | – – | P(F) | [6, 14–16] | – Natural SWE not considered<br>– Signals must be known as more precise function for considering parasitic effects<br>– Algorithm for separation not known |
| **Extended synchronous sampled average models—κ ≥ 1** | | | | | | |
| dSPAM | – | + | + | P(F) | [11, 17] | – Parasitic behavior can be considered by more iteration for the switch time determination (higher computation effort) |
| SPAM | – – | – | – | P | [7, 8, 18, 19] | – Implicit equations have to be solved (high computation effort)<br>– Parasitic behavior can be considered by the Equal-Area-Method (high computation effort) |
| ICAM | – – | ? | o | P | [5, 20–23] | – Implicit equations have to be solved<br>– The behavior of all signals has to be known<br>– Algorithm for separation not known<br>– Separation of fast and slow signals necessary |

dSPAM are negligible. However, the modeling approaches have different properties, which are discussed and evaluated in the following section by the criteria, mentioned in Sect. 1. The evaluation is concluded in Table 2.

## 4.1 Computation Effort

The computation effort for the presented methods depends on the assumption, simplification and implementation on the real-time system. However, a rough relative distinction can be given by considering the sequence of the simulation algorithms in Figs. 6 and 10. For the continuous conduction mode, when no natural switching events have to be considered and no implicit equation occur for ICAM and SPAM as well as no equation has to be solved for determining ratios consequently, the sequence for ICAM and SPAM is shorter than the sequence for dSPAM models. However, the coefficients of ICAM or SPAM models are functions of the ratios, while the continuous time state space representations of the dSPAM are constant and the ratios determine the discretization time, only. Therefore, the computation effort for the conventional methods is quiet small. For the discontinuous conduction mode, when natural switching events must be considered, the computation effort is more or less defined by the numerical method and its convergence speed. The function for determining the ratios in the dSPAM algorithm is always a scalar, while the implicit equation of the ICAM and SPAM must be solved at the same time for all ratios, referring to natural switching events. Moreover, the implicit equation of the ICAM and SPAM is usually nonlinear, caused by the discretization, while the dSPAM model can be a linear problem, when the time behavior is linear. Therefore the computation effort for the extended dSPAM is quiet high, but however lower than for the ICAM and SPAM.

## 4.2 Algorithmizability

The ICAM needs expert knowledge of the circuit, to separate the fast and slow subsystem. By the author's best knowledge a systematic approach, which does not need expert's knowledge, is unknown.

For the conventional SPAM (assuming that correction is not necessary) and conventional dSPAM, the continuous state space representations have to be known, only, which can be systematically determined, e.g., by the modified nodal analysis. The extended SPAM requires expert knowledge for applying suitable assumptions and simplifications. For applying the extended dSPAM, the characteristic voltage or current has to be known to determine the ratios. This can also be determined by circuit analysis methods and does not represent an issue.

### 4.3 *Generalizability*

The conventional average models are limited to the continuous conduction mode and limit the generalizability significantly. Without additional modifications, the extended SPAM is limited to circuits, whose average values do not become zero. Notwithstanding some simplifications and assumptions, the extended ICAM solves the integrals directly and it is well generalizable. However, the extended dSPAM does not use simplifications or assumptions, anyway.

### 4.4 Platform Usability *(Processor, FPGA)*

The conventional average models do not use any divisions in the state space representations. However, the discretization might require divisions or even calculation of inverse matrices. Thus, they are favored for processor-based simulation. However, FPGA-based real-time simulation is possible, but special measures for the discretization are necessary to guarantee stable simulation. For the extended SPAM and the extended ICAM, the ratio calculation usually contains divisions and limits their platform usability to processors. When linear time behavior can be assumed and the discretization can be simplified by interpolation [12], the extended dSPAM is applicable for FPGA-based real-time systems. However, the calculation of the time, when a switching event occurs, requires only few divisions. Thus, for dSPAM an implementation on a processor is preferred, but for high switching frequencies the FPGA-based real-time system can be a better choice, due to its parallel calculation.

## 5 Conclusion

This contribution investigates different approaches for average models and their usability in the Hardware-in-the-Loop simulation of power electronic circuits. After an introduction and a classification of models for Hardware-in-the-Loop simulation, the modeling methods were described and illustrated by the application example of a buck converter. Afterwards, the properties were discussed to evaluate the methods according to computation effort, algorithmizability, generalizability and platform usability.

**Acknowledgements** The contribution is accomplished within the project "Echtzeit-Schaltungssimulation", funded by the Federal Ministry of Education and Research (BMBF) of Germany under grant number 17N1909. The authors would like to thank the company dSPACE for supporting the research project, too.

## References

1. Kiffe, A., Riediger, W., Schulte, T.: Advanced preprocessing and correction-methods for automated generation of FPGA-based simulation of power electronics. In: Proceedings of 16th International Power Electronics and Applications (EPE-ECCE 2013 Europe) (2013)
2. Wagener, A., Schulte, T., Waeltermann, P., Schuette, H.: Hardware-in-the-loop test systems for electric machines in advanced powertrain applications. In: Proceedings of SAE (2007)
3. Belanger, J., Venne, P., Paquin, J.-N.: The what, where and why of real-time simulation. Planet RT **1**, 1 (2010)
4. Graf, C., Maas, J., Schulte, T., Weise-Emden, J.: Real-time HIL-simulation of power electronics. In Proceedings of Industrial Electronics Conference, pp. 2829–2834 (2008)
5. Keong, N.P.: Small signal modeling of DC-DC power converters based on separation of variables. Master thesis, University of Kentucky, USA (2003)
6. Sun, J., Grotstollen, H.: Symbolic analysis methods for averaged modeling of switching power converters. IEEE Trans. Power Electron. **12**, 537–546 (1997)
7. Sun, J., Mitchell, D.M., Greuel, M.F., Krein, P.T., Bass, R.M.: Averaged modeling of PWM converters operating in discontinuous conduction mode. IEEE Trans. Power Electron **16**, 482–492 (2001)
8. Davoudi, A., Jatskevich, J.: Parasitics realization in state-space average-value modeling of PWM DC-DC converters using an equal area method. IEEE Trans. Circuits Syst. **54**, 1960–1967 (2007)
9. Middlebrook, R., Cuk, S.: A general unified approach to modelling switching-converter power stages. In: Proceedings of the IEEE Power Electronics Specialists Conference (1976)
10. Al-Numay, M.: A new discrete-time simulation method for switched systems using averaging. Int. J. Modell. Simul. IASTED **21**, 288–291 (2001)
11. Kiffe, A., Geng, S., Schulte, T., Maas, J.: Real-time simulation of power electronic circuits based on discrete averaging method. In: Proceedings of 37th Annual Conference on IEEE Industrial Electronics Society, IECON 2011, Melbourne, Australia, 7–10 Nov 2011
12. Kiffe, A., Schulte, T.: HIL simulation of a rectifier with power factor correction. In: 6th Conference on Simulation and Testing, Berlin, Germany, 15–16 May 2014
13. Davoudi, A., Jatskevich, J., De Rybel, T.: Numerical state-space average-value modeling of PWM DC-DC converters operating in DCM and CCM. IEEE Trans. Power Electron. **21**, 1003–1012 (2006)
14. Vorperian, V.: Simplified analysis of PWM converters using model of pwm switch. Part I: continuous conduction mode. IEEE Trans. Aerosp. Electron. Syst. **26**, 490–496 (1990)
15. van Dijk, E., Spruijt, H.J.N., O'Sullivan, D.M., Klaassens, J.B.: PWM-Switch modeling of DC-DC converters. IEEE Trans. Power Electron. **10**, 659–665 (1995)
16. Sanders, S.R., Noworolski, J.M., Liu, X.Z., Verghese, G.C.: Generalized averaging method for power conversion circuits. IEEE Trans. Power Electron. **6**, 333–340 (1991)
17. Al-Numay, M.S.: Discrete-time model for PWM converters in discontinuous conduction mode. In: Proceedings of 12th International Power Electronics and Motion Control Conference (2006)
18. Davoudi, A., Jatskevich, J.: State-space averaging of switched-inductor-cell for PWM Dc-Dc converters considering conduction losses in both operational modes. In: Proceedings on Circuits and Systems, pp. 827–830 (2006)
19. Davoudi, A., Jatskevich, J., Chapman, P.L.: Computer-aided dynamic characterization of fourth-Order PWM DC–DC converters. IEEE Trans. Circuits Syst. **55**, 1021–1025 (2008)
20. Sun, J., Mitchell, D.M., Greuel, M.F., Krein, P.T., Bass, R.M.: Averaged modeling of PWM converters operating in discontinuous conduction mode. IEEE Trans. Power Electron. **16**, 482–492 (2001)
21. Vorperian, V.: Simplified analysis of PWM converters using model of PWM switch. Part I: continuous conduction mode. IEEE Trans Aerosp. Electron. Syst. **26**, 490–496 (1990)

22. Kiffe, A., Formann, S., Schulte, T., Maas, J.: Hardware-in-the-loop capable state-space-averaging models for power converters in discontinuous conduction mode considering parasitic component behavior. In: Proceedings of 14th European Conference on Power Electronics and Application (2011)
23. Vorperian, V.: Simplified analysis of PWM converters using model of PWM switch. Part II: discontinuous conduction mode. IEEE Trans. Aerosp. Electron. Syst. **26**, 490–496 (1990)

# Hardware-in-the-Loop Test Process for Modern E/E Systems

**Tino Schulze and Jann-Eve Stavesand**

**Abstract** Automobiles are highly interconnected mechatronic systems. Over the last years, the software of these systems has played an increasingly important role in the value chain. Current production vehicles contain more than 80 electronic control units (ECUs) with over one hundred million lines of software code. Advanced driver assistance systems and functions for autonomous driving will further increase the requirements for the E/E systems. Furthermore, these systems will become even more complex due to the use of sensors to capture the vehicle environment, the interconnection of the ECU functions, and the influence of external communication. This paper outlines an ECU test process example in which an advanced driver assistance function is tested on a hardware-in-the-loop (HIL) test system. The necessary process steps and interdependencies to ISO 26262 are identified. The paper also explains how to simulate an ideal test process under consideration of the version and variant flows, and how to use different validation systems to perform different validation steps throughout the development process. The paper aims to analyze the various questions arising in the E/E validation process to identify the advantages the users have throughout the entire development cycle.

**Keywords** Hardware-in-the-loop (HIL) · Virtual validation · Test process · ISO 26262 · E/E validation

## 1 Challenges for the Validation of Modern E/E Systems

It is clear that the functionalities of E/E systems will be even more interconnected in the future. The most striking examples are advanced driver assistance systems (ADAS), x-by-wire, active safety systems and, particularly, highly automated and

T. Schulze (✉) · J.-E. Stavesand
dSPACE GmbH, Paderborn, Germany
e-mail: tschulze@dspace.de

J.-E. Stavesand
e-mail: jstavesand@dspace.de



C. Gühmann et al. (eds.), *Simulation and Testing for Vehicle Technology*,
DOI 10.1007/978-3-319-32345-9_23

autonomous vehicles. To achieve the desired functionality in these fields, various mechatronic actuators, sensors, and ECUs work together. This means that the ECUs in a vehicle, the vehicles themselves, and the vehicles and infrastructure will be more and more interconnected in the future. This trend towards interconnectedness can already be seen at the ECU level where different software components interact, and also in ECU networks that are connected via bus systems [1]. The bus systems used to establish ECU networks range from the classic CAN, LIN, and FlexRay buses in the automotive industry, to the new, Ethernet-based networks such as SOME/IP and AVB, up to systems such as WLAN, LTE, etc., which enable communication with the infrastructure.

The increasing interconnectedness of the systems is accompanied by the high amount of variants and versions of vehicles, which primarily affects the ECUs and embedded software. ECUs are used in different vehicle variants and production series. Very often, the only difference between the ECUs is their software version, which is designed for the features and functional range of a specific vehicle. However, the increasing complexity of the E/E network stemming from new, highly interdependent functions, makes the system more susceptible to errors [2]. Because safety requirements are particularly important for advanced driver assistance systems, validation tasks are also playing an ever more important role. For example, for vehicles that are moving towards ADAS and autonomous driving, the combination of versions and variants has to be observed even more due to the increasing complexity of the E/E system.

At the same time, developers want to find and solve errors in a product and the entire system as early as possible to prevent potential recall campaigns and costly hotfixes in later development phases or in the field. But in these early development phases, usually not all components of an E/E system are available, because they are still being developed. New validation concepts and platforms will therefore gain importance in the future to cover the stronger need for testing. These new concept allow to cope with the many versions and variants as well as the vehicle environment influencing the validation effort of advanced driver assistance systems in early development stages. In specific, the high number of possible driving situations, which needs to be tested, can be addressed [3].

Another factor influencing for the development and test processes are standards and norms, which are becoming increasingly relevant for E/E validation. These documents standardize the processes for developing safety-related E/E systems in vehicles (ISO 26262) and aircraft (DO-178C) [4]. In the automotive industry, the syllabus of the International Software Testing Qualifications Board (ISTQB) is also crucial, as it aims to integrate the standards for software testing in the development of E/E systems. ISO 29119 may play a role in the future as well. The standard was published as a new standard for software testing [5].

In summary, the increasing interconnectedness, the high amount of versions and variants, the influence of standards and product liability, the influence of the vehicle environment on the E/E systems, and the growing cost pressures force developers to find innovative solutions for the development and validation tasks of mechatronic systems. To master these challenges, new technical aspects have to be considered and, more importantly, questions about the process need to be answered. The following sections use the example of validating an ADAS function on a HIL test bench to present possible solutions.

## 2 What Needs to Be Tested?

Section 2 shows the typical tasks involved in E/E validation, from software components to interconnected ECU networks, and describes the available test systems and methods.

### *2.1 Overview of the Relevant Test Objects and Test Phases for Validating E/E Systems*

First, the term "system" needs to be defined. Typically, OEMs use the term when referring to a complete vehicle system that consists of various ECUs. For suppliers, on the other hand, a system is usually the ECU, which consists of a combination of different software modules, basic software components, and hardware components. These individual ECUs are then connected via a bus to form an ECU network, which is responsible for the E/E system of the vehicle.

Figure 1 shows the setup of a vehicle's E/E system, consisting of the software components, the basic software components and operation system parts as well as the hardware layers at the individual ECU level. At the vehicle level, it also illustrates the entire ECU network connected via the required bus system.

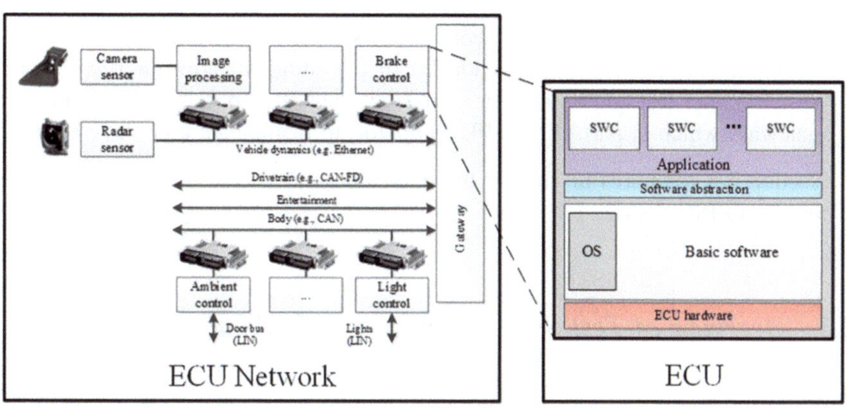


**Fig. 1** ECU software and ECU network

During validation, this setup and all its individual parts have to be tested one after the other, starting with the single software component and ending with the integration of the entire network in the vehicle.

Each integration phase of the E/E system comprises different tests that observe and verify the characteristics of each ECU and the ECU network.

- The **software component test** checks whether the individual software components are implemented correctly according to the specified requirements. Typically, PC-based simulations and some hardware-in-the-loop (HIL) simulations are used for this.
- The **ECU function test** checks the interaction of the software components. Here, a wide range of validation methods can be used as well, from PC-based simulation up to HIL simulation. These tests are typically performed by the suppliers or, if the OEMs integrate their own software components in the overall system, by the OEMs.
- The **software integration test** is performed according to the hierarchical structure of the software architecture. After all software components have been integrated, their interplay in the completely embedded software has to be verified as well. In this validation step, hardware issues and especially the validation of bus communication play an important role. Therefore, HIL test benches are the ideal validation tool.
- The last milestone in the software development is the **acceptance and release test**, where the entire embedded software is tested. In contrast to the software integration test, this test phase is performed only when the software is being executed on the target hardware. It therefore requires HIL tests.
- In the subsequent test phase, **testing distributed functions**, the individual system components are successively combined into vehicle units. These units are typically partial networks used to realize specific functionalities that are distributed across several ECUs. This also makes it necessary to validate the bus communication between the ECUs.
- During the **testing networked functions** phase, the functional behavior and interplay of all system components is assessed at the system level. The ISO 26262 standard calls for the validation of the communication both within the system under test and with other vehicle systems. It is possible to use an error gateway, which is part of the HIL simulator, to manipulate bus messages [6].

Section 3.3 compares in Table 1 the phases described here with the test phases according to ISO 26262.

## 2.2 *Overview of Test Systems in the E/E Validation Process*

Today, validating E/E systems predominantly involves simulation-based methods, such as hardware-in-the-loop simulation. These validation platforms have been joined in recent years by PC-based simulation systems, which can be used in earlier

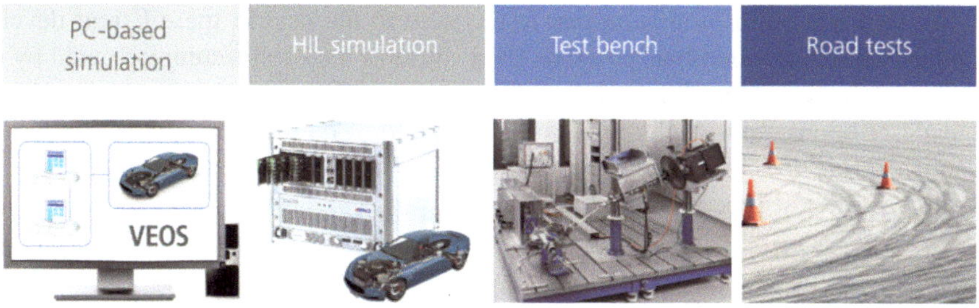


**Fig. 2** Test systems for E/E validation

development phases [7]. Then there are tests performed on test benches and in the real vehicle to release the entire system for the final production. A dedicated selection of validation platforms is used, for example, for validating airbag ECUs in "virtual crash tests" at BMW [8]. Figure 2 divides the test systems into four parts.

PC-based offline simulation lets developers perform the first tests on a PC. This can usually be done on a standard PC, making it possible to use existing PC hardware. PC-based simulation therefore complements the validation process, particularly in the early development phases [9]. Tests can be executed in parallel, allowing for parallel editing of version—and variant-dependent tests, which results in significant time benefits.

In addition to the traditional tests in a vehicle and on the test bench, hardware-in-the-loop simulation is an efficient and industry-proven step for reliably managing the pending test tasks. There are HIL systems of different sizes, from small to large. **Component HILs** are designed for testing individual ECUs. They require the plant model for the ECU and restbus simulation for the missing ECUs. **System HILs** or "cluster HILs" are used especially for tightly connected ECUs or for validating partial networks in the vehicle. **Networked HILs** or "virtual vehicles" contain all the real ECUs. This lets developers test a vehicle's entire E/E architecture [10].

Highly dynamic test benches for mechatronic components and systems that are connected to HIL simulators are used when it is not possible to access the ECU via its electrical interfaces alone: For example, when connecting mechanical loads to an electric drive in an electric power steering system or when stimulating the integrated sensors for mechatronic ECUs, e.g., for the ESP. These test benches can also be used to measure the influence of the real parts on the ECUs.

Tests in real vehicles are an integral part of a validation process. They are detailed and consider a high number of environmental factors, material characteristics, and construction details, which are difficult to simulate. However, in-vehicle testing is expensive and cannot always be adjusted to new challenges. The advanced driver assistance systems in particular are the reason why not all test cases can be performed reproducibly in the vehicle, because it would be increasingly difficult or even impossible to execute the diverse driving situations.

To make the transition from one test system to the next in the different development phases as smooth as possible, these systems have to be complemented by a seamless tool chain. Such a tool chain lets the users reuse test descriptions, models, bus descriptions, and much more from one platform on the next. Validation tasks can therefore be distributed across the development process, which further improves the test process efficiency.

## 2.3 Technical Challenges for the Test Systems

The challenges for E/E validation mentioned in Sect. 1, require not only process-related answers but also new technical solutions for new E/E functionalities. This section takes a closer look at two pronounced challenges that are and will be particularly important for validating especially ADAS applications.

The main characteristics of ADAS applications are the tight interconnection of ECUs, the capturing of the vehicle environment via sensors, as well as the communication with other vehicles (V2V) and the environment (V2X). To master the increased amount of communication, the automotive industry is starting to use new, Ethernet-based bus systems, which create new network structures in the vehicles. The first Ethernet-based applications will be used in the following areas: (camera-based) driver assistance, comfort and entertainment functions, and ECU flashing. The different protocols and/or combined mechanisms used in this context are DoIP, AVB (IEEE1722), IEEE802.1AS, ISO 15118 and SOME/IP. The latter is currently undergoing standardization by ASAM and AUTOSAR. In particular, the use of service-based communication opens up new possibilities, which have to be reflected in the test systems. ECU communication is no longer restricted to point-to-point communication but can, as usual in networks, be implemented with various architectures. Network participants are found automatically so even different equipment variants of the vehicles can be simulated at the network level. Test systems also have to be able to handle the large amount of data that is exchanged via Ethernet-based networks. This means that the test systems need the performance required for restbus simulation in order to send and receive data, and also manipulate it for specific test scenarios. High-performance data networks between the real-time processor and the bus simulation boards are crucial especially for HIL simulation. The bus simulation tools have to integrate the bus description standards, which evolves dynamically, in time to be able simulate given network descriptions.

Another technical aspect that needs to be considered in the context of advanced driver assistance systems is the inclusion of environment conditions in vehicle ECU validation. Vehicles with advanced driver assistance systems or autonomous driving functions have communication and satellite interfaces that detect the vehicle's environment and communicate with it. Modern test systems have to be able to simulate these systems' interfaces in real time and generate scenarios that simulate the vehicle's movements in realistic driving scenarios and environments. Here, too, the computational power of the systems plays a decisive role, because the environment simulations are becoming more and more realistic and therefore require more computational power. The test systems have to calculate sensor models, simulate fellow vehicles and roads, etc. All of these simulations must be computed in real time to provide a consistent simulation of the environment. This generates a high demand for computational power, which requires up-to-date real-time computation nodes and, for increasing demands, multiprocessor options.

# 3 The Process Context of Validation Tasks

The previous section presented the relevant test objects (software components, ECUs, ECU networks, etc.) and the test phases (software component test, software integration test, etc.) for validating an E/E system and provided an overview of the most commonly used test systems. This section describes a generic test process based on ISO 26262 [11], which can be used in a seamless validation tool chain from PC-based simulation, to HIL testing, to mechatronic test benches [12]. A close look at the processes used by OEMs and suppliers naturally reveals different processes for validation and integration. Each validation process is unique. Based on experience from the automotive industry and with a view to ISO 26262, the subsequent sections therefore attempt to describe a validation process that is as generic as possible.

## *3.1 ISO 26262 Road Vehicles—Functional Safety*

The ISO 26262 standard is specifically designed for the automotive industry and is intended to be applied to safety-related systems that are based on E/E functionalities. The 'ISO 26262 Road vehicles—Functional safety' is an adaptation of IEC 61508, written to satisfy the needs specific to the application sector of E/E systems within road vehicles up to 3.5 tons gross-weight. This covers all activities during the development life cycle of systems composed of electrical, electronic, and software elements that provide safety-related functions.

The standard consists of 9 normative parts, from the concept phase over product development to the production and operation of the affected E/E systems, and a guide-line for ISO 26262 as the 10th part.

### 3.1.1 Relation to Product Liability

Manufacturers should ensure that their development processes and also development tools are state-of-the-art, meaning that they are the highest level of development of a system at this particular time [13]. This ensures that a tool or a product is developed under the best possible technologies.

State-of-the-art development processes and tools are important for manufacturers of safety-related automotive systems since there is a close relationship to the manufacturer's liability for these products.

Under German law for example, manufacturers are generally liable for personal damage caused by the malfunction of a product [14]. But if the malfunction could not have been detected by the technological state of the art, the liability is excluded [15].

Thus, development in accordance to ISO 26262 is definitely a significant factor for manufacturers of safety-critical systems [16].

## *3.2 Test Processes in the Automotive Industry*

In the automotive industry, the V-cycle (Fig. 4) has long been established as a model for development and test processes. The right branch of the V-cycle contains the individual test phases and integration levels for testing automotive ECUs. The integration levels are based on the integration levels for ECUs as discussed in Sect. 2.1. Typically, integration gradually moves from single software components to the entire vehicle. This means that many test steps are involved, which differ in test object, test phase, test system and the related work packages. The main challenge is to coordinate these steps and use them consistently and reliably throughout the entire verification process so the high number of tests can be distributed across the available development time.

For software development and particularly software testing, the syllabus of ISTQB [17] contains recommendations on how to organize such a test process. The syllabus calls this the fundamental test process. This fundamental test process is described in the following paragraphs, connecting the organizational recommendations of the ISTQB syllabus to the roles known from traditional software development [18] (Fig. 3).

According to the syllabus, planning the verification process includes identifying the number and scope of the individual test phases in the V-cycle: i.e., how many and which integration steps exist, the test objects, the units under test (UUT) and systems under test (SUT) for each level of integration. The test goals of each test phase also have to be specified. Developers also have to plan which information, tools, and tests have to be available in each phase.

The test process begins with the **Planning phase**. The main tasks in this phase are resource allocation and assigning roles for performing the tests. During this phase, a test strategy is developed that considers the prioritization of tests determined via risk analysis and the test intensity for individual system parts. The test strategy comprises the test phases in the right branch of the V-cycle and defines the levels of integration as well as the test methods (SIL, MIL, PIL, HIL, etc.). During the **Analysis and Design** phase, test cases are specified and simulation models are created. Analyzing whether the requirements and specifications that lay the foundation for test design (test creation) can be tested is another main task in this phase. This analysis in turn leads to requirements for test case creation. The **Implementation and Execution**

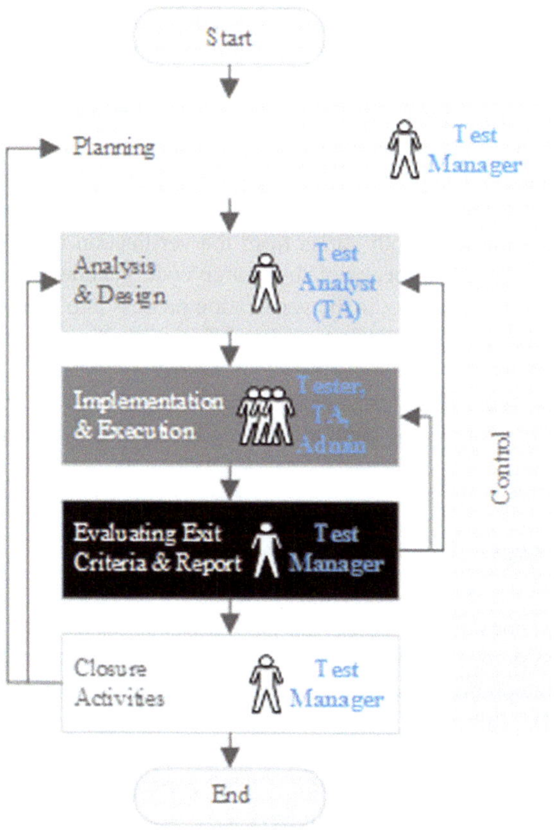


**Fig. 3** Test process

phase starts with concretizing the logical test cases, i.e., determining the input values. One logical test case is often the basis for several concrete test cases [19], which are defined by different parameter sets. The parameter sets link the test cases to specific variants and versions, which makes it possible to manage them in the overall verification process. This separation of implementation and parameterization is also used in model development to reuse generic models for different test phases and systems. The next step, Evaluating Exit Criteria and Report, involves the evaluation of the test results on the basis of the previous risk analysis. Against this backdrop and considering the additional effort for executing new test cases, developers have to decide whether a test can be completed or the test object has to be modified in one iteration so that further testing activities have to follow. Retrospective analysis and evaluation are the final test activities. They identify the best practices for future projects. In the Control phase, running test activities are checked against the previously created test plan, giving developers the possibility to intervene at any time. Figure 3 shows the sequence of the various phases.

A look at the validation processes of different OEMs shows that they perform similar planning tasks and also define test phases, test objects, and test systems. For example, a simulation-supported validation process usually involves the following phases: module tests, software integration tests, tests of individual ECUs, tests of distributed functions, and in-vehicle tests [20]. However, different OEMs use different test systems and test processes for each integration step. For example, module tests are carried out by model-in-the-loop (MIL) testing or software-in-the-loop (SIL) testing, and the test depth varies from the verification of function specifications at the model level to checks of the source code generated from the models. ISO 26262 takes this variability in the verification process into account and includes it in a schema that can be adjusted to the different verification processes of the OEMs and suppliers.

The following section analyzes the test steps according to ISO 26262 and presents the simulation-supported methods ISO 26262 names for each test level.

## 3.3 *Test Process According to ISO 26262*

ISO 26262 describes recommendations and methods to increase the functional safety of E/E systems. To ensure functional safety, the standard recommends a validation and integration process that is also represented by the right branch of the V-cycle (Fig. 4). This shows that the standard was developed with a view to practicality and considers many aspects that are already practiced in reality. Table 1 shows how the test phases defined in ISO 26262 correspond to the integration steps known from industry and described in Sect. 2.1.

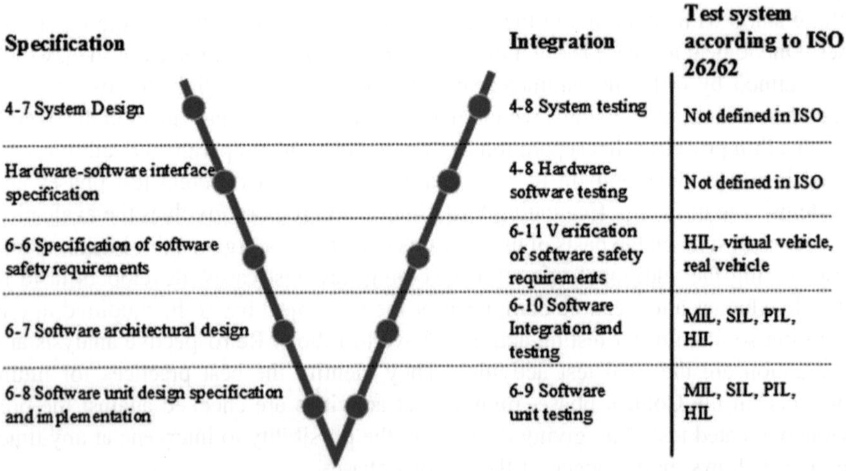


**Fig. 4** ISO 26262 process description and test systems

**Table 1** Test phase mapping

| Test phases Acc. to Sect 2.1 | Test phases Acc. to ISO 26262 |
|---|---|
| Software component test | Software unit test |
| ECU function test | Software integration and test |
| Software integration test | Software integration and test |
| Acceptance and release test | Verification of SW safety requirements |
| Distributed functions | Hardware-software integration (HIL, test bench, vehicle)/software integration and testing at the functional level (offline simulation) |
| Networked functions | System integration |

The ISO 26262 standard thus includes mechanisms that are already used by most OEMs and suppliers. Figure 4 shows that the standard describes the integration levels and appropriate test benches in the same way as they are already being used in the automotive industry. From the individual software components up to the complete vehicle, it describes the successive integration of individual components. For the software unit test, for example, this means that different test methods are recommended for SIL-based validation and HIL-based validation.

The following section outlines a test process according to ISO 26262 using the example of the validation of an ADAS function on a HIL test bench.

# 4 Illustrating the Generic Test Process Using the Example of Validating an ADAS Function on a HIL Test Bench

This section uses a specific tool chain and test suite to show the process for validating an ADAS function on a HIL test bench. This serves to illustrate what the integration steps described in ISO 26262 could look like, and how to implement the planning data, parameters and tasks from the test strategies and the work packages identified by ISTQB. This section shows how parameters, models, test cases and system configurations from a central data and test management tool are used in a specific level of integration on a dedicated test system. It then shows how to manage the different versions and variants of the artifacts in a database system so they can be traced and reused throughout the entire development process.

Figure 5 [21] indicates which information has to be available during the **Implementation and Execution** phase to execute an automated HIL test. The **Planning and Control** phase is used to set the test goals and define which test objects to test with which test system. The models, model parameters, test cases, and test case parameters are created and centrally managed in the **Analysis and Design** phase. In the **Implementation and Execution** phase, all of this information is combined and executed on the HIL system by using test suites. This modular method decouples some of the work steps so the complexity of the entire validation process is not as noticeable in the individual parts of the development process.

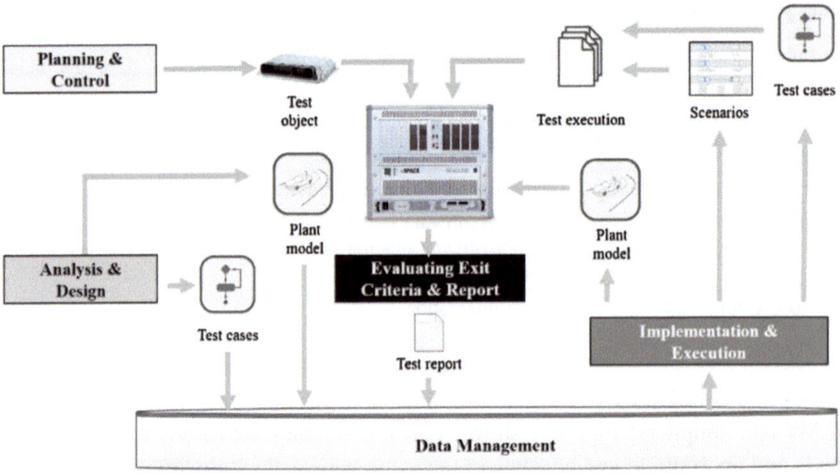


**Fig. 5** HIL testing process

To make this process more tangible, the dSPACE tool chain and the Euro NCAP test suite [22] are used to describe the test of an autonomous emergency braking (AEB) system and a forward collision warning (FCW) system [23]. The Euro NCAP test protocol states which tests and test scenarios have to be executed to validate the respective functionalities of the AEB and FCW. This information is used as requirements for the test cases and is managed in the data management tool SYNECT (Fig. 6).

In the **Analysis and Design** phase, logical test cases are derived for each of these requirements and then implemented in expert tools such as dSPACE AutomationDesk, which is used for test automation. In addition, parameter sets are

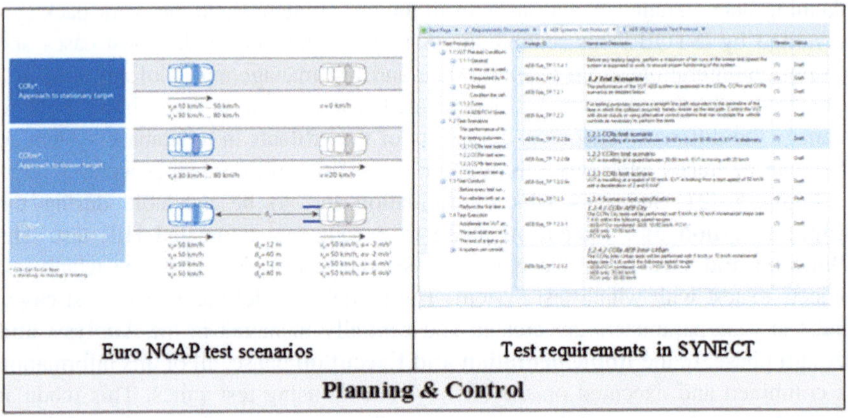


**Fig. 6** Planning and control phase for the Euro NCAP testing example

created. These parameter sets link the test cases to specific versions and variants and make the test cases available for various **test execution plans** (variant-dependent compilation of concrete test cases) and test cases. The variant dependency also lets developers link the test cases to the requirements. This is usually done in the **Analysis and Design** phase and managed centrally in the data management tool. The models and model parameters are also handled in this central data management tool. Because this data management tool is a database and provides company-wide access, various test process steps can be combined at the start of a test execution and to configure the overall system. This standardization makes it possible to use a high number of artifacts at different times in the development process.

In the present example, the data management tool dSPACE SYNECT and the dSPACE Workflow Management (WFM) solution manage all parameters, test cases, test results, plant models and automate processes for configuring the dSPACE HIL component test systems so that all necessary information can be combined and executed on the test system in the **Implementation and Execution** phase. This makes it possible to manage the flood of variants. These superordinate tools ensure that all the software and hardware involved in the test process is automatically configured and put into operation. From the test-phase-dependent configuration of the test system up to the final test report, all necessary steps are covered. The involved tools are configured accordingly and the results are evaluated automatically. This shows that the central data management tool can manage and evaluate the entire test process. The variant management at the hardware and software level is therefore completely traceable, as demanded by the **Planning and Control** phase of the fundamental test process.

For Euro NCAP, this means that the concrete test cases from the **Implementation and Execution** phase are combined to a **test execution plan** that is used to validate a dedicated functionality, e.g., the AEB function (Fig. 7).

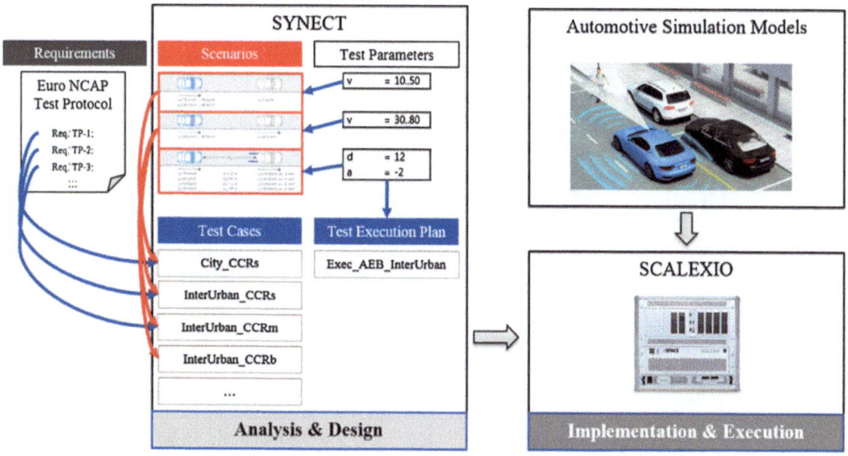


**Fig. 7** The phases analysis and design and implementation and execution for the Euro NCAP testing example

From the perspective of ISO 26262, this could mean that in the **Planning and Control** phase, it was determined that during **software integration and testing** (refer to ISO 26262 part 6-10) the distributed functions of the AEB or FCW are to be tested. According to Sect. 2.1, for example, this would correspond to the **testing distributed functions** phase, i.e., the functional test of one distributed function. The test object is therefore a distributed function. In this example, the test is executed on a HIL simulator [4]. But it could also be executed on a PC-based simulation platform. This means that this integration step involves a number of test methods. ISO 26262 suggests the following methods:

- **Requirements-based test**—To validate the software architecture.
- **Interface test**—To validate the functional dependencies of the software components, which are usually distributed across several ECUs for AEB and FCW functions.
- **Fault injection test**—To inject faults by changing variable values, producing electrical faults, etc.
- **Resource usage test**—To capture the execution times and the performance of software on the target hardware or ECU prototypes, to measure latencies in bus communication, etc.

The data management tool can provide the parameters associated with these methods so the HIL test can be executed automatically. The **test execution plan** includes all test cases that were associated in SYNECT Test Management with the relevant test phase, test object and executable tests, according to their version and variant configuration.

The data management tool then automatically captures and saves the test results and adds the correct version and variant configuration information, which complies with ISO 26262 part 8-7, **configuration management**. The configuration of the test execution is now stored in the database and can be reused. The test results are checked against the requirements and evaluated (Fig. 8).

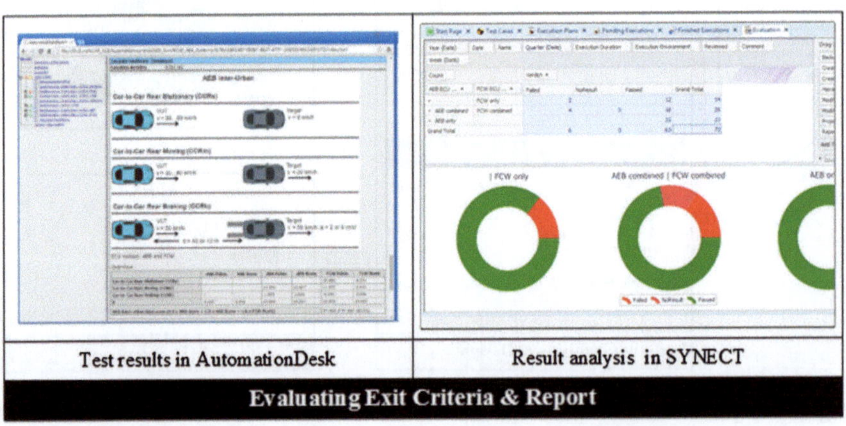


**Fig. 8** Evaluating exit criteria and report phase for the Euro NCAP testing example

The central access to the data, configurations and tests of the entire verification process makes it possible to handle the tests in different development phases, test systems and test objects, so the entire ECU testing process can be managed. This enables experts to work with the tools specific to their field of work. A central management tool is then used to associate the data to the appropriate test scenarios.

## 5 Example of a Seamless Test Process for E/E Systems

Section 4 illustrates how to plan and execute the validation of an ECU function in a specific test phase in consideration of ISO 26262, on a specific test system, with a generic test process. What would a validation tool chain for E/E systems have to look like to master the challenges presented in Sect. 1? Combining different methods in different development phases is a worthwhile approach to finding and eliminating errors early on. PC-based offline simulation will play a central role, because it is easy to use and can be distributed across many PCs where for parallel use. This means that a high number of scenarios can be executed, particularly for the **software component test** (refer to Sect. 2.1). HIL simulation will further strengthen its position and importance, because the overall systems must be tested together with their real hardware components. This is a clear requirement for the **verification of software safety requirements** in ISO 26262. Bus systems can also be tested under realistic conditions. The HIL test benches provide reproducible testing conditions and automated operation. For the final release, mechatronic test benches and real in-vehicle tests will remain an integral part if the physical properties of the vehicles have to be considered. However, in the future many more tests will have to be simulation-based. Because a vehicle's environment and the traffic situations influence the behavior of advanced driver assistance systems, the amount of test cases that need to be considered will increase to the point where in-vehicle tests can no longer cover every situation.

Figure 9 shows a suggestion for which test phases (Sect. 2.1) the individual test systems (Sect. 2.2) could be used.

An additional testing option comes with PC-Based Simulation for Testing Distributed Functions. Testing distributed functions in general requires real ECUs and thererfore HIL testing is necessary. Nevertheless distributed functions could be tested earlier in the development process on software component level with parts of the basic software with PC-based test systems such as dSPACE VEOS [9].

To use the validation methods efficiently in the development process, there must be a seamless process tool chain that can be used by different departments and in different development phases to make the use of tests, test results, user interfaces, and models traceable. A database-type solution as described for one phase of the generic test process in Sect. 1 is useful so all artifacts of the test process can be provided to various users and departments from a central location and under consideration of versions and variants. This type of solution also makes the entire process traceable, from the requirements to the test evaluation. Another key to

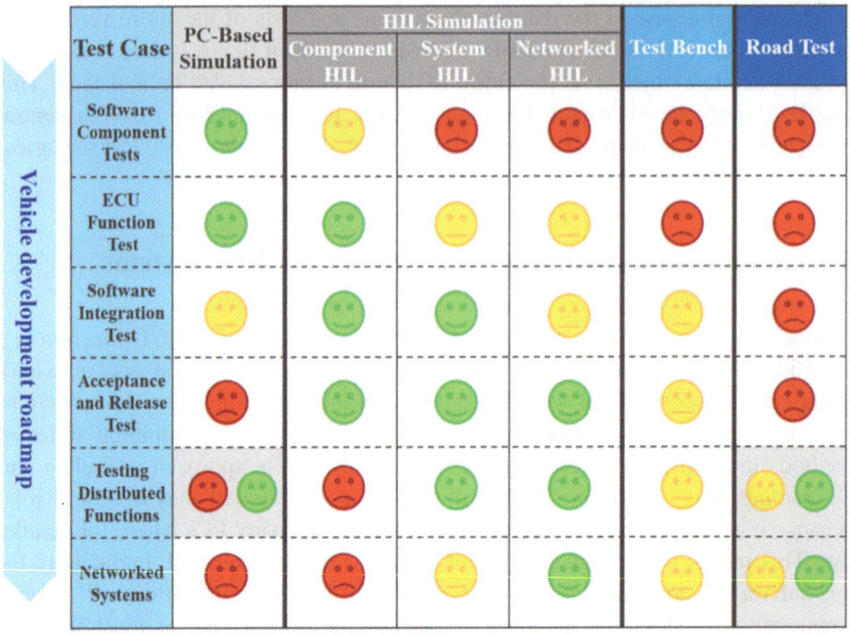


**Fig. 9** Use of test systems in the vehicle development process

efficient processes is a tight integration of PC-based simulation and HIL-based simulation, as provided by dSPACE VEOS and SCALEXIO. This makes it possible to reuse simulation models, environment models, user interfaces, and test frameworks [3].

## 6 Summary and Outlook

This paper has shown that E/E systems are becoming more complex, creating greater challenges for the validation processes. Development engineers should be able to use the same tools in all development and validation steps so the systems can be used efficiently and traceably in the different development phases. The transitions between the individual systems must be seamless, especially from PC-based simulation to HIL simulation, so overall models that are composed of model parts can be used flexibly, and virtual ECUs can be used in early development phases with an increasing share of real hardware as the development progresses. It must also be easy to combine HIL test benches, from simple component HILs to complex networked HILs (virtual vehicles), to have the flexibility needed to adjust the validation tasks to the project status in the vehicle development process.

An example for the validation of an ADAS function on a HIL test bench based on the Euro NCAP test suite has shown what a tool environment that is integrated into a data management solution could look like. Databases support the user by making the use of tests, models, parameters, and test environments traceable throughout multiple development stages. Since test cases are linked to a version and variant management system, it is always clear which validation environment a test case is linked to. This is in line with the relevant standards.

Models play a particularly important role for ADAS applications. To validate the systems by means of simulation, developers have to overcome challenges such as simulating vehicle dynamics and using environment models as well as sensor models. The stricter requirements for computing the environment simulation in real time also increase the requirements for HIL simulation hardware, such as dSPACE SCALEXIO.

The requirements will continue to grow. The connection of vehicles and infrastructure via a data network will increase the need for new simulation methods, such as simulating WLAN communication in HIL scenarios.

## References


1. Siebenpfeiffer, W.: Vernetztes Automobil: Sicherheit - Car IT - Konzepte. Springer – Vieweg (2014)
2. Weber, J.: An aspect driven approach for the analysis, evaluation and optimization of safety within the automotive industry. SAE 2010 World Congress and Exhibition (2010)
3. Schulze, T., Krügel, K.: Continuous validation of driver assistance functions simulation and testing automotive electronics. IAV Conference on Simulation and Testing for Vehicle Technology, Berlin (2014)
4. Himmler, A., Lamberg, K., Beine, M.: Hardware-in-the-Loop Testing in the Context of ISO 26262. SAE Technical Paper (2012)
5. Reid, S.: ISO/IEC/IEE 29119: the new international software testing standards (2014)
6. Wallentowitz, H., Reif, K.: Handbuch der Kraftfahrzeugelektronik. Vieweg & Sohn, Wiesbaden (2006)
7. Miegler, M.: Warum virtuelles Testen besser ist als reales und umgekehrt. Stuttgart: Autotest (2014)
8. Kranzer, C., Stavesand, J.-E.: Vollautomatisiertes Crash-Simulationssystem. Hanser Automotive, pp. 34–37, 7–8 2015
9. Getos, S., Krügel, K.: Die virtuelle Absicherung in der Praxis. Simulation und Erprobung in der Fahrzeugentwicklung, Baden-Baden (2014)
10. SCANIA, dSPACE.: A moving variety. dSPACE Magazin, March 2015
11. ISO 26262:2011. (2011). Road Vehicles - Functional Safety. International Standard, International Organization for Standardization
12. Schulze, T., Stavesand, J.-E.: Wann virtuell und wann real? - Ein Skizze der E/E-Testsysteme der Zukunft. Graz: Grazer Symposium Virtual Vehicle (2015)
13. Duisberg, A.: ISO 26262—a reliable standard from a product liability viewpoint? 4. Euroforum Jahrestagung. Stuttgart, 12–14 Sept 2012
14. Gesetz über die Haftung für fehlerhafte Produkte (Produkthaftungsgesetz - ProdHafG). (1989)
15. Klindt, T.: Haftungsrechtliche Wirkung von technischen Normen. Euroforum-Konferenz ISO 26262 (2014)
16. Himmler, A., Lamberg, K., Schulze, T., Stavesand, J.-E.: Testing of real-time criteria in ISO 26262 related projects—maximizing productivity using a certified COTS test automation tool. SAE World Congress. Detroit: SAE Technical Paper 05 Apr 2016

17. ISTQB.: Foundation Level Syllabus (2011)
18. Spillner, A., Linz, T.: Basiswissen Softwaretest. Heidelberg: dpunkt.verlag (2003)
19. Roßner, T., Brandes, C., Goetz, H., Winter, M.: Basiswissen Modellbasierter Test. dpunkt. verlag (2010)
20. Morciniec, T.: Test suite minimierung durch Vermeidung mehrfacher Fehlernachweise. Stuttgart: Autotest (2014)
21. Stavesand, J.-E.: Vehicles, versions, variants. Automotive Testing Expo. Stuttgart (2015)
22. Euro NCAP.: Retrieved 23 Oct 2015 from http://www.euroncap.com/en
23. Krügel, K., Hordys, G., Deutenberg, G., Sänger, N.: Innovative stragetgies for validation and testing in function development. IAV Symposium on Automotive Powertrain Control Systems (2014)

# Optimization of Modular Wiring Harnesses by Means of Regression Models for Temperature Prediction of Wire Bundles

**A. Rius, A. Garcia and M.A. Díaz**

**Abstract** Automotive wiring harnesses have become heavier and more complex due to their increasing number of electrical components. It is now desired to reduce their mass of copper. For this purpose, experimentation can be partially replaced by simulation, but it is still impossible to exhaustively simulate all of the combinations of modular wiring harness. This proposed approach consists of carrying out simulations using the FEM method and using their results to create regression models. Polynomial formulae can give the same information as simulations within a clearly reduced time and satisfying accuracy. An optimization algorithm introduced in this study will use them to assign new cable cross-sections of harnesses considering their currents and the ambient temperature.

**Keywords** Wire harness · Optimization · Weight reduction · Regression · Models · Temperature prediction

## 1 Introduction

The number and overall weight of electric components commercial vehicles has risen across the recent decades. Some European Automotive brands offer wide possibilities of customization for their vehicles, so that customers can choose features of their cars among a large variety. Furthermore, worldwide car sales demand

A. Rius (✉) · A. Garcia
Universitat Politècnica de Catalunya (UPC), Barcelona, Spain
e-mail: extern.armand.rius@seat.es; armand.rius.rueda@gmail.com

A. Garcia
e-mail: garciae@ee.upc.edu

A. Rius
RückerLypsa, SL, Barcelona, Spain

M.A. Díaz
Centro Técnico de Seat, SA, Barcelona, Spain
e-mail: manuel-alberto.diaz@seat.es



C. Gühmann et al. (eds.), *Simulation and Testing for Vehicle Technology*,
DOI 10.1007/978-3-319-32345-9_24

extra customization capabilities for car manufactures to offer specific features of their models adapted for each continent. In terms of magnitude, the overall number of possible combinations of features of a car model can reach $10^{10}$.

Different simulators have been developed to predict the temperature of wires or wire bundles in vehicle applications, with the aim of optimization. Nevertheless, the huge number of combinations makes it unaffordable to simulate all of the feature combinations due to a matter of computation time.

Wires must be dimensioned in accordance to their worst case. This involves the fact that wires in bundles transmit heat amongst them. Therefore, a wire must be dimensioned observing other wires in all of the segments of the wiring harness throughout it extends.

The task of optimizing the mass of copper maintaining high reliability in such complex systems demands extremely fast ways of predicting temperature. This document proposes using regression models with the form of multidimensional polynomials providing the necessary information to carry out this task.

## 1.1 Scope

Wiring harnesses are generally assembled manually, so that all of its segments contain sets of undistributed wires forming random layouts. Thus, it is mandatory to design wire harnesses in such a way that the internal distribution of the bundles do not affect their maximum temperature. This latter requirement demands that all bundles have homogeneity in their heat generation, viewed as the $f$ parameter ($W/m^3$) in the heat dissipation equation, which must be roughly uniform all over the cross-section of any random bundle. It must be assumed that $f$ equals to zero in all regions of a bundle except in conductor regions (generally copper), whose Joule effect heat depends on their current density $J$ and resistivity $\rho$ responds to the following equation:

$$f = J^2 \cdot \rho(T) \tag{1}$$

The latter equation presents the resistivity of copper as a function of the temperature, $T$. Since the resistivity $\rho$ is related to the conductor material, if we want to keep $f$ uniform over a cross-section of a bundle containing wires with different conductor materials, $J$ will have to be adapted to each of them so that $f$ is approximately uniform. The different dependency of $\rho$ on the temperature for different materials complicates this, but it is still possible to find all the values of $J$ assuming a given temperature, which could be a certain point of temperature between the ambient temperature and the maximum expected temperature of the bundle. This document only explores the case of bundles with unique composition of materials (copper and polyvinylchloride).

Nonetheless, power wire bundles will frequently include signal wires carrying negligible currents, which prevent us to maintain this desired unique Joule heat per unit volume $f$ for all of the wires. These signal wires will have nearly zero heat dissipation in their cores, since the wire manufacturers cannot provide wires with unlimitedly thin diameters. No matter how low the current might be, these wires will always be cold, i.e. providing nearly $f = 0$ heat to the system. Their presence enlarges de bundle and provides negligible Joule heat at the same time. The resulting bundle dissipates slightly more heat by means of convection and radiation, due to a greater surface of contact with the environment, when compared with the same bundle lacking these signal wires.

These aspects make us understand that these cold wires can only benefit bundles by lowering their maximum expectable temperature for given currents. Since wire dimensioning must be done according to a reasonable worst case, this document deals solely with bundles containing just power wires with approximately uniform current density, $J$.

## *1.2 Critical Dimensions of a Homogeneous Conductive Cylinder with Uniform Current Density*

If all of the wires in the bundle are made of the same materials and they have the same current density, the temperature of the center will increase with the radius of the bundle. Therefore, it is possible to find a certain critical area in the bundle $A_{crit}$ corresponding to the area that makes the temperature of the bundle reach certain maximum acceptable temperature $T_{max}$ for a given current density, $J$. Assuming for a moment that the bundle is a cylinder of conductive material with a uniform current density $J$, the heat equation in steady state might be written as follows:

$$-\lambda \cdot 2\pi rL \cdot \frac{dT}{dr} = f \cdot \pi r^2 L = F \tag{2}$$

where $\lambda$ is the thermal conductivity of the cylinder material, and $r$ and $L$ are respectively its radius and its length. Integrating from $r = 0$ where the maximum temperature $T_c$ is found, up to $r = R$, where the temperature is $T_s$:

$$T_c - T_s = \frac{fR^2}{4\lambda} = \frac{fA}{4\pi\lambda} \tag{3}$$

Now, the temperature of the external surface of the cylinder can be found by equaling the entire generated heat $F$ with all the dissipated heat due to radiation and convection:

$$2\pi RL \cdot \alpha(T_s) \cdot (T_s - T_a) = \pi R^2 Lf = F \tag{4}$$

$$\rightarrow T_s = \frac{fR}{2\tilde{\alpha}} + T_a \tag{5}$$

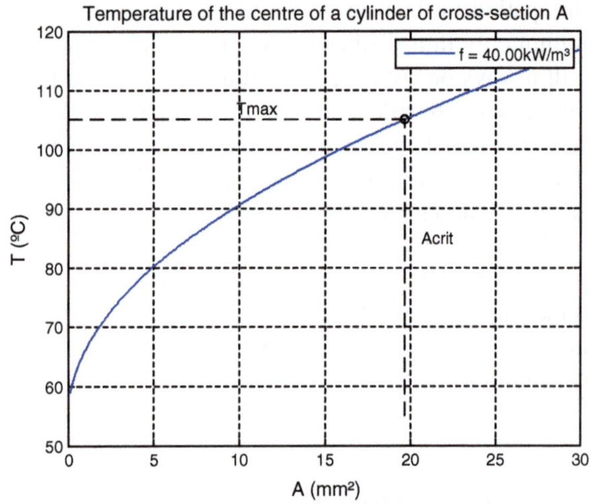


**Fig. 1** Center temperature in a cylinder of cross-section *A* and uniform heat generation *f*

Here, an averaged and constant value of $\alpha(T)$ (W/m$^2$K) is used. Now, the maximum temperature in the center and its dependence with the cross-section of the cylinder can be found:

$$T_c = T_a + \frac{fR^2}{4\lambda} + \frac{fR}{2\tilde{\alpha}} \tag{6}$$

$$T_c = f\left(\frac{A}{4\pi\lambda} + \frac{\sqrt{A}}{2\tilde{\alpha}\sqrt{\pi}}\right) + T_a \tag{7}$$

Equation (1) is represented graphically in Fig. 1. Equaling $T_c$ to a maximum value of temperature $T_{max}$, the relation between $f$ and its critical radius $R_{max}$ can be found, and this way $A_{max} \equiv A_{crit}$ is obtained:

$$R_{max} = -\frac{\lambda}{\tilde{\alpha}} + \sqrt{\frac{\lambda^2}{\tilde{\alpha}^2} + \frac{4\lambda(T_0 - T_{amb})}{f}} \tag{8}$$

$$A_{max} = 4\pi\lambda^2\left(-\frac{1}{2\tilde{\alpha}} + \sqrt{\frac{1}{4\tilde{\alpha}^2} + \frac{T_0 - T_{amb}}{\lambda J^2 \rho}}\right)^2 \tag{9}$$

This can be seen graphically in Fig. 2.

Conversely, it is useful to consider the minimum area $A_{min}$ as a function of the total current $I$ rather than the current density, Fig. 3.

It is also convenient to consider the maximum current $I_{max}$ as a function of the total cross-sectional area of the cylinder, Fig. 4.

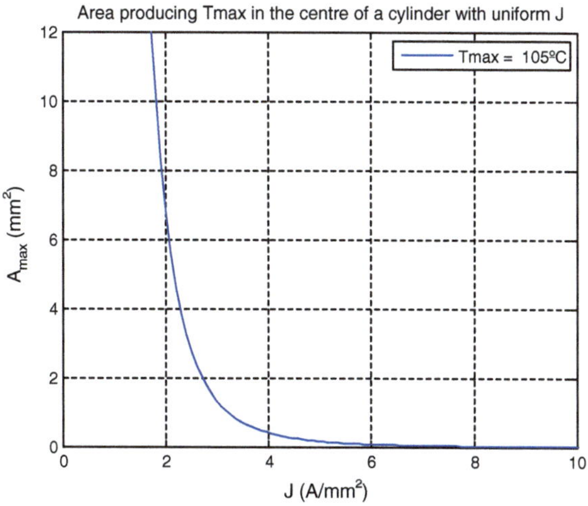


**Fig. 2** Maximum area of a cylinder conductor set to uniform current density $J$

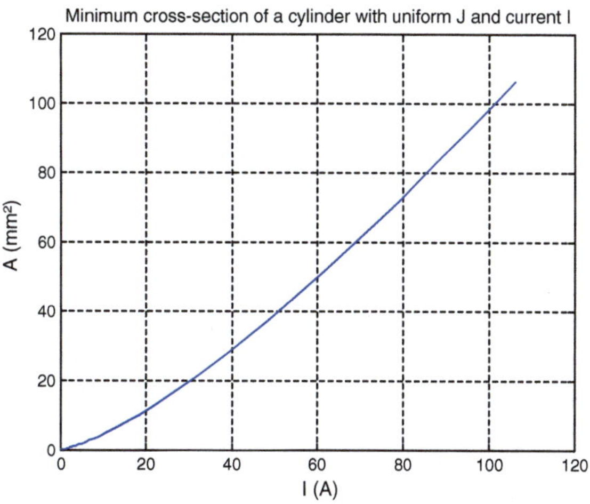


**Fig. 3** Minimum cross-sectional area of a cylindrical conductor subjected to a current $I$ and uniform current density. Smaller areas would produce temperatures above $T_{max}$

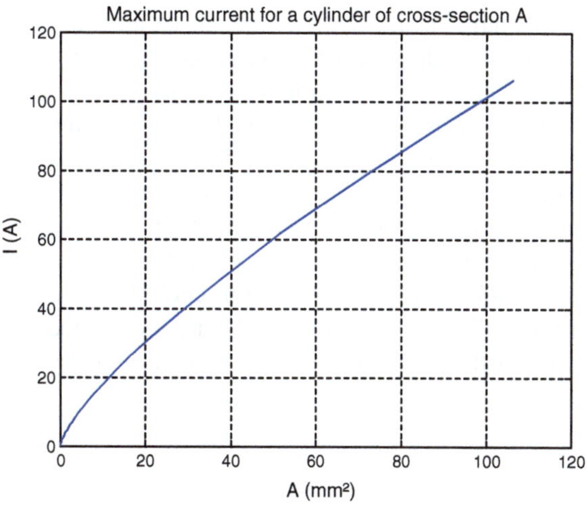


**Fig. 4** Maximum current for a cylindrical conductor with cross-sectional area $A$. Greater currents would produce temperatures above the maximum acceptable value

These latter relations are interesting, since they allow for estimating the maximum area of an arbitrary bundle, given the sum of its currents or its area. This will be used later on in this document.

## 2 Critical Currents of Wire Bundles Found by FEM Simulations

Actual wire bundles can be regarded as a heterogeneous mix of different materials, which are reduced here to copper, still dry air, polyvinylchloride (PVC) and poly-ethylene terephthalate (PET) fleece. The behavior of this system is similar to the case of the prior plotted homogeneous cylinder, yet not as simple to express by means of its parameters, such as $\lambda$ or $\rho$.

In order to observe the corresponding curves for actual heterogeneous bundles, finite-element simulations have been carried out. In order to allow radial symmetry, simulated bundles contain specific amounts of wires forming radial symmetry around a central wire ($N = 6n + 1 = 1, 7, 13, 19, 25, 31, 37\ldots$), all of them having the same diameter. This ensures all bundles have approximately round shapes. Round shapes with uniform current density are capable of producing the worst-case hot spot in the center of the bundle, and they are the typical shape of manufactured segments of wiring harnesses.

The critical value of current density of each bundle is found by means of iterative simulations, which try to raise the temperature at the center to exactly $T_{max} = 105$ °C, which is one of the standard values of acceptable steady temperature for PVC in vehicle wiring applications.

Founding the critical current $I_{crit}$ of the bundle for a given area instead of the opposite has the advantage that the geometry and the mesh of the finite elements are generated only once and reused for each new iterated simulation, which speeds up the process. The resulting list of area-current pairs can be swapped afterwards to either obtain the maximum current for a given area or the minimum area for a given current.

We have maintained an ambient temperature of 70 °C for the first set of simulations in order to explore the possibilities of these hypotheses. The x-axis is the sum of all the cross-section areas of the wires—areas of their conductors—Fig. 5.

We can observe a well-defined curve with the same characteristics as the one of the homogeneous cylinder, Fig. 5. Again, we plot the critical area of the bundle (the sum of all of the copper areas) versus the sum of all of the currents, and we can fit a third degree polynomial to it, which provides tight prediction bounds, Fig. 6.

This polynomial expression plotted in Fig. 6 can be used to estimate the critical area of a bundle, given its sum of currents. Wires might be dimensioned assigning areas so that the sum of all of the areas of copper equals this value of critical area. The fact that available cross-sections in industry are discrete values turns this one

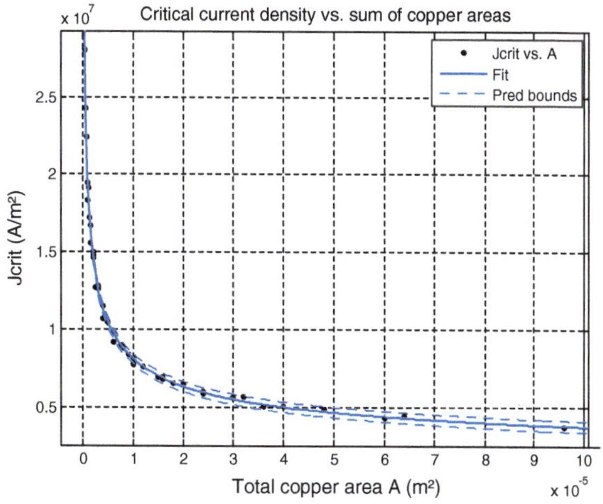


**Fig. 5** Results of critical current densities of wire bundles versus their areas of copper obtained by FEM simulations

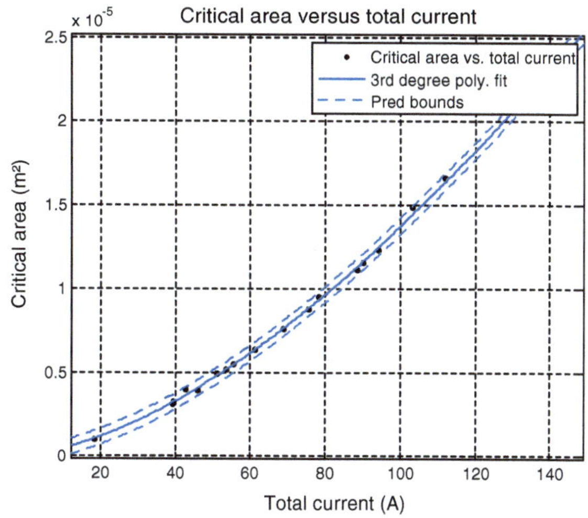


**Fig. 6** Polynomial fit for minimum area versus total current of bundle, with prediction bounds

into a simple discrete optimization problem: minimize the area of copper always remaining above its minimum value. This problem will be discussed and solved later in this document.

## 3 Regression Models for Wire Bundle Dimensioning

Linear regressions are carried out with the aim of obtaining general expressions for the critical area of arbitrary wire bundles. With 171 simulated cases and using the stepwise method, we obtained expressions for the logarithms of the critical area and the critical current.

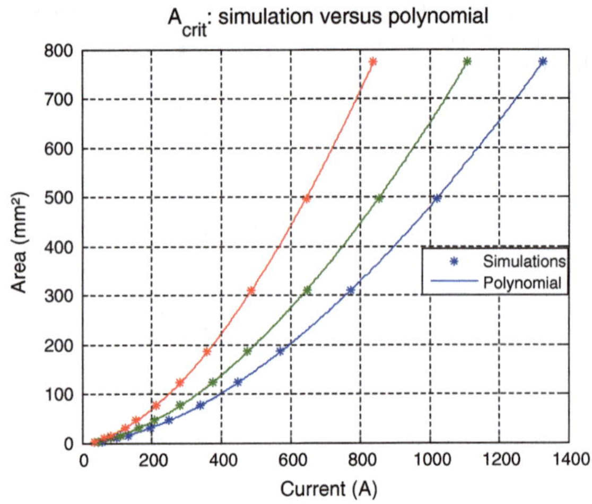


**Fig. 7** Regression model: minimum area of the bundle versus its sum of currents

Since what we want to predict in the first case is the critical area, the predictors are reduced to the number of wires $n$, the sum of their currents, $I$ and the ambient temperature $T_a$. Moreover, the ratio between copper and PVC can be used as an additional predictor for the response variable $I$. These predictors are included in the regression with pertinent operations such as square root or logarithm to coincide better with the theoretical formulae for cylindrical conductors discussed in this document.

Both regressions have obtained satisfactory results with $R^2 \approx 1$, especially the one for the critical current, due to the extra predictor related to the mass of PVC versus the mass of copper present in the bundle. This can be observed in Figs. 7 and 8.

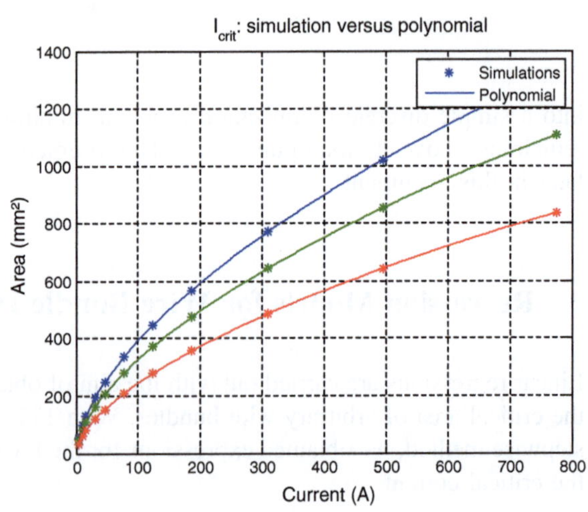


**Fig. 8** Regression model: maximum current of the bundle versus its sum of cross-sections

# 4 Integer Optimization for Cross-Section Assignment

As previously discussed, once the critical area is obtained, the cross-sections of the wires must be chosen so that the final area is minimum and greater than its critical value. If it were possible to buy wires with unconstrained cross-sections—i.e. continuous and infinite availability of sizes—then the problem would be easily solved by multiplying each current by the obtained value of current density $J$, obtained by dividing the total current of the bundle by the sum of the areas of the wires ($J = I/A$). However, cross-sections can be only discrete values, and it is not completely trivial to assign them in an optimum way.

## 4.1 Integer Linear Programming

Let $N$ be the number of wires of the bundle, indexed by $k = 1 \ldots N$. Let $Q_k$ be a set of available cross-sections for the wire $k$ with $S$ different values of area $a_{ki}$, $i = 1, 2 \ldots S$:

$$Q_k = [a_{k1}, a_{k2}, \ldots, a_{kS}] \tag{10}$$

Given $N$ wires in a bundle, for each wire there will be $S$ Boolean variables $b_{ki}$ determining whether the wire $k$ is associated to the cross-section $i$ in $Q_k$.

$$x = \begin{bmatrix} b_{11} \\ \vdots \\ b_{1S} \\ b_{21} \\ \vdots \\ b_{2S} \\ \vdots \\ \vdots \\ b_{NS} \end{bmatrix} \quad x_{min} = \begin{bmatrix} 0 \\ \vdots \\ 0 \end{bmatrix} \quad x_{max} = \begin{bmatrix} 1 \\ \vdots \\ 1 \end{bmatrix} \tag{11}$$

Since these variables are Booleans, it must be imposed as constraints that they are integers with lower and upper values respectively 0 and 1. Additionally, since one wire can only have one cross-section, it must be imposed that only one variable for the wire $k$, $b_{ki}$, $i = 1 \ldots S$ can be 1, and the rest must be 0. Specifically, this is achieved by imposing these Booleans to sum exactly 1:

$$\sum_{i=1}^{S} b_{ki} = 1 \quad k = 1, \ldots, N \tag{12}$$

Expressed in a matrix with $N$ rows and $S{\cdot}N$ columns:

$$A_{eq} \cdot x = b_{eq} \tag{13}$$

$$\begin{bmatrix} 11\ldots1 & \ldots & 0 \\ \vdots & 11\ldots1 & \vdots \\ 0 & \ldots & 11\ldots1 \end{bmatrix} \cdot x = \begin{bmatrix} 1 \\ \vdots \\ 1 \end{bmatrix} \tag{14}$$

The main constraint of the problem is that the sum of all of the areas must not exceed the given critical or maximum area.

$$\sum_{k=1}^{N} \sum_{i=1}^{S} b_{ki} \cdot a_i \geq A_{crit} \tag{15}$$

This can be expressed in matrix form as follows:

$$A = [Q_1 \quad \ldots \quad Q_k] \tag{16}$$

$$A \cdot x \geq A_{crit} \tag{17}$$

The objective function of the problem is the total mass of copper, which is not merely proportional to the sum of cross-section areas of the bundle. Instead, we must minimize the volume of copper, for which we need to multiply all of the areas by the length of each wire, $l_k$. Additionally, we can multiply each of these volumes $k$ by a relative weight expressing the importance of the wire $k$ in the system. Typically, these relative weights $w_k$ will express the expected ratio of manufactured instances of the wire $k$ over the total number of manufactured cars, which is a useful parameter in modular systems. In other terms, we express the volume of copper as the overall sold copper including all of the vehicles of a certain model, rather than the volume of copper of just one car.

$$V = [Q_1 \cdot l_1 \cdot w_1 \quad \ldots \quad Q_k \cdot l_k \cdot w_k] \cdot x \tag{18}$$

The problem is therefore defined as follows:

$$\min V \begin{cases} A \cdot x \geq A_{crit} \\ A_{eq} \cdot x = b_{eq} \\ 0 \leq x \leq 1 \end{cases} \tag{19}$$

Particularly, this method is implemented by giving each wire two possible cross-sections: one with the maximum current density below the critical value, and the other with the minimum current density above the critical value. This defines $Q_k = [a_{k1}\ a_{k2}]$. Then the algorithm chooses the cross-section for each wire so that the $V$ parameter of the set is minimized, keeping the total area greater than $A_{crit}$.

**Table 1** Results of optimization of random bundles

| $T_{amb}$ (°C) | $N$ | $\Sigma I$ (A) | $A_{crit}$ (mm$^2$) | $A_{total}$ (mm$^2$) | $T_{max}$ (°C) |
|---|---|---|---|---|---|
| 62.80 | 8 | 114.99 | 15.63 | 16.00 | 105.17 |
| 79.00 | 24 | 292.78 | 103.62 | 105.00 | 105.26 |
| 67.94 | 12 | 130.87 | 23.45 | 24.00 | 101.47 |
| 82.32 | 25 | 322.06 | 136.15 | 137.00 | 105.72 |
| 60.46 | 27 | 351.90 | 88.10 | 89.00 | 106.55 |
| 62.91 | 29 | 353.80 | 94.37 | 95.50 | 105.75 |
| 59.37 | 18 | 206.56 | 49.36 | 50.50 | 91.66 |
| 59.08 | 5 | 70.03 | 6.28 | 7.25 | 102.12 |
| 81.08 | 11 | 152.34 | 43.36 | 44.00 | 103.56 |
| 72.39 | 34 | 413.26 | 161.46 | 162.50 | 103.60 |
| 71.50 | 8 | 100.22 | 15.33 | 15.50 | 103.89 |
| 59.35 | 31 | 354.96 | 89.88 | 90.00 | 106.13 |
| 80.59 | 21 | 276.46 | 99.52 | 101.00 | 105.02 |
| 73.66 | 37 | 453.54 | 203.45 | 205.00 | 101.25 |
| 65.53 | 5 | 53.08 | 4.53 | 5.00 | 101.66 |
| 70.40 | 18 | 242.48 | 82.53 | 84.50 | 93.57 |
| 67.05 | 6 | 84.28 | 10.08 | 11.00 | 102.74 |
| 57.28 | 36 | 438.63 | 130.62 | 132.50 | 102.74 |
| 62.20 | 3 | 41.73 | 2.87 | 3.00 | 105.91 |
| 58.70 | 30 | 397.17 | 106.47 | 107.00 | 103.26 |
| 60.52 | 31 | 407.05 | 115.65 | 117.00 | 103.31 |
| 62.20 | 33 | 428.68 | 132.92 | 133.00 | 104.19 |
| 67.52 | 5 | 35.60 | 2.46 | 2.50 | 103.78 |

The optimization algorithms assumes that the critical area obtained by means of the regression model provides a bundle with a maximum temperature of $T_{max}$, in this case $T_{max}$ = 105 °C. In order to validate the system, some simulations of random bundles have been performed (see Table 1).

## 5 Conclusions

We obtained expressions that provide the minimum sum of cross-sections of a wire bundle with given currents and algorithms capable of assigning feasible cross-sections to them. Results show great accordance between the models and the finite-element simulations for both the expressions of minimal area for given currents and maximum current for given area. This satisfactory result ought to be now complemented with experimentation, which is expected to be in accordance to the obtained results due to the already good accordance of the finite elements simulations and reality.

As for the cross-section assignment algorithm, the linear method provides the best results regarding maximum temperature reached in the bundle, which is rarely above 105 °C. It is now necessary to validate the result of these expressions and dimensioning algorithms experimentally in order to find imperfections in the actual maximum temperature of the bundles.

## 6 Future Work

Actual wire harnesses include in many cases mixed types of wires, with different materials. That is why throughout the elaboration of this study it has come to our minds the necessity of including the presence of different materials to the bundles. It seems crucial to extend this study finding regression models that include the proportion of different typical materials in the bundle, such as other isolation materials or aluminium.

In a similar way, the purpose of optimizing the used mass of copper in the wiring harnesses, it appears useful to consider how can these so-called cold signal wires inside the bundle influence the maximum temperature. It could be done by including in these further regression models not only the ratio of different materials but also information about wires with negligible currents. It is expected that the presence of these wires will provide more favourable results, in such a way that some copper could be saved. Nevertheless, this must be validated with results.

## References

1. Ilgevičius, A.: Analytical and numerical analysis and simulation of heat transfer in electrical conductors and fuses. Universität der Bundeswehr München (2004)
2. Čiegis, R., Ilgevičius, A., Ließ, H., Meilūnas, M., Suboč, O.: Numerical simulation of the heat conduction in the electrical wires. Math Model Anal **12**(4), 425–439 (2007)
3. Frei, S., Diebig, M.: Simulation-based optimization of multi voltage automotive power supply systems. In: Vehicle Power and Propulsion Conference (VPPC), IEEE, pp. 1–6. 15–18 Oct 2013
4. Diebig, M., Frei, S.: Modeling of the automotive power supply network with VHDL-AMS. In: Vehicle power and propulsion conference (VPPC), IEEE, pp. 1–3 (2010)
5. Čiegis, R., Meilūnas, M., Jankevičiūtė, G., Ilgevičius, A.: Determination of heat conductivity coefficient of a wire bundle by inverse problem solution method. Electron. Electr. Eng. ISSN 1392–1215. **2**(90) (2009)
6. Loos, F., Dvorsky, K., Ließ, H.: Two approaches for heat transfer simulation of current carrying multicables. Math Comp. Simul. **101**, 13–30 (2014)
7. Benthem, R.C. van, Grave, W. de, Doctor, F., Nuyten, K., Taylor, S., Jacques, P.A.: Thermal analysis of wiring for weight reduction and improved safety. National Aerospace Laboratory NLR. NLR-TP. AIAA/ICES Conference, Portland, Oregan, USA 18–21 July 2011
8. Harbrecht, H., Loos, F.: Optimization of current carrying multicables. Comput. Optim. Appl. (2013)

9. Ließ, H.: Optimal dimensions for electrical conductors in mobile systems. In: Electric drives production conference (EDPC), 1st international. IEEE, pp. 139–142 (2011)
10. Olsen, R.S., Holboll, J., Gudmundsdóttir, U.S.: Dynamic temperature estimation and real time emergency rating of transmission wires. In: Power and energy society general meeting IEEE, pp. 1–8, 22–26 July 2012
11. Loos, F., Ließ, H., Dvorsky, K.: Simulation methods for heat transfer processes in mechanical and electrical connections. In: Electric drives production conference (EDPC), 1st international. IEEE (2011)
12. Short, T.A.: Electric power distribution handbook. CRC Press LLC, USA (2004)
13. Lewis, R. W., Nithiarasu, P., Seetharamu, K. N.: Fundamentals of the finite element method for heat an fluid flow. Wiley, New York (2004)
14. Blomberg, T.: Heat conduction in two and three dimensions: computer modelling of building physics applications. Lund University, Sweden (1996)
15. VDI-Heat atlas, 2nd edn. Springer, Berlin (2010)
16. Hussain, K.M., Hussain, D.: Telecommunications and networks. Butterworth-Heinemann, UK (1997)
17. Brabetz, L., Ayeb, M., Jilwan, G., Gräbel, P., et al.: A new approach to the test, assessment and optimization of robust electrical distribution systems. SAE Int. J. Mat. Manuf. (2013)
18. Combettes, S., Sontheimer, T., Rougemaille, S., Glize, P.: Weight optimization of aircraft harnesses. In: Advances on practical applications of agents and multi-agent systems. Advances in intelligent and soft computing, vol. 155, pp. 229–232 (2012)
19. Lin, C., Rao, L., D'Ambrosio, J., Sangiovanni-Vincentelli, A.: Electrical architecture optimization and selection—cost minimization via wire routing and wire sizing. SAE Int. J. Passeng. Cars—Electron. Electr. Syst. (2014)

# Part XI
# Test Driver Assistance Systems

# Validation of Automated Driving Functions

**Ruben Schilling and Torsten Schultz**

**Abstract** The validation of the development of driver assistance functions over millions of driven test kilometers finds its limits already today. The systems evolve rapidly from pure assistants over automated driving maneuvers towards fully autonomous driving. In the process an explosion of complexity is generated through interconnection and addition of sensors as well as strongly increased model complexity to represent the environmental aspects relevant for driving dynamics better and timelier. Here we would like to present an idea how an effort reduction of the validation can be realized. We propose to start the validation of intelligent systems before the road test: Features that intelligent systems use to classify the environment can selectively be varied in regards of test stimuli. The sensor state space is far easier to handle here than in the road test. Variations can be generated noticeably more efficient and goal oriented. In conclusion relevant parts of intelligent systems can be validated in driving dynamics relevant scenarios with less effort.

## 1 Introduction

A major source of innovation for today's vehicles is the addition of advanced driver assistance systems (ADAS) and their improvement over generations. In the big picture these systems develop from pure assistants over automated driving maneuvers towards fully automated driving. Fully automated driving is a scenario that provides whole new opportunities and drastic change for people's daily lives.

In the validation of driver assistance systems often 100,000 km and more of road tests scenes are collected for a single system. Driving maneuvers and situations are usually specified at the start of the development and the tests are recorded to

R. Schilling (✉) · T. Schultz
Berner & Mattner Systemtechnik GmbH, Gutenbergstr. 15, Berlin 10587, Germany
e-mail: Ruben.Schilling@berner-mattner.com

T. Schultz
e-mail: Torsten.Schultz@berner-mattner.com

 
C. Gühmann et al. (eds.), *Simulation and Testing for Vehicle Technology*,
DOI 10.1007/978-3-319-32345-9_25

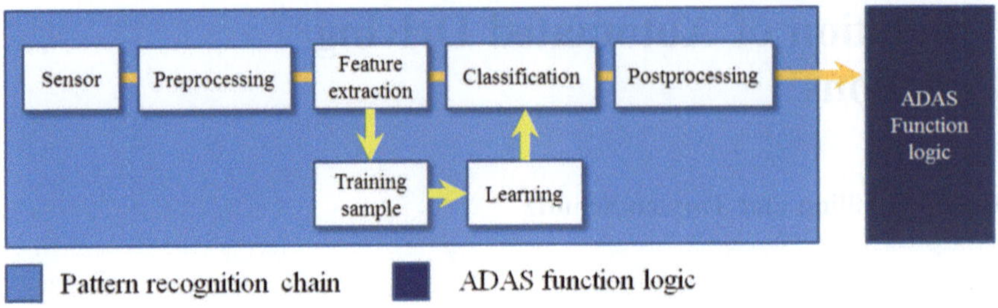


**Fig. 1** The general architecture of ADAS or intelligent system can often be depicted by the combination of a pattern recognition chain with an ADAS specific function logic such as zones, warning or concepts utilizing actuating elements

produce a repeatable benchmark suite for the monitoring of the maturity grade of the system as it is developed (Fig. 1).

Machine learning techniques are methods that pave the way towards autonomous driving. The ability to learn from limited data generalized recognition concepts, that work in an otherwise unexplored world is what makes these technique very powerful concepts for intelligent or even autonomous vehicles. Yet the concept of generalization is typically phrased in terms of an error. We give an introduction to errors of machine learning models in the next section.

The general architecture of ADAS or intelligent system can often be depicted by the combination of a pattern recognition chain [1] that sends its recognized classes to the ADAS function logic. The pattern recognition chain starts with the raw sensor data. This is followed by a typically sensor specific preprocessing. From preprocessed data features are extracted. These can then be used to train the system offline or in the live system to feed a trained classifier. The classes of the classifier are then often post processed using additional heuristic knowledge to filter false positives. The final classes are then passed to the ADAS function logic. Typical example of ADAS function logics are warning zones, warn concepts (timings and warning levels) combined with specific conditions.

In practical development projects often the test focus lies on vehicle maneuvers and scenarios relevant for the ADAS function, such as e.g. take over maneuvers. Systematic testing of the intelligent system's internals is rarely employed. This leads to a high risk, as the developers do not know if the system generalizes well, once it is deployed to customers. Typically this puts a lot of pressure on people who participate in road tests to manually recognize all abnormal behavior during the road tests. When driving 100,000 km of road tests or more, keeping track of all issues and their potential interactions becomes an error prone process. Errors are usually put into a database and fixed in the progress. If the encountered errors are relevant in the sense of, that they are representative of situations that will occur in the future use of the vehicle is hard to estimate. Furthermore during typical road tests often the same routes are driven. Each drive gives rise to variations of course, as every test drive is somehow specific. But the opportunities to stress specific aspects of the intelligent system are strongly limited as only the system as a whole can be stressed from the outside.

Hence the abstraction layer of features for intelligent systems is inefficient to stimulate when executing testing according to traditional test concepts that typically comprise hierarchical levels such as unit tests, integration tests, system tests and road tests. The level of indirection to stimulate this abstraction layer is too high to efficiently test intelligent systems here. However the variations of features that would appear under relevant driving conditions are one of the most critical factors for the performance of intelligent systems.

## 2 Validation of ADAS Functions Through Simulation of Features

In real ADAS series development projects the data to train systems can only be collected piece by piece, as the required kilometers of road driving are time consuming.

Collecting data to train and validate machine learning models is also generally a time consuming activity. For intelligent systems this means, that although we usually have a lot of data in total, it has to serve so many different training purposes, that for each recognition problem we do not end up in a data rich situation. This has consequences for the assessment of faults of the intelligent systems.

Generally the error of a machine learning model can be defined as in [2]: During training a machine learning model is adapted to predict the training data as good as possible. The training error is the error that remains between the models prediction and its training data. The generalization error of a machine learning model is the performance of that model on independent test data, i.e. data that has not been utilized to train the model. The generalization error is what actually matters in practice. This error reflects the promise that the system will behave well during many years of ownership of a customer and while being exposed to unseen conditions and environments.

The textbook picture (see [2]) is to split data into training, validation and if possible test data. The training data is then used to fit the models, the validation data to compare alternative models to each other before a final model is chosen and the test data to predict the final chosen models generalization capability [2]. This is an ideal case for data rich situations that unfortunately do not happen often in ADAS development projects.

Therefore we propose in this article to augment the data for testing purposes through simulation (see Fig. 2). For many scenarios it is possible to come up with simulations how certain features would behave. This enables goal specific, direct stimulation of feature variations and helps to validate the system in less time. Also in sparse data situations (e.g. when learning rare events, such as “20” speed signs in traffic sign recognition), it may prove practically impossible to stimulate the feature space through road tests. Here simulation can help to explore real world variations and test the intelligent systems performance. A low fault level is always the goal as

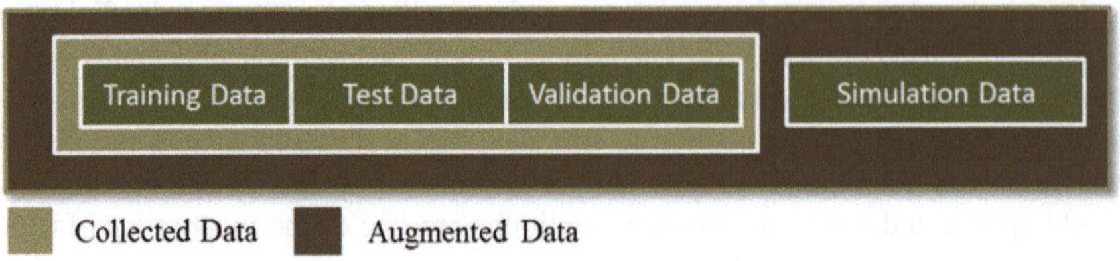


**Fig. 2** Testing purposes through simulation

todays intelligent systems are either safety relevant or pure comfort functions, that shouldn't negatively affect customer experience.

As an example consider said "20" speed sign. Instead of trying to sample this sign under various conditions we can simulate the sign (see Fig. 3). Here we take the original sign (left), apply aspect ratio changes, motion blur, white balance adjustments or a combination of these effects (right). This corresponds directly to real world scenarios that could be encountered. The classification tree method [3] is one typical, systematic approach to generate test cases and structure the problem. The sensor space of a gray image sensor has $256^{1280*720}$ states. The classification tree in (Fig. 4) requires two test cases for minimal coverage. The classifications in the classification tree correspond directly the relevant scenarios and can be simulated as in Fig. 3.

The simulation data can be rich due to the cheap cost to produce them. Testing with simulated data enables predictions if the developed intelligent system can generalize from the sparse training data in richer, realistic scenarios.

As described above usually the individual components of the pattern recognition chain of intelligent systems are not systematically tested for faulty behavior. With this approach it is possible to do so in a systematic way.

It is often possible to find realistic scenarios and write a sufficient simulation to replicate their effects on relevant features. Above we show exemplary a rare data for traffic sign recognition ("20" speed limit), that is hard to sample in the real world. In the above example we augment the test data by simulating aspect ratios, motion blur, white balance issues or a combination of these effects. If we were to require the sampling of all these scenarios on the road this would prove impossible given the time constraints of real projects. The simulation itself on the other side was straightforward to do and could be easily extended and refined to cover more

**Fig. 3** Simulation of relevant scenarios; from left to right: Original rare sample speed sign; Change of aspect ratio; (Motion) blur; White balance change; Aspect ratio, blur and white balance changes combined

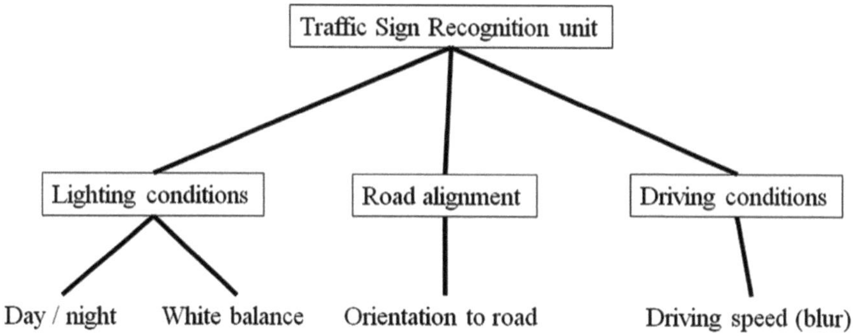


**Fig. 4** A classification tree (without concrete test cases) for some realistic scenarios considered here

scenarios. Test cases can then be generated by varying the simulation parameters using standard test methodology, such as limit value checking or generating representatives.

## 3 Conclusion

In this article we proposed an alternative way to test intelligent systems in real world development projects. We exemplified how simulations are often possible with low effort and how they cover a vast set of test cases. We showed, that this can be a helpful technique to explore an otherwise complex sensor space and to provide an assessment method for situations that are hard to sample. Furthermore we showed, that it is possible to naturally employ standard testing techniques to help generating test cases, e.g. to cover combinatorial testing needs. We believe this is an opportunity already for today's projects to improve the real world quality of intelligent systems and provides a method to help dealing with increasing complexity of these systems in their evolution towards autonomous driving. We believe, that for future development, when complexity of the systems rises and their decision playground is largely increased it will be a necessity to come up with validation approaches similar to the one we outlined here.

## References


1. Theodoridis, S., Koutrumbas, K.: Pattern Recognition, 3rd edn. Academic Press, London (2006)
2. Hastie, T., Tibshirani, R., Friedman, J.: The Elements of Statistical Learning, 2nd edn. Springer Series in Statistics (2009)
3. Grochtmann, M., Grimm, K.: Classification trees for partition testing. Softw. Test. Verification Reliab. **3**(2), 63–82 (1993)

# Systematic Use Case Driven Environmental Modeling for Early Validation of Automated Driving Functionalities

**Sebastian Siegl and Martin Russer**

**Abstract** In the automotive industry, current activities focus heavily on the development of automated driving systems (ADS). ADS process environmental data from different sensors, which are fused to generate a model of the surrounding world. Actors in the generated model are objects, which are e.g., classified as vehicles or pedestrians. The actors run in parallel, as in the real world actions from traffic participants can be taken independently and asynchronously from each other. Different actors may have impose different use cases in the interaction with the system. For verification and validation of these functionalities a method is required, that allows for a realistic and hence parallel modeling of the system under test's environment. Additionally, the method should allow for structured testing in compliance with international norms such as the ISO 26262 and the first international standard for software testing ISO/IEC/IEEE 29119, published in 2013. In this contribution we present an approach for creating environmental models for structured testing of automated driving systems with a constructive method. One step is the enumeration of all possible sequences, but we first decompose the task into manageable units by input/output dependency analysis. The expected behavior is formalized in temporal logic. In doing so, the effort for the creation of the model is feasible in industry. On the other hand, the test model guarantees the representation of all possible scenarios of use, making it a stable basis to derive significant test cases. We applied the method on an embedded system functionality. The functionality's components were architectured using the AUTOSAR 3.2 standard and implemented with Matlab Simulink. The proposed method allows for a use case driven and feasible creation of environmental models for the early validation of automated driving functionalities.

S. Siegl (✉)
AUDI AG, I/EF-563, 85045 Ingolstadt, Germany
e-mail: Sebastian.siegl@audi.de

M. Russer
AUDI AG, I/EF-84, 85045 Ingolstadt, Germany
e-mail: Martin.russer@audi.de



C. Gühmann et al. (eds.), *Simulation and Testing for Vehicle Technology*,
DOI 10.1007/978-3-319-32345-9_26

**Keywords** Automated driving functionalities · Validation · Verification · Requirements · Model based testing · Environmental modeling

## 1 Introduction

Testing is an essential activity for validation and verification in the development of automated driving functionalities. By observing the execution of the function-under test, one judges, whether the functionality behaves as expected. Misbehavior and malfunctions can be identified. As testing provides realistic feedback of the behavior, it is a key activity in industry before releasing a product on the market.

Requirements definition is the first main activity after the decision for the development of a functionality is made. It also constitutes the first activity in which errors can be made. It is even the most critical activity with regard to faults and defects, because defects discovered in late development phases might have their origin in the initial requirements. In 2013 the first release of the international standard for software testing ISO/IEC/IEEE 29119 [1] was released. The ISO 29119 complements the ISO 26262 [2] w.r.t. the activities for software testing. It states the main purpose of requirements based testing as 'to determine whether the test item meets end-user requirements'. It should be noted, that in the same section it is highlighted that testing may suffer if the requirements are incomplete or not consistently specified. Following Boris Beizer's bug taxonomy, the main defects in requirements can be classified into incorrectness, inconsistency, incompleteness, and obsolescence. In this paper we present an approach, to base the testing activities on a validated, consistent and complete environmental model.

## 2 Use Case Oriented Constructive Requirements Modeling

In this section we introduce foundations of our approach: The modeling elements of the environmental model, and the method to analyze and transform the requirements into the model.

### *2.1 Modeling Elements*

The environmental modeling is done by creating Time Usage Models (TUM) [3]. The representation by a set of TUMs is the output of the constructive requirements modeling process, which is described in more detail in the following section.

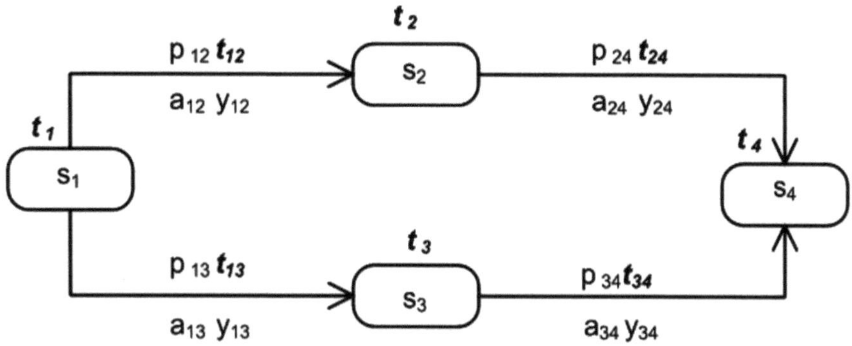


**Fig. 1** Underlying graph abstraction of Time Usage Model

A Time Usage Model consists of:

- A set of states $S = \{s_1;\ldots;s_n\}$, that represent possible states of use.
- A set of arcs A, representing state transitions. An arc from state $s_i$ to state $s_j$ denoted by $a_{ij}$.
- A set of stimuli $Y = \{y_1;\ldots;y_m\}$. A stimulus $y_j$ is assigned to each arc.
- The transition probability from state i to state j, denoted by $p_{ij}$ for the transition $a_{ij}$.
- A probability density function (pdf) $t_i$ to specify possible state sojourn timest o each state $s_i$.
- A pdf of state transition time $t_{ij}$, which is assigned to each arc $a_{ij.}$ This pdf describes e.g. the possible stimulus execution times for an arc $a_{ij.}$

Test cases can be sampled from the model statistically using the probabilities. Another possibility to derive test cases is the graph abstraction on the model, and to use deterministic algorithms to achieve e.g. a coverage level of states or transitions.

In Fig. 1. the elements of a TUM are presented in their representation in a graph.

## 2.2 *Basic Principles of Constructive Requirements Modeling*

The method for creating a TUM is the key activity, since the following activities for validation and determination of estimators about the reliability depend on the quality of the model. The method for creating a TUM as the test model follows the principles of sequence-based specification (SBS). This includes the identification of the system's boundaries as well as the enumeration of all sequences of stimuli, considering their responses over time.

Following this procedure one ensures, that the final dynamic model represents the environment completely, consistently, and traceably correct.

Previously uncovered system behavior is identified by this method. Requirements must be derived for it and formulated. The stakeholders who are responsible are invoked to manage this task. They have to derive a new requirement and add it to the existing set of requirements, usually in a document based form. So this procedure is a technique for analyzing and clarifying the requirements.

# 3 Leveraging Environmental Modeling

Applying the sequence-based specification in practice showed some significant improvements in the analysis and modeling of requirements and therefore led to better test suites and more reliable software. With the increase of the complexity of functionalities, the feasibility of applying this process is limited to functions with a manageable number of inputs and level of complexity. A safety relevant function with four inputs and simple timing conditions resulted in a TUM containing 35 states. In order to identify the approximately 230 state transitions, more than 600 sequences of stimuli had to be analyzed.

To address this problem of complexity and feasibility, the analysis of the requirements can help to reduce the necessary effort. These analyses concentrate on dependences between input signals and are influenced by our experiences with sequence-based specification in practice. We introduce these methods in the following sections.

## *3.1 Conditions and Prioritization*

Often, the processing of a function is active only if a certain trigger condition is met. This can be for many reasons such as saving processing time or power. If such an activation stimulus exists and the output of the respective functionality is independent from the rest of the system, it can be treated in a separate enumeration process. This step helps to reduce the enumeration effort significantly.

The same principle can be applied to hierarchically structured functionalities as well. Figure 2 shows the processing hierarchy with dependences on various conditions. If and only if the trigger condition is met, one of the available functions will be called and the specific return value will be returned. Otherwise, a default or error behavior is expected. The graph shows the hierarchy of the functionalities activation conditions, with function A as the highest order function. If function A is called, its return value is generated and the module step is finished. Depending on the input conditions, only one function is called in one step. The mutual exclusion resulting from the hierarchy allows to simplify the sequence-based enumeration process by dividing it according to the functions. Hence an enumeration process is done for all functions separately. This leads to a larger number of enumerations, yet they are less complex and less time consuming.

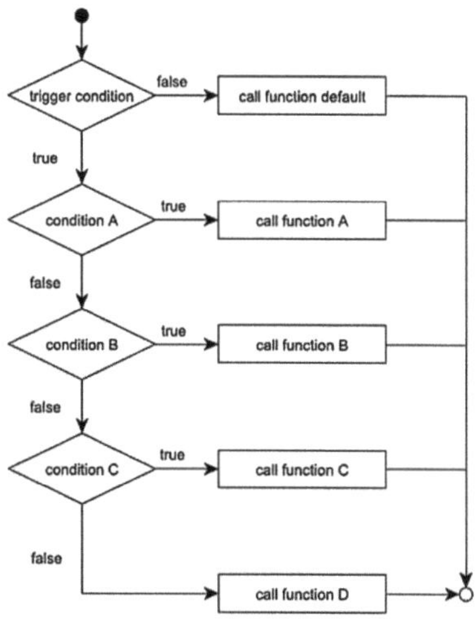


**Fig. 2** Function call graph

## 3.2 Input-Output Dependency Analysis

Another way to reduce the overall complexity of the enumeration process is to identify independent input stimuli. After defining the systems boundaries, the lists of input and output stimuli are defined. If the focus is on low level functionality, it is hard to find independent input values because often one is at a level in which one module implements one feature. Having a look at higher levels of abstraction, it's more likely to find independent input signals as these functionalities encapsulate and coordinate the underlying functionalities.

Figure 3 shows a software component containing six input signals and three output signals. The subsystem represents the boundaries of the functionality under consideration. The result of the Input-Output Dependency Analysis is depicted in Fig. 4 in order to provide a simpler view on the result instead of listing the requirements of the example. The analysis of dependent input variables has to certainly be done on the requirements. Figure 4 shows that input one (notation: In1) to four are used to calculate output one (Out1) and two. Output three depends only on input five and six. Thus, In5 and In6 do not affect the outputs calculated by In1 to In4 and vice versa. We propose to split up the enumeration process for the two mentioned groups of inputs. If supported by the testing tool, the test cases derived by the two different enumerations can be run in parallel as the inputs cannot affect each other according to the specification.

Sometimes, safety critical functions require the testing of every possible input combination. In this case, it is still possible to separate the enumeration processes. This leads to two different test suites. Testsuite A runs all the valid sequences developed by enumerating In1 to In4, while the independent stimuli can be stimulated in parallel in the background. The same principle can be applied to

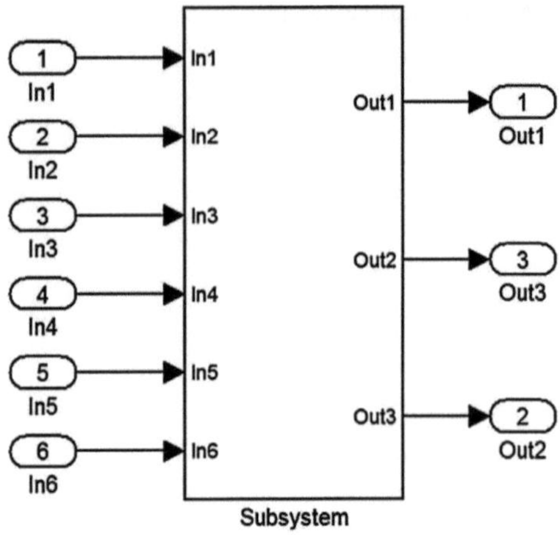


**Fig. 3** Software subsystem

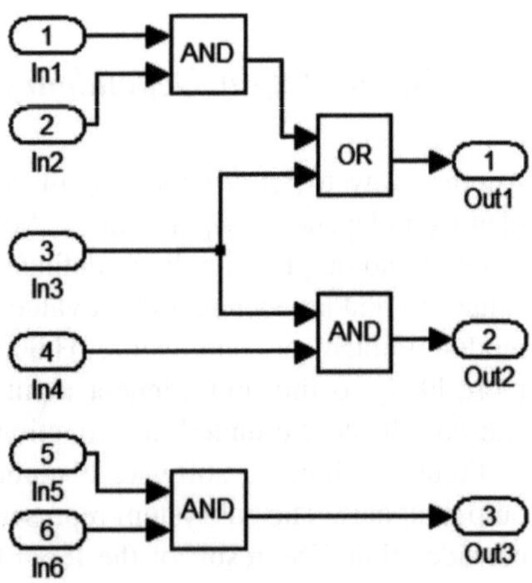


**Fig. 4** Input output dependency analysis

Testsuite B, where the enumerated sequences of In5 and In6 can be evaluated together with the combinations of In1 to In4.

## 3.3 *Requirement Assessment Using Rules in Temporal Logic*

Instead of defining local expected returns and assessments, as in the classic sequence-based specification, we define global assessments [4]. These are defined in temporal logic. Global assessments can be seen as a global observer checking the output of the SUT. For each enumeration step at least one global assessment and the corresponding requirement must be defined.

## 4 Model Composition

After decomposing the input streams and applying the enumeration separately, the final model consists of different regions. Figure 5 depicts an example of a model. Completely parallel regions are on the top, below are regions with enumerations, which can be run in parallel, but must be synchronized. These regions with synchronization points result from dependencies on trigger conditions.

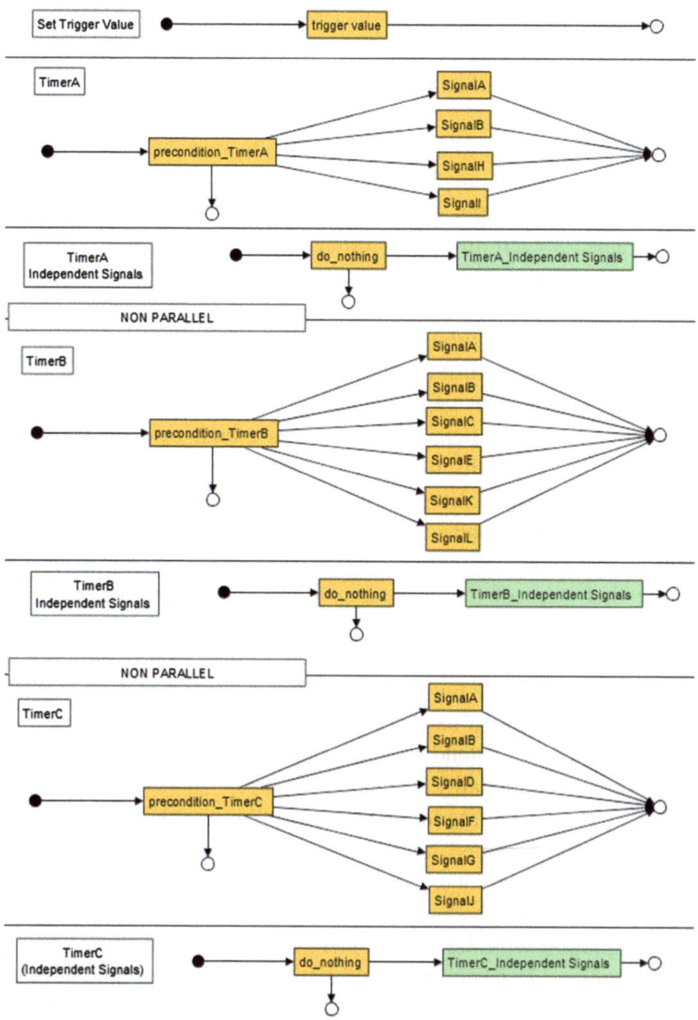


**Fig. 5** Composed environmental model

## 5 Application on Functionality

After case studies and the elaboration of the theoretical aspects, which were introduced in Sects. 2 and 3, we applied the method in practice on a real implementation. In this section we give a brief introduction to the embedded function and illustrate the methods and processes by means of the example. This example and the previous sections can be read in more detail in [5].

The functionality controls the timing of various sub-functionalities in the interior of a car. The software architecture was specified in AUTOSAR, [6] as it is done in the development of automated driving systems. Depending on the interaction of a human with the car like opening or closing doors as well as the car's environment, the function decides whether to trigger modules and checks the activation, the duration of an activation or the deactivation after specific timers.

The selected module of the functionality has a total input of 14 signals, it processes all information in a single step and provides the results by six output signals. Most input signals are described by boolean values and define a certain state of the car. The six output signals give information about the three available timers, two signals for each timer. One describes the running state, a second is true whenever a timer elapses. None of the output signals are returned and used as an input of the module.

After defining the test environment, the proposed methods were put into practice. On the top layer, the environmental consists of 52 transitions and 26 states and covers all 615 legal sequences gained from three enumerations, that could be done separately after the analysis steps. This provides a nice view on the system compared to the overall complexity of the system and model. Furthermore, the visual way of building up the model with preconditions plus stimulus provided a good basis for understanding and discussion between the stakeholders of the functionality.

During the first steps of the sequence-based enumeration, the method quickly discovered the incompleteness of the specification. It was notable, that many requirements define an expected result under specific conditions, but make no statement of the expected behavior if the condition is not met. Consequently, there were sequences for which no requirement specified the output of the function. It was therefore necessary to extend six requirements. This conspicuousness occurred at all three different functionality dependent timers.

In addition, inconsistent requirements were discovered. For some sequences, multiple requirements applied and consequently resulted in a contradictory behavior. It is remarkable, that these inconsistencies were discovered only by considering the timing dependent behavior. Additional requirements had to be defined in consultation with the stakeholders of the functionality.

All in all, 12 of the 17 existing requirements had to be altered and three additional requirements had to be added. The derived requirements in combination with the enumerated sequences were the basis of the environmental model. It served as test model. Executing the test cases showed no errors as requirements and implementation had already been corrected during the enumeration process. This allows for an early validation of the requirements and implementation before the final testing activities.

## 6 Conclusions

The development of automated driving systems that process parallel and complex inputs from the environment poses not only a challenge for the test environments, such as virtual simulation, but also for the methods of determining the significant and required test scenarios.

The methods for verification and validation of software were substantiated in 2013, when the first, second, and third parts of the international standard for software testing ISO/IEC/IEEE 29119 were released. It states the main purpose of requirements based testing as 'to determine whether the test item meets end-user. In the same section it remarks that testing may suffer if the requirements are incomplete or not consistently specified.

In this contribution we presented our approach to facilitate the creation of a complete, traceably correct, and consistent environmental model in industry. Its foundation lies in a separate modeling of the environmental actors with a constructive method.

By this approach, the subsequent activities for quality assurance, such as validation and verification, measurement of coverage criteria, and estimators e.g. of the reliability profit from this approach.

The application of constructive requirements modeling can be hardly feasible for complex functionalities, because it can result in a rapid growth of the state space. The state space explosion results from the fact, that the method itself requires all possible stimulations to be sequentialized on a certain level of abstraction. To deal with this, we presented techniques, that allow for an early reduction of the modeled state space.

Input streams are classified in two categories: completely independent and parallel, and those which can be run in parallel between synchronization points. The input streams are separately enumerated, which reduces tremendously the state growth. The final complete environmental model is a composition of all models. In doing so, the creation of the model is de-composed into manageable pieces: the creation of the model is feasible for real applications in industry, for the purpose to test automated driving functionalities.

## References

1. ISO/IEC/IEEE 29119 Software testing (2013)
2. ISO 26262.: Road Vehicles Functional Safety International Standard, 1st edn. (2011)

3. Siegl, S., Hielscher, K.-S., German, R.: Introduction of time dependencies in usage model based testing of complex systems. In: IEEE Systems Conference 2010, 4th Annual IEEE, S, pp. 622–627 (2010)
4. Müller, T.C., Lochau, M., Detering, S., Saust, F., Garbers, H., Märtin, L., Form, T., Goltz, U.: Umsetzung eines modellbasierten durchgängigen Entwicklungsprozesses für AUTOSAR-Systeme mit integrierter Qualitätssicherung. Informatikbericht 2009-06, Technische Universität Braunschweig, Dezember 2009
5. Siegl, S., Russer, M.: Constructive modelling of parallelized environmental models for structured testing of automated driving systems, cyber physical systems. In: Design, Modeling, and Evaluation: 5th International Workshop, CyPhy 2015, Amsterdam (2015)
6. Bringmann, E., Krämer, A.: Model-based testing of automotive systems. In: Proceedings of the 2008 International Conference on Software Testing, Verification, and Validation (2008)

MIX
Papier aus verantwortungsvollen Quellen
Paper from responsible sources
FSC® C105338

If you have any concerns about our products,
you can contact us on
**ProductSafety@springernature.com**

In case Publisher is established outside the EU,
the EU authorized representative is:
**Springer Nature Customer Service Center GmbH**
**Europaplatz 3, 69115 Heidelberg, Germany**

Printed by Libri Plureos GmbH
in Hamburg, Germany